T0329727

**Spectrum Sharing in
Cognitive Radio Networks**

Spectrum Sharing in Cognitive Radio Networks

Towards Highly Connected Environments

Prabhat Thakur

and

Ghanshyam Singh
University of Johannesburg
Auckland Park, Johannesburg, South Africa

Registered Office
John Wiley & Sons, Inc., 111 River Street, Hoboken, NJ 07030, USA

Editorial Office
111 River Street, Hoboken, NJ 07030, USA

For details of our global editorial offices, customer services, and more information about Wiley products visit us at www.wiley.com.

Wiley also publishes its books in a variety of electronic formats and by print-on-demand. Some content that appears in standard print versions of this book may not be available in other formats.

Library of Congress Cataloging-in-Publication Data
Names: Thakur, Prabhat, author. | Singh, Ghanshyam, author.
Title: Spectrum sharing in cognitive radio networks : towards highly
  connected environments / Prabhat Thakur and Ghanshyam Singh, University
  of Johannesburg Auckland Park, Johannesburg, South Africa.
Description: First edition. | Hoboken : Wiley, 2021. | Includes
  bibliographical references and index.
Identifiers: LCCN 2021015272 (print) | LCCN 2021015273 (ebook) | ISBN
  9781119665427 (hardback) | ISBN 9781119665434 (adobe pdf) | ISBN
  9781119665441 (epub)
Subjects: LCSH: Radio resource management (Wireless communications) |
  Cognitive radio networks.
Classification: LCC TK5103.4873 .T44 2021 (print) | LCC TK5103.4873
  (ebook) | DDC 621.384–dc23
LC record available at https://lccn.loc.gov/2021015272
LC ebook record available at https://lccn.loc.gov/2021015273

Cover Design: Wiley
Cover Image: © Pobytov/Getty Images

Set in 9.5/12.5pt STIXTwoText by Straive, Pondicherry, India

10 9 8 7 6 5 4 3 2 1

Contents

Preface

Technology is rapidly transforming our daily live, local community, as well as entire globe and everyone should be able to participate in this transformation. Currently, we rely on the wireless connected devices and systems to not only enable on-demand, pervasive communications for a large proportion of the population, but also other critical application areas such as scientific and medical research, industrial control and automation, and public safety. Thus, the communication technologies are crafting new corporate social responsibility initiatives which address global problems, support innovative ideas, and enable opportunities for all in recent increasingly digital world. As the wireless communication technologies and its applications continue to flourish, the demand for precious spectrum resources, which is an essential foundation block to support the wireless communications in the globally connected world, will continue to grow. Recently, due to exponential increase in the number of wireless connected devices with other bandwidth hungry services have exploded huge communication traffic and we expect that the demand for spectrum will continue to increase as new wireless technologies and applications requirements continue to emerge in the foreseeable future which results in the spectrum scarcity. This voracious enthusiasm for additional spectrum resources cannot be met by simply allocating new spectrum. Therefore, the usable capacity of spectrum must be expanded with innovative technologies, regulatory reforms, and removal of market barriers. The cognitive radio is one of the innovative technologies which has the potential to effectively address the spectrum scarcity problem and radically change the way we utilize spectrum. Due to its potential impact, various stakeholders – including regulatory policymakers, wireless device manufacturers, telecommunication service provider, and academic researchers – have shown strong interest in it, especially with respect to research and development.

The cognitive radio technology has emerged as a prone candidate for exploiting the increasingly flexible licensing (dynamic spectrum access) of the spectrum for the wireless communication system. The regulatory bodies have come to realize that most of the time, a large portion of certain licensed frequency band remain empty/unused. Therefore, to redress this, a new regulation would allow for devices which are able to sense and adapt to their spectral environment, such as cognitive radio to become secondary user and such users are wireless devices that opportunistically employ the spectrum already licensed to the primary users. The primary users generally associate with the primary spectral licensed holder and thus have higher priority right to the spectrum. The intuitive objective behind secondary spectrum licensing is to improve the spectral efficiency of the network, whereas depending on the type of licensing without affecting higher priority users.

In the cognitive radio network, the spectrum prediction, spectrum monitoring, and medium access control protocols play an important role to exploit the spectrum opportunities, manage the interference to the primary users, and coordinate. The dynamic leasing, in which some wireless devices opportunistically employ the spectrum rather than choose for a long-term sub-lease. In order to exploit the spectrum, we require a device which is able to sense the communication opportunity and then take actions based on the sensed environment. The cognitive radio offers a novel way of solving spectrum underutilization problems. The emergence of Federal Communication Commission's secondary market initiative has been brought both the obvious desire for spectral efficiency as well as empirical measurements showing that most of the time certain licensed frequency remain unused. The goal of secondary market initiative is to remove unnecessary regulatory barriers to new secondary market-oriented policies. The key points of the book that benefits the readers are as follows:

- Provides decent background about the fundamentals of the spectrum sharing techniques.
- Explores the advanced frame structures with spectrum accessing techniques.
- Role of spectrum prediction and spectrum monitoring techniques for interference-free spectrum sharing as well as for effective spectrum mobility is analyzed.
- Due to the demand of distributed architectures in various applications, the energy and spectral efficient frameworks are presented by using the self-scheduled medium access control (SMC-MAC) protocol in the CRNs.
- The frameworks of CR-NOMA for further improvement in the spectral efficiency as compared to CR are also the unique contributions of this book.
- The interference management schemes for the spectrum sharing are illustrated.
- The exploitation of CR for the internet-of-vehicles, that is CR-inspired internet-of-vehicles, adds as a novel contribution in the book.

Thus, it is perceived that this book enables readers about the recent advances in the field of spectrum sharing, strategies of mathematical modeling, theories of spectrum sharing in addition to the primary activities of cognitive radio network.

This book puts together a rich set of research articles featuring recent advances in theory, design, and analysis of cognitive radio networks with its connected environments. The book consists of 13 chapters, which cover a wide range of topics related to the cognitive radio technology, in particular, the topics covered in this book include fundamental challenges and issues in designing cognitive radio systems, information-theoretic analysis of cognitive radio systems, spectrum sharing, spectrum sensing and coexistence issues, adaptive physical layer protocols and link adaptation techniques for cognitive radio, different techniques for spectrum access by distributed cognitive radio, and cognitive medium access control protocols. The book is organized as follows. Chapter 1 starts with the connected environments, evolution of the wireless communication as well as technical perspectives by using the Third Generation Partnership Project (3GPP) with state-of-the-art spectrum sharing/access techniques and the fundamental issues related to cognitive radio networks with its several connected parameters and the major research challenges mostly from a signal processing and communication-theoretic perspective are well presented. The potential advantages, limiting factors, and characteristic features of the existing cognitive radio spectrum sharing domains are thoroughly discussed. The comparison of various spectrum accessing techniques such as interweave spectrum access, underlay spectrum access, overlay spectrum access, and hybrid spectrum access is presented. As the complexities of wireless technologies increase, novel multidisciplinary approaches for the spectrum sharing/management are required with inputs from the technology, economics, and regulations. To identify the available spectrum resource, decision on the optimal sensing and transmission time with proper coordination among the users for spectrum access are the important characteristics of spectrum sharing methods.

Chapter 2 describes a novel hybrid-cum-improved spectrum access technique to improve the throughput and data loss of the cognitive radio networks. The hybrid-cum-improved spectrum access technique consists of two advanced frame structures which explore the hybrid spectrum accessing strategy to utilize the channel in the presence of primary user. The closed-form expressions of the throughput and data loss for the proposed cognitive radio networks are derived numerically and the simulation results are the witness of superior performance with reference to the throughput and data loss.

Chapter 3 explores the cognitive radio networks in the high traffic environments where the spectrum prediction plays an important role to select a particular channel for spectrum sensing from the pool of channels on the basis of idle prediction probability. The considered frame structure has spectrum prediction phase before the spectrum sensing and data transmission phase. We have exploited the concept of hybrid spectrum access strategy to improve the throughput of the considered frame structure. The probability of primary users' emergence in the data transmission period is very significant which needs to be detected to avoid data loss and interference with primary user; however, there is no mechanism to perform this function. The spectrum monitoring technique has been used simultaneously with the data transmission period which is an advanced technique that exploits the received signal characteristics to detect the emergence of primary user. The use of spectrum monitoring improves the performance of high-traffic cognitive radio networks in terms of data loss, power loss, and interference-at-primary user.

Chapter 4 discusses the potential issues concerning the random selection of spectrum sensing channel after the spectrum prediction phase in the cognitive radio networks. A novel approach of improved channel selection is proposed which relies on the probabilities of channels by which predicted idle and the closed-form expressions for the throughput of cognitive user are derived. To achieve the prediction probabilities, the pre-channel-state-information is prerequisite, which may be unavailable for particular scenarios; therefore, a modified selection method is introduced to avoid the sense and stuck problem. For additional improvement in the throughput of cognitive user, a new frame structure is introduced, in which the spectrum prediction and spectrum sensing periods are exploited for simultaneous transmission of data via the underlay spectrum access technique.

Chapter 5 introduces the concept of imperfect spectrum monitoring error and has analyzed the effect of imperfections on the data loss, power wastage, interference efficiency, and energy efficiency in the cognitive radio network. The hardware impairments and channel random nature result in the imperfections in the spectrum monitoring process which are presented through the probability of spectrum monitoring error. The imperfection in spectrum monitoring degrades the performance of cognitive radio networks when analyzed for the different scenarios of the traffic intensity and probability of spectrum monitoring error.

Chapter 6 explores the cooperation among cognitive users for the homogeneous and heterogeneous cognitive radio networks for spectrum monitoring. The Binomial and Poisson-Binomial distribution functions are used to compute the probability of spectrum monitoring error after cooperation in the homogeneous and heterogeneous cognitive radio networks, respectively. The cooperation among cognitive users for spectrum monitoring improves the performance of cognitive radio networks in terms of the data loss, interference efficiency, and energy efficiency.

Chapter 7 presents the concept of spectrum mobility by using the spectrum prediction and spectrum monitoring techniques simultaneously, to detect the emergence of primary users. In this strategy, the decision results of the spectrum prediction and monitoring techniques are fused using AND and OR fusion rules, for the detection of emergence of primary user during the data transmission. Further, the closed-form expressions of the resource wastage, achieved throughput, interference power at primary user, and data loss for the proposed approaches are derived. In a special case, when the prediction error is zero, the graphical characteristics of all metric values overlies the spectrum monitoring approach, which further support the proposed approach.

Chapter 8 discusses a hybrid framework in the distributed cognitive radio networks with a novel frame structure and the self-scheduled multichannel-medium access control protocol is developed for the proposed frame structure. The distributed network architecture is a suitable option to overcome the limitations of the centralized architecture. Each cognitive user performs all the functions of spectrum accessing, individually in the distributed network architecture; therefore, the self-scheduled multichannel-medium access control protocol plays a key role. The hybrid spectrum access technique is used to exploit the active channels with constrained power transmission. The proposed framework is also analyzed for the perfect and imperfect spectrum sensing scenarios. It is perceived that the proposed framework outperforms the conventional self-scheduled multichannel-medium access control protocols as reported in the literature with reference to the spectral efficiency/utilization as well as interference efficiency.

Chapter 9 exploits a unique and key enabler technique of future generation communication which improves spectral and energy efficiency while satisfying the constraints on users' quality-of-service requirements, that is the non-orthogonal multiple access (NOMA) technique. The potential frameworks of NOMA implementation over cognitive radio (CR) as well as the feasibility of proposed frameworks are presented as CR-NOMA framework. Further, the differences between proposed CR-NOMA and conventional CR frameworks are discussed and the potential issues regarding the implementation of CR-NOMA frameworks are explored.

Chapter 10 explores a spectral-and-interference-efficient framework named as MIMO-based CR–NOMA communication system using collectively three spectral efficient techniques such as cognitive radio, NOMA, and multiple-input-multiple-output (MIMO). The proposed framework is analyzed and the closed-form expressions for throughput at each user due to the number of transmitting-and receiving-antennas are derived numerically for the downlink and uplink scenarios. In addition to this, the total/sum throughput for different frameworks such as CR–NOMA, CR–MIMO, and MIMO-based CR–NOMA systems is also derived for both the downlink and uplink scenarios. In order to satisfy the interference constraints at the PU due to cell-edge/far-user transmission in the uplink scenario, a new metric known as interference efficiency is derived. Furthermore, the proposed frameworks are simulated for downlink and uplink scenarios and the relationship between throughput of cell-centered/near-user and cell-edge/far-user are presented. The presented results reveal that the proposed MIMO-based CR–NOMA system outperforms the existing MIMO–NOMA, CR–NOMA, and CR–OMA systems in terms of the cognitive users individual throughput, total throughput, and interference efficiency of the system.

Chapter 11 discusses the interference management between the cognitive radio networks to enhance the spectral efficiency. The cognitive radio networks are classified as interfering and non-interfering interference scenarios, particularly the self-interference and user-to-user interference, which can be managed by frequency channel and power allocation techniques. Further, the interference cancellation techniques in the cognitive radio networks are explored and have also proposed the cross-layer interference mitigation in the networks. Moreover, the interference avoidance techniques by advancing the different constituents of cognitive cycle that are the spectrum sensing, spectrum accessing, spectrum monitoring, and spectrum mobility are illustrated.

Chapter 12 emphasizes over the potential applications of the cognitive radio such as the internet-of-things and internet-of-vehicles or vehicular networks. It started with the digitization techniques of the fourth industrial revolution (4IR), where IoTs is a very popular technique that is subdivided into the industrial IoTs (IIoTs) and consumer (CIoTs). Further, another potential application, the connected vehicles/internet-of-vehicles is a prominent part whose potential constituents that are vehicle-to-vehicle, vehicle-to-infrastructure, infrastructure-to-vehicle, vehicle-to-pedestrians are illustrated. The potential simulation frameworks in order to simulate the vehicular networks are discussed and a comparison of those simulation techniques and open research challenges is illustrated.

Chapter 13 explores the radio resource management perspectives in the internet-of-vehicles (IoVs) networks with different wireless access technologies such as vehicular communication, cellular communication, and short-range static communication. The vehicular communication techniques comprise the dedicated short-range communication (DSRC), wireless access for vehicular communication (WAVE), and communication architecture and land mobile (CALM). The cellular communication techniques are 3G, 4G, and 5G, wireless access for microwave (WiMAX), and satellite communication where LTE, LTE-Advanced, and NR are popular techniques. Further, the spectrum sharing perspectives in the IoV networks with CR frameworks are explored and the potential research challenges for particular constituents of the cognitive cycle are illustrated.

In summary, the book provides a unified view of the state-of-the-art of cognitive radio wireless communications and networking technology, which should be accessible to a readership with basic knowledge about wireless communications and telecommunications networking. The readership may find the rich set of references in each of the chapters very useful. The authors have performed a good job by providing a concise summary of all the chapters at the preface of the book. I would strongly recommend the book to graduate students and researchers and engineers working or intending to work in the area of cognitive radio networks and its connected environments. Although numerous journal/conference publications, tutorials, and books on cognitive radio have been published in the last few years, the vast majority of them focus on the various physical-layer attributes of the technology. More importantly, these technical publications discuss the cognitive radio in isolation, essentially as a stand-alone system or network, with little regard for how it may interact with legacy wireless systems or how heterogeneous cognitive radio systems may collaborate with each other. Although this book's main theme is efficient spectrum sharing in cognitive radio networks, its specific focus areas are quite different from the existing literature. The prime intend of this book is to provide a comprehensive discussion on how cognitive radio technologies can be employed to enable spectral efficient wireless communication system. In other words, the discussions in this book revolve around how cognitive radio technologies can be used to enable various wireless networks to coexist and efficiently share spectrum. The intended readership of this book includes wireless communications industry researchers and practitioners as well as researchers in academia. The readership is assumed to have background knowledge in wireless communications and networking, although they may have no in-depth knowledge of cognitive radio technologies. The intention of this book is to introduce communication generalists to the technical challenges of the various coexistence techniques and mechanisms as well as solution approaches which are enabled by cognitive radio networks with connected environments.

This book distinguishes itself from the existing prosperous literature of cognitive radio networks. The existing literature presents a self-contained introduction of the emerging cognitive radio networking paradigm outlining the theoretical fundamentals and requirements for enabling such a technology. The emphasis of such books is on the theoretical design, optimization, and performance evaluation of opportunistic spectrum access in cognitive radio networks. The main challenge of existing distributed opportunistic spectrum management schemes is that they do not consider the unavoidable practical limitations of today's cognitive radio networks such as the inability to measure the interference at the primary receivers. Consequently, optimizing the constrained cognitive radio network performance based only on the local interference measurements at the cognitive radio senders does not lead to truly optimal performance due to the existence of hidden or exposed primary senders. More specifically, the existing schemes have a cognitive radio sender decide its transmission strategy based on its local interference measurement – while such decisions should have been made based on the interference measurement at the nearby primary receivers to be interfered with its transmission. However, there does not exist a practical mechanism that enables a cognitive radio to determine the interference at nearby primary receivers. Furthermore, the existing transceiver technologies and spectrum measurement techniques are incapable of accurately assessing the spectrum usage over a wide frequency range due to the limitations imposed by the transceiver hardware.

This book targets a wide range of readers including but not limited to researchers, industry experts, and senior undergraduate as well as graduate students from academia. On the one hand, the readers with theoretical interests will experience an unprecedented treatment of the conventional cognitive radio network performance optimization problem that takes into account the practical limitations of recent technologies. Further, the readers interested in real-life distributed cognitive radio network realization will be exposed to a first-of-its-kind clean-slate implementation approach that demonstrates the significant multi-faceted performance improvement. This book offers the reader a range of interesting topics portraying the current state-of-the-art in cognitive radio technologies. In simple terms, while several existing opportunistic spectrum access approaches have been developed and theoretically optimized, they are challenged by the inherent constraints of practical implementation technologies. Analyzing these constraints and proposing an attractive and practical solution to counter these limitations are the basic aims of this book.

This book is an extension of the PhD thesis of Dr. Prabhat Thakur submitted to the Jaypee University of Information Technology, Solan, India, 2018, under the supervision of Prof. Ghanshyam Singh. The authors are indebted to numerous colleagues for the valuable suggestions during the entire period of manuscript preparation. The authors are especially thankful to the Professor B N Basu and Professor S K Kak, IIT (BHU), India, for motivation. We would also like to thank publishers at John Wiley & Sons, Inc., in particular Brett Kurzman, Victoria Bradshaw, and Sarah Lemore, for their helpful guidance and encouragement during the creation of this book. The authors would not justify their work without showing the gratitude to their family members who have always been the source of strength to tirelessly work to accomplish this assignment. We would like to acknowledge and thank several colleagues as well as Masters students and PhD scholars who have made this book possible. The first author would not justify his work without showing gratitude to his family members who have always been the source of strength for working tirelessly to accomplish the assignment. I owe my deepest gratitude toward my mother Shrimati Samundri Devi and father Shri Gian Chand Thakur for their continuous support and understanding of my goals and aspirations. I give my greatest gratitude to my parents for offering all-around support during the period of my studies and research. Their patience and sacrifices will remain my inspiration throughout my life. I am thankful to my brother Abhishek Thakur and grandmother Shrimati Banto Devi for loving me and not complaining for their share of time. Moreover, I am thankful to all the family members and relatives for loving me and encouraging at every stage of life. I am grateful to Prof. Ghanshyam Singh, University of Johannesburg, Johannesburg, South Africa, for his valuable guidance and encouragement. His vast experience and deep understanding of the subject proved to be an immense help to me. The 2nd author, Prof. Ghanshyam Singh, is also thankful to his wife, Swati Singh; daughter, Jhanvi; and son, Shivam, for sparing their time for this work.

Special Acknowledgements

We sincerely thank the authorities of University of Johannesburg, Johannesburg, South Africa, especially, Prof. Saurabh Sinha, Deputy Vice Chancellor: Research and Internationalization and Prof. Khmaies Ouahada, Head, Department of Electrical and Electronic Engineering Sciences for their kind support to come up with this book.

University of Johannesburg   Prabhat Thakur

Ghanshyam Singh

List of Acronyms

Acronym	Meaning
3GPP	Third generation partnership project
ACS	Adaptive channel sensing
ASAF	Antenna sub array formation
AS	Antenna selection
ACI	Adjacent channel interference
ADS	Automated deriving system
AV 2.0	Automated Vehicles 2.0
AV 3.0	Automated Vehicles 3.0
AWGN	Additive white Gaussian noise
BANs	Body area networks
BS	Base station
CCH	Common control channel
CSMA	Carrier sense multiple access
CA	Collision avoidance
CSI	Channel state information
CM	Cooperative spectrum monitoring
CNR	Channel-to-noise ratio
CoMP	Coordinated multipoint
CIoTs	Consumer IoTs
CoIoTs	Cognitive IoTs
CALM	Communication architecture for land mobile
CR	Cognitive radio
CRN	Cognitive radio network
CC	Cooperative communication
CCC	Common control channel
CCRN	Cooperative cognitive radio network
CRAHN	Cognitive radio ad-hoc networks
CRCN	Cognitive radio cellular networks

CWLAN	Cognitive wireless local area networks
CWMN	Cognitive wireless mesh networks
CRSN	Cognitive radio sensor networks
CCU	Centralized/controlling cognitive user
CDM	Code domain multiplexing
Conv	Conventional
Conv-Rand-Sel	Conventional random selection method
Conv-Pro-Sel	Conventional proper selection method
Conv-1st-F	Conventional spectrum access with first advanced frame structure
Conv-2nd-F	Conventional spectrum access with second improved frame structure
DFT	Discrete Fourier transform
DoS	Denial-of-service attack
DL	Deep learning
DVB	Digital-video broadcasting
DSRC	Dedicated short-range communication
DTIUM	Data transmission in underlay mode
EDGE	Enhanced Data rates for GSM Evolution
FCS	Fixed channel sensing
FMC-MAC	Flexible multi-channel coordination medium access control
FC	Fusion center
FBMC	Filter-bank-based multicarrier
GPRS	General Packet Radio Service
GFDM	Generalized frequency division multiplexing
HCRN	Homogeneous/Heterogeneous cognitive radio network
HTCRNs	High traffic cognitive radio networks
HC-MAC	Hardware-constrained medium access control
HMM	Hidden Markov model
HSA	Hybrid spectrum access
Hybrid-Conv-F	Hybrid spectrum access with conventional frame structure
Hybrid-1st-prop-F	Hybrid spectrum access with 1st proposed frame structure
Hybrid-2nd-prop-F	Hybrid spectrum access with 2nd proposed frame structure
HSPA	High-Speed Packet Access
ITU	International Telecommunication Union
I4.0	Industry 4.0
IV	In-vehicle communication
IIoTs	Industrial IoTs

ISO	International Standard Organization
LDS-CDMA	Low-density spreading – code division multiple access
LICSPA	Low-interference channel status prediction algorithm
LORA	LOss differentiation rate adaptation
MLP	Multilayer perceptron
MOON	M-out-of-N
MIMO	Multiple-input-multiple-output
MMC	Millimeter-wave communication
MA	Multiple access
MAC	Medium access control
M2H	Machine-to-human
MUSA	Multiuser shared access
MUSIC	Multiple signal classification
NCM	Non-cooperative spectrum monitoring
NP	Neyman–Pearson
NC	Number of channels
NN	Neural network
NHTSA	National Highway Traffic Safety Administration
NTIA	National Telecommunications and Information Administration
NOMA	Non-orthogonal multiple access
NFV	Network function virtualization
OMC-MAC	Opportunistic multi-channel medium access control
OFDMA	Orthogonal frequency division multiple access
OSI	Open system interconnection
PUEA	Primary user emulation attack
PO-MAC	Pre-emptive opportunistic medium access control
Prop-Pro-Sel	Proposed proper selection method
Prop-Rand-Sel	Proposed random selection method
PDM	Power domain multiplexing
PTDMA	Pattern division multiple access
PD-NOMA	Power domain-NOMA
PUDM	PU-first-decoding mode
SFDM	CU/SU-first-decoding mode
QoS	Quality-of-service
QoE	Quality of experience
QPSK	Quadrate phase-shift keying
REC	Receiver error count
RSMA	Resource shared multiple access
RFID	Radio-frequency identification

SUMO	Simulation of urban mobility
SM	Spectrum monitoring
SSA	Static spectrum access
SCMA	Sparse code multiple access
SC	Superposition coding
SIC	Successive interference cancellation
SNR	Signal-to-noise ratio
SJAS	Subset-based joint AS
SINR	Signal-to- interference-plus-noise ratio
SWIPT	Simultaneous wireless information and power transfer
SDWN	Software-defined wireless networking
STFT	Short-time Fourier transform
TCS	True channel states
TV	Television
TRAI	Telecommunication Regulatory Authority of India
UFMC	Universal filtered multi-carrier
UMTS	Ultra mobile telecommunication services
V2V	Vehicle-to-vehicle
V2B	Vehicle-to-broadband
V2I	Vehicle-to-infrastructure
V2P	Vehicle-to-pedestrians
VANET	Vehicular ad-hoc network
WRAN	Wireless regional area network
WBANs	Wireless body area networks
WSNs	Wireless sensors networks
WAVE	Wireless access for vehicular communication
WiMAX	Worldwide Interoperability for Microwave Access
Wi-Fi	Wireless Fidelity

List of Figures

List of Tables

List of Symbols

Symbol	Notation
B_D	Bandwidth of the downlink channel
B	Bandwidth of PU signal
$C(t)$	Received signal at CU receiver due to CU transmission
DL_c	Data loss in the conventional approach
DL_1	Data loss in the first proposed approach
DL_2	Data loss in the second proposed approach
DL	Data loss
EL	Energy loss
EST	Effective switching time
f_s	Sampling frequency
IE	Interference efficiency
IF	Interference to the PU communication due to CU transmission
IP	Interference power at PU
h_{ss}	Channel gain coefficient between CU transmitter and CU receiver
h_{ps}	Channel gain coefficient from PU transmitter to CU receiver
h_{sp}	Channel gain from CU transmitter to PU receiver
h_{cci}	Channel gain coefficient between the CUs' in a base station and ith CU in the network
h_{pci}	Channel gain coefficient between the PU-base station and ith CU in the network
h_{Di}	Channel gains from the base station to the ith users
$H_D i$	The channel matrix from BS to the ith CU in the downlink scenario
$H_D P$	The channel matrix from the PU transmitter to the CU-4 in the downlink scenario

H_Ui	The channel matrix from ith CU to BS in the uplink scenario
H_UP	The channel matrix from the CU-4 to PU transmitter in the downlink scenario
H_0	Hypothesis for the absence of PU
H_1	Hypothesis for the presence of PU
IPU_{CUi_NCM}	The interference power introduced at PU due to ith CU in MIMO-CR-NOMA framework.
IEU_{CU4_NCM}	The interference efficiency of the ith CU in MIMO-CR-NOMA framework
K	Path loss exponent
k_{avg}	Average number of packets lost due to monitoring error
k_{an}	Average number of packets lost in the network without considering the effect of traffic intensity
k_{TAEPU}	Total number of packets to be transmitted after the emergence of PU
k_{comp}	Complete data loss in the proposed network
N	Number of channels
N_0	Number of packets in the data transmission period
N_{CUi}	Number of antennas on the ith CU
N_{PUT}	Number of antennas on the PU transmitter
n_UBS_i	Noise power at the ith antenna of BS in uplink scenario
N_p	Noise power at CU receiver
N_Di	The noise vectors at the ith CU in the downlink scenario
N_UBS	The noise vectors at the BS in the uplink scenario
N_UPU	The noise vectors at the PU in the uplink scenario
N_{PPU}	Noise power at PU receiver
N_{PCU}	Noise power at CU receiver
N_{BS}	Number of antennas on the BS
NC	Number of channels
N_{0i}	N_{0i} is the noise power at user-1
N_{PPU}	Noise power at PU receiver
N_{PCU}	Noise power at CU receiver
N_{BS}	Number of antennas on the BS
P_d	Probability of detection
P_f	Probability of false alarm
P_1/P_s	CU transmission power with interweave mode
P_2	CU transmission power with underlay mode
P_DP_i	Power assigned for the ith antenna of PU transmitter in the downlink scenario

$P(H_0)$	Probability of the channel being idle
$P(H_1)$	Probability of the channel being active
P_{pe}	Probability of wrong prediction/probability of spectrum prediction error
P	Traffic intensity of PU
P_{sn}	Probability of success in the nth event
P_m	Probability of misdetection
P_{me}	Probability of spectrum monitoring error
PE	Packet energy
PT_s	Starting time of packet
PT_E	Ending time of packet
PC	Power consumption
P_{Pr}	Powers desired for the spectrum prediction process
P_S	Powers desired for the spectrum sensing process
P_U	Maximum power that can be transmitted from the base station in order to avoid the interference with the adjacent cells
P_{PU}	The power transmitted by the PU base station
P_{in}	Probability of interference
P_i	Power transmitted by the ith user
PW	Power wastage
P_P	Power required by one packet for its complete process such as transmission, channel passing, reception, etc.
Q_{me}	Probability of error after cooperation
$Q_D i$	Input covariance matrix of input vector $X_D i$
$Q_U i$	Input covariance matrix of input vector $X_U i$
RA	Achieved throughput
$RARDL$	Ratio of achieved throughput to data loss
$r(t)$	Received signal at CU receiver at time t
RW	Resource wastage
R_{DNi}	Throughput of the ith NOMA user in downlink scenario
R_{DOi}	Throughput of the ith OMA user in downlink scenario
R_{UNi}	Throughput of the ith OMA user in uplink scenario
R_{UOi}	Throughput of the ith OMA user in uplink scenario
R_{Hpi}	The number of nonzero singular values of the channel matrix $H_D i$
RD_{CUi_MCN}	Throughput achieved at the ith CU in the downlink scenario for the MIMO-CR-NOMA communication system

RD_{CUi_MC0}	The throughput of ith CU when the interference at CU-4 is zero
RU_{CUi_MCN}	The throughput of ith CU in the uplink scenario for the MIMO-CR-NOMA communication system
RU_{CUi_CN}	Throughput achieved at the ith CU in the uplink scenario for the CR-NOMA communication system
S_i	ith signal
SU	Spectrum utilization
S_{Nd}	Decoded signal for the Nth user
ST_{ssh}	Switching time as move from sensing to sharing interval for data transmission
ST_{tm}	Transmission mode switching time
ST_{ch}	Channel's switching time
$s(t)$	PUs' transmitted signal at time t
SNR_{s1}/SNR_s	Signal-to-noise ratio at CU receiver in interweave transmission mode
SNR_{s2}	Signal-to-noise ratio at CU receiver in underlay transmission mode
SNR_p	SNR at CU receiver due to the primary transmitted power
$SNRP_s$	SNR at the PU receiver due to CU transmission
$SNRP_p$	SNR at the PU receiver due to PU transmission
$SNRP_{ps}$	Total SNR at PU receiver due to both the PU and CU transmission
SRD_{CR-OMA}	Sum-throughput of the CR-OMA framework in the downlink scenario
$SRD_{CR-NOMA}$	Sum-throughput of the CR-NOMA framework in the downlink scenario
$SRD_{CR-MIMO}$	Sum-throughput of the CR-MIMO framework in the downlink scenario
$SRD_{MIMO-CR-NOMA}$	Sum-throughput of the MIMO-CR-NOMA framework in the downlink scenario
SRU_{CR-OMA}	Sum-throughput of the CR-OMA framework in the uplink scenario
$SRU_{CR-NOMA}$	Sum-throughput of the CR-NOMA framework in the uplink scenario
$SRU_{CR-MIMO}$	Sum-throughput of the CR-MIMO framework in the uplink scenario
$SRU_{MIMO-CR-NOMA}$	Sum-throughput of the MIMO-CR-NOMA framework in the uplink scenario

$SNR_D i$	Power assigned to the ith CU in terms of SNR per unit channel gain
τ_s/τ	Sensing time
τ_p/T_p	Prediction time
T_{sh}	Sharing period
T_{ct}	Contention period
T_{tr}	Data transmission period
T	Frame duration
TI	Throughput efficiency
TP	Throughput of CU
TP_c	Throughput of CU in the conventional approach
TP_1	Throughput of CU in the first proposed approach
TP_2	Throughput of CU in the second proposed approach
w_i	AWGN including intercell interference at the ith user receiver
$w(t)$	Noise signal at CU receiver at time t
$X_D i$	Input signal vectors assigned for the ith CU in the downlink scenario
$X_U i$	Input signal vectors at the ith CU in the uplink scenario
$Y_U i$	Output signal vector available at the ith CU in the downlink scenario
y_i	Received signal at the ith user
y_U	Complete received signal at the base station
$Y_D i$	Output signal vector available at CU-1 in the downlink scenario
$Y_U BS$	The output vectors at the BS
$Y_U PU$	The output vectors at the PU
Γ	Signal-to-noise ratio of CU signal at the PU
Λ	Parameter of Poison distribution
μ	Parameter of Binomial distribution
$\sigma_D 1_i$	Parallel channel gain coefficients from the ith antenna of the BS to the ith antenna of CU-1
$\sigma_D 4_i$	Parallel channel gain coefficients from the ith antenna of the BS to the ith antenna of CU-4
$\sigma_D P_i$	Parallel channel gain coefficients from the ith antenna of PU to the ith antenna of CU-4
$\sigma_U 1_i,$	Parallel channel gain coefficients from the ith antenna of the CU-1 to ith antenna of BS
$\sigma_U 4_i$	Parallel channel gain coefficients from the ith antenna of the CU-4 to ith antenna of BS

1

Introduction

1.1 Introduction

The connected environment is a reality which we are going to achieve in the near future and internet-of-things (IoTs) have contributed to rapid progress in this journey. The IoTs have played a key role to establish the meaning of word "Connected" because the idea of enabling the things/devices with the Internet eases the process of communication with other things/devices in the nearby or far places. Parallelly, we are progressing immensely in the wireless communication generations where we are in the implementation stage of fifth generation and research is in progress for sixth or next generation (6G/XG). This progress in the wireless communication generations leads to improved capacity, reliability, and latency that motivate the end users to use the mobile phones as an integral part of their daily life. The rapid progress in the direction of connected environments and wireless communication generations has increased the number of wireless connected devices exponentially in recent years and as per the various survey articles, in the upcoming years, this increase will be surprising. As per Martech Advisor, the average number of wireless connected devices to each person will be 15. M. Hatton [1] has revealed that the number of IoT connected devices will increase to 24.1 billion in 2030 which was 7.6 billion in 2019 and will increase the revenue to USD 1.5 trillion which will be almost three times when compared with USD 465 billion in 2019 [2]. As per the Tankovska, the number of wireless connected devices in the 2018 was 22 billion, and are predicted to be 38.6 and 50 billion in 2025 and 2030, respectively [3]. This explosive increase in the number of wireless connected devices and network service requirement for these devices poses potential challenges to the future wireless communication systems in which the most prominent confront is the scarcity of radio resources/spectrum. A report from the Federal Communications Commission (FCC) [4] reveals a fact that the emerging spectrum scarcity comes from

Spectrum Sharing in Cognitive Radio Networks: Towards Highly Connected Environments,
First Edition. Prabhat Thakur and Ghanshyam Singh.
© 2021 John Wiley & Sons, Inc. Published 2021 by John Wiley & Sons, Inc.

the inefficient spectrum usage due to the fixed spectrum allocation scheme rather than the real spectrum scarcity. The dynamic spectrum access (DSA) [5–15] is proposed as a promising approach to overcome the aforementioned issues. The prominent framework which enables the transformation of DSA approach into physical implementation is the cognitive radio (CR). The CR allows cognition in wireless networks and enables the concerned wireless communication devices to adapt to real-time wireless environments. The CR can enhance the spectrum utilization if the primary network providers/licensed users/primary users (PUs) allow the unlicensed users/secondary users/cognitive users (CUs) to access the licensed spectrum, provided that the PUs are protected from the interference of CUs [5]. Further, the recent advancement about the connected environments, generations of wireless communication, and third generation partnership project (3GPP) has been briefed as follows.

1.1.1 Connected Environments

It is worth mentioning that the connected world comprises the connected environments where we live and are connected with other people/things – socially, politically, economically, and environmentally from the small scale to the extended large scale as shown in Figure 1.1a and b. The information and communication is a key component of the digitization and connecting various things and persons across the world. The body area network (BAN) is a small connected environment over the human body which collects the data from human body and either human takes the responsive step or report data to the doctor at remote place via different communication technologies such as third/fourth/fifth/sixth generation (3G/4G/5G/6G), via wireless fidelity (WiFi), or Bluetooth relies on the distance. Further, humans live in home and continuously try to ease their life where connected home with the help of IoT devices enables them to operate the home appliances remotely either using hand gestures or voice processing. In a smart home, most of the appliances such as lighting bulbs, fans, air-conditioners, refrigerators, televisions, locking systems, etc., are controlled remotely and even from thousands of kilometers because of IoTs and this eases the human life significantly [16–21]. There are a number of connected homes in a city and other connected environments such as smart parking systems, smart hospitals, smart drainage systems, smart offices, smart monitoring systems, etc., which lead to formulate a smart city or connected city environments. The popular communication technology for connected vehicles is the dedicated short-range communication (DSRC). As per various proposed definitions, "A smart city is a city that integrates its systems – from local labour markets to financial markets, from local government to education, healthcare, transportation and utilities – to drive efficiency

(a)

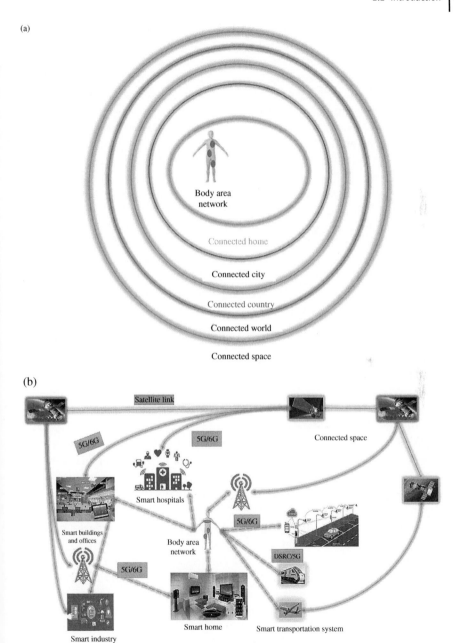

(b)

Figure 1.1 (a) The evolution of the connected environments from human body to the space. (b) The scenario of the connected environments.

for the benefit of its inhabitants." A smarter city is a more collaborative, connected, and responsive city. It enables greater data innovation to meet citizens' needs [22, 23]. Further, multiple connected cities form the connected district and then to connected states, and connected counties, and then connected world. The next step after connected world is connected universe where initially we need to explore the connected space and some efforts are in progress in this direction with microscale satellite devices [24, 25]. Some recent works and literature in the direction of connected environments are presented as follows.

Hofman et al. [26] have illustrated the 5G NetMobil project which is a progressive step toward the implementation of connected vehicles. The authors have presented an analysis of future automotive applications and their needs in the 5G NetMobil. Further, the authors have provided a description of several technical solutions developed within the project to respond to these arising communications demands in terms of reliability and latency. Y. Nakamoto [27] has highlighted some technological issues in the connected world that is assumed to consist of embedded systems and connects the physical and cloud systems by using internet services. The author has emphasized over two areas which are as follows: (i) implementation of connected world from the embedded system viewpoint where the research directions and challenges of security are discussed and (ii) the exploration of a design method for integrating functional safety with security in a vehicle which comprises a behavioral and development process model. Chen et al. [28] have presented the inception of 5G till 2020 and further plans for next generation of wireless communication, that is, 6G till 2030. The authors have perceived that when internet-of-everything (IoE) is an important constituent of 5G, the internet-of-intelligence (IoI) with connected people, connected things, connected intelligence in order to reduce the human intervention in most of the activities, and self-decision-making will play a vital role in the 6G. It is also revealed that 6G is not an extended version of 5G in terms of improved data rate; however, completely a new generation with emphasis on the emerging applications, new design philosophy and hardware capabilities toward the higher spectrum, artificial intelligence, sensing, privacy, and security perspectives. Further, the higher capacity and efficiency will enable the collaboration among machines, robots, vehicles, and other devices to achieve the fully connected environment. The authors in [29] illustrated the 5G radio access technology (RAT) comprising the data rate evolution from 3G to LTE and 5G. The potential discussed topics are radio access technologies, waveform designs, massive multiple-input-multiple-output (MIMO), channel coding, etc. Further, the detailed physical layer channel structures with necessary features and functionalities are described and the complete radio protocol design of the 5G RAT is also presented.

1.1.2 Evolution of Wireless Communication

The wireless communication has revolutionized the communication perspective and applications in the society. The first generation (1G) of wireless communication was introduced in 1980 with prime technology advanced mobile phone systems (AMPS), and progressed periodically with global system for mobile (GSM) communication; that is, second generation (2G) in 1992, third generation (3G) in 2000, fourth generation (4G) with prime technologies of long-term evolution (LTE) and LTE-Advanced in the 2010, and fifth generation in 2020. The evolution of aforementioned generations with prime modulation, accessing techniques, data rate, bandwidth, and data applications is depicted in Figure 1.2. The previously undefined acronyms in Figure 1.2 are defined in Table 1.1.

The Internet is an integral and potential constituent to provide the always and reliable wireless connectivity where the existing and upcoming wireless communication generations have a significant role. Today, the research of the fifth generation of wireless communication is completed and we are in the implementation stage. Further, the communication professionals and research community are looking for the sixth generation (6G) and beyond. Some potential research efforts toward the next generation communication technologies are as follows. Rapaport et al. [30] have presented a review where they explored various opportunities and challenges for 6G and beyond where key emphasis is on above 100 GHz. The authors have emphasized on the recent regulatory and standard body rulings that are anticipating wireless products and services above 100 GHz as well as illustrate the viability of wireless cognition, hyper-accurate position location, sensing, and imaging. Further, the methods in order to reduce the computational complexity and simplify the signal processing used in adaptive antenna arrays by exploiting the special theory of relativity to create a cone of silence in over-sampled antenna arrays that improve performance for digital phased array antennas are illustrated. Akyildiz et al. [31] have presented the potentials of the next generation wireless communication such as 6G and beyond that will fulfill the requirements of a fully connected world and provide ubiquitous wireless connectivity for all. Transformative solutions are expected to drive the surge for accommodating a rapidly growing number of intelligent devices and services. Major enabling technologies to achieve connectivity goals within 6G networks include: (i) a network operating at the THz band with much wider spectrum resources, (ii) intelligent communication environments that enable a wireless propagation environment with active signal transmission and reception, (iii) pervasive artificial intelligence, (iv) large-scale network automation, (v) an all-spectrum reconfigurable front-end for DSA, (vi) ambient backscatter communications for energy savings, (vii) the internet of space things enabled by Cube-Sats and UAVs, and (viii) cell-free massive MIMO communication networks. The authors have explored the various devices in

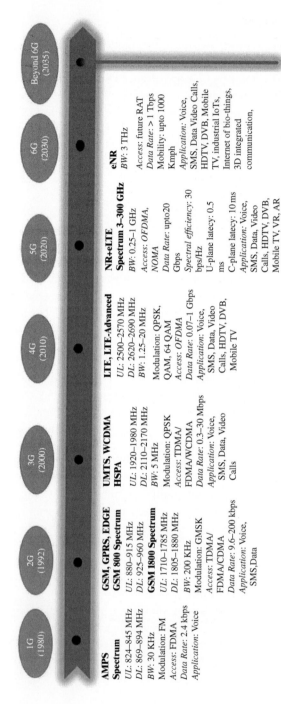

1G (1980)

AMPS
Spectrum
UL: 824–845 MHz
DL: 869–894 MHz
BW: 30 KHz
Modulation: FM
Access: FDMA
Data Rate: 2.4 kbps
Application: Voice

2G (1992)

GSM, GPRS, EDGE
GSM 800 Spectrum
UL: 880–915 MHz
DL: 925–960 MHz
GSM 1800 Spectrum
UL: 1710–1785 MHz
DL: 1805–1880 MHz
BW: 200 KHz
Modulation: GMSK
Access: TDMA/
FDMA/CDMA
Data Rate: 9.6–200 kbps
Application: Voice,
SMS,Data

3G (2000)

UMTS, WCDMA
HSPA
UL: 1920–1980 MHz
*DL:*2110–2170 MHz
BW: 5 MHz
Modulation: QPSK
Access: TDMA/
FDMA/WCDMA
Data Rate: 0.3–30 Mbps
Application: Voice,
SMS, Data, Video
Calls

4G (2010)

LTE, LTE-Advanced
UL: 2500–2570 MHz
DL: 2620–2690 MHz
BW: 1.25–20 MHz
Modulation: QPSK,
QAM, 64 QAM
Access: OFDMA
Data Rate: 0.07–1 Gbps
Application: Voice,
SMS, Data, Video
Calls, HDTV, DVB,
Mobile TV

5G (2020)

NR+eLTE
Spectrum 3–300 GHz
BW: 0.25–1 GHz
Access: OFDMA,
NOMA
Data Rate: upto20
Gbps
Spectral efficiency: 30
bps/Hz
U-plane latecy: 0.5
ms
C-plane latecy: 10 ms
Application: Voice,
SMS, Data, Video
Calls, HDTV, DVB,
Mobile TV, VR, AR
IoTS, Wearable
networks, Smart
home, telemedicine

6G (2030)

eNR
BW: 3 THz
Access: future RAT
Data Rate: > 1 Tbps
Mobility: upto 1000
Kmph
Application: Voice,
SMS, Data Video Calls,
HDTV, DVB, Mobile
TV, industrial IoTs,
Internet of bio-things,
3D integrated
communication,

Beyond 6G (2035)

Application: Voice, SMS, Data,
Video Calls, HDTV, DVB,
Mobile TV, industrial IoTs,
Internet of bio-nano things, 3D
integrated communication,
Satellite communication,
Space communication, IoNT,
IoBNT, Quantum
Communication

Figure 1.2 The evolution of wireless communication generations.

Table 1.1   List of acronyms in Figure 1.2.

AMPS	Advanced mobile phone systems
AR	Augmented reality
BW	Bandwidth
DL	Downlink
DVB	Digital video broadcasting
eLTE	enhanced LTE
EDGE	Enhanced data rate for GSM evolution
FM	Frequency modulation
FDMA	Frequency division multiple access
GSM	Global system for mobile communication
GPRS	Global packet radio services
GMSK	Gaussian minimum shift keying
HSPA	High-speed packet access
HDTV	High-definition television
IoBnTs	Internet-of-bio-nano-things
IoBTs	Internet-of-bio-things
LTE	Long-term evolution
LTE-Advanced	Long-term evolution-advanced
NR	New radio
NOMA	Non-orthogonal multiple access
OFDMA	Orthogonal FDMA
TDMA	Time division multiple access
UL	Uplink
UMTS	Ultra-mobile telecommunication services
QPSK	Quadrature phase shift keying
QAM	Quadrature amplitude modulation
WCDMA	Wideband CDMA

addition to the physical and MAC layer perspective and challenges for terahertz (THz) communication systems. Further, the authors have presented the fundamentals of the intelligent communication environments in addition to the functionalities of intelligent environments. Moreover, a layered architecture that comprises the metamaterial plane, sensing and actuation plane, computing plane, and computing plane and communication plane of the intelligent communication environment is presented. Chowdhury et al. [32] have presented and illustrated a potential vision for the next/sixth generation (XG/6G) of wireless communication as well as its network architecture. The key emerging technologies, such as artificial intelligence, terahertz communications, wireless optical technology, free-space optical network, block-chain, three-dimensional networking, quantum communications (QC), unmanned aerial vehicles (UAVs), cell-free communications, integration of wireless information and energy transfer, integrated sensing and communication, integrated access-backhaul networks, dynamic network slicing, holographic beamforming, backscatter communication, intelligent reflecting surface, proactive caching, and big data analytics, will play a significant role in order to achieve the standards suggested or defined for the 6G. The enabling techniques for the above-mentioned technologies and their implementations are discussed in detail. Further, the expected constituents of the potential technologies, such as internet-of-nano-things (IoNTs), the internet-of-bio-things (IoBTs), and QC, are explored in detail. Zhu et al. [33] have presented the vision for 6G and beyond by exploiting the currently emerging techniques that are artificial intelligence (AI), Big Data (BD), and cloud computing (CC), which are further recognized as ABC and their role in 6G wireless communication is compared with the role of vitamin ABC in the human body. Moreover, the authors have developed a learning approach for wireless artificial intelligence defined as knowledge plus data-driven deep learning (KD-DL) method. The proposed KD-DL structure is an open controllable AI model which exploits the resolution ability and predictive ability of wireless BD (WBD), which is described with the help of an example of resource optimization of the ultradense networks. In addition to this, a layered framework of mobile networks using cloud/edge/terminal computing is proposed and their achievable efficiency is also illustrated. Ziegler and Yrjola [34] have enlightened on the 6G indicators with their significance and performance where the potential attributes of the 6G wireless communication systems such as Terahertz spectrum-based radio access, network as sensor, edge centric and flow-based architecture, are described. Further, the authors have presented a heat-map of key performance attributes and value impact categories of growth, efficiency, and sustainability. Moreover, briefs about the key inferences for business model transformation are given and suggested the building of the next generation business models on the basis of upcoming business opportunity, scalability, replicability, and sustainability. It is believed that as per the previous 3GPP releases, every 10 years witnessed a new generation of wireless communication and implementation of that generation relies on the ability of the researchers to address the technical problems and defined the standards for the upcoming generation. Brito et al. [35] have addressed the upcoming challenges for standardization of the next generation wireless communication systems in Brazil. The authors have presented the 6G Brazil project whose key aim is to structure the research efforts and to influence the standardization process by emphasizing over the Brazilian society and economy. Elmeadaw and Shubair [36] have presented recent works to fulfill some of the potential demands of the next generation communication systems such as data rate, capacity, latency, reliability, resources sharing, energy/bit, etc. In addition to this, the authors have described the opportunities and research challenges for the next generation wireless communication systems. Yang et al. [37] have revealed that the 6G will demand very high spectral and energy-efficient wireless communication designs and by keeping this in mind, the authors have outlined various technical challenges as well as their probable solutions on the physical and network layer design perspectives. In addition to this, the authors have explored various security concerns and progress in the test bed developments. Alsharif et al. [38] have illustrated various perspectives of 6G such as vision and key features, challenges and potential solutions, and research activities and the authors tried to highlight the new horizons for future research directions. The areas where authors have emphasized are the energy efficiency, intelligence, spectral efficiency, security, secrecy, privacy, affordability, and customization. Further, the authors have presented challenges and potential solutions for future 6G. Zhang et al. [39] have revealed that the interesting and key point that eases the customer life is 1000 times reduction in the price from consumer perspective. The authors have classified the current candidate technologies in a well-organized manner as well selected the AI-assisted intelligent communication. Blue et al. [40] have explored the potentials and challenges of the spectrum management in the next generation wireless communication system because of very high bandwidth required for one particular channel while at the same time, the number of wireless connected devices will increase tremendously. The authors have perceived that the spectrum management in the XG will be more complex because of the variety of spectrum bands and spectrum sharing models. In [40], the authors have overviewed the spectrum management and decision perspectives in 5G and have explored the role of spectrum sharing in the 6G and other enabling techniques. Zhang et al. [41] have presented a vision for the 6G where the foundation technologies are inherited from the 5G. The authors have presented the state-of-the-art about the 5G potential technologies such as mobile broadband, IoTs, and AI as well as specify the necessity to explore the discussed techniques for 6G. The three major techniques considered for 6G are mobile ultra-broadband, Super IoTs, and AIs where it is revealed that THz is the prominent technology that can serve as a milestone to achieve ultra-broadband. Conversely, the symbiotic radio and satellite IoTs are the concepts to transform the IoTs into Super IoTs; however, the deep learning and reinforcement learning are the potentials of AI. Further, the authors have illustrated the fundamental principle of each technology, key challenges, and state-of-the-art as well as potential solutions.

1.1.3 Third Generation Partnership Project

The 3GPP is a group of multiple telecommunication bodies such as Association of Radio Industries and Businesses (ARIB), Automatic Terminal Information Services (ATIS), China Communications Standards Association (CCSA), European Telecommunications Standards Institute (ETSI), Telecommunications Standards Development Society, India (TSDSI), Telecommunications Technology Association (TTA), etc., which is responsible to target and define the standard for telecommunication technologies, radio access techniques, core network, and service availability which covers the complete system mobile configuration. The working wings of the 3GPP are as follows: (i) Radio access networks (RAN), (ii) Services and Systems Aspects, and (iii) Core networks and Terminals.

In 1998, the 3GPP was formulated in order to define the standards for third generation of wireless communication with key emphasis on the universal terrestrial radio access (UTRA) and frequency division duplexing (FDD). Further, this group proposes the standards and norms for upcoming advances in the wireless communication and now has become a well-recognized group for standardizing the various specifications for particular release. Till now, the 16 release is defined and Release 17 is under progress. The highlights of prospective releases of 3GPP are described in Figure 1.3. It is worth mentioning that the dynamic spectrum sharing (DSS) and enhanced DSS (eDSS) are the prime focus areas in the release 16 and 17 because of the nature of efficient exploitation of the spectrum. Further, the fundamentals of CR technology are presented in Section 1.2.

1.2 Cognitive Radio Technology

The CR is an advanced version of the reconfigurable device which supports multiple technologies and is known as software defined radio (SDR) coined by Mitola in 1991 [42]. However, the prominent limitation of SDR is the absence of intelligent nature whereas the CR is an intelligent device which means it senses its electromagnetic environment and dynamically/autonomously adjusts its radio operating parameters to adapt system operation such as mitigate interference, assist interoperability, and access secondary markets [5]. The general framework of CR communication system is classified mainly into four classes namely, spectrum sensing, spectrum analysis and decision, spectrum access, and spectrum mobility as shown in Figure 1.4. The spectrum sensing is the basic step of CR communication systems in which the cognitive user (CU) has to sense the spectrum status and the activities of PU periodically, in order to perceive the idle channels. The spectrum sensing approach is broadly of two types such as the centralized and distributed approaches. There is a fusion center in the centralized spectrum

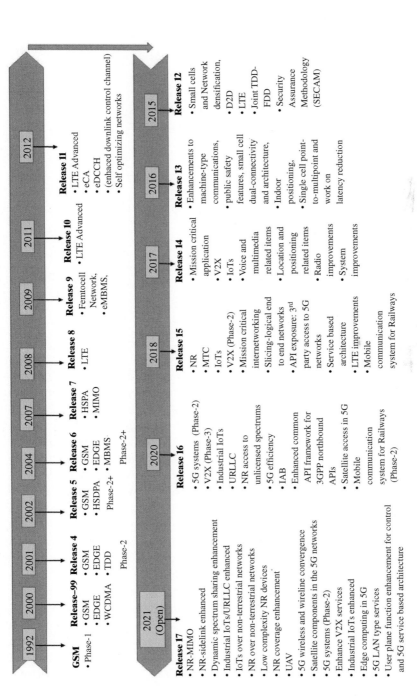

Figure 1.3 The evolution of the 3GPP releases for wireless communication.

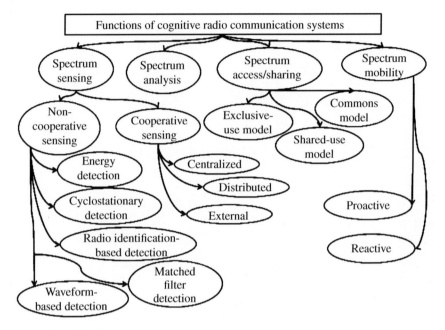

Figure 1.4 The general framework of cognitive radio communication systems.

sensing of the CRN where all the CUs in a network sense the status of spectrum individually and then transmit their sensed information to the fusion center. On the other hand, in the distributed spectrum sensing, all the CUs sense the information and exchange within themselves to take the decision. There are various spectrum sensing techniques which are further categorized as: noncooperative spectrum sensing, cooperative spectrum sensing, and interference-based spectrum sensing as shown in Figure 1.4 and well explored by Yucek and Arslan [43–54]. The next step after spectrum sensing is the spectrum analysis in which the sensed information is analyzed to yield the information about the spectrum holes and then the decision is taken by optimizing the system performance metrics [55]. The third step of the CR system is the spectrum access; after taking the decision on the spectrum holes, now CU accesses the spectrum holes to establish the communication [56–62]. However, in order to avoid the collision between the licensed and unlicensed user, the accessing of spectrum should be based on the medium access control (MAC) layer protocols and finally the CR system performs the function of spectrum mobility, in which it emphasizes on the handoff of the operating frequency in the CR system [63–65]. For the spectrum mobility, the CU must have the ability to detect the emergence of PU during the CUs' data transmission and two significant techniques for this are the spectrum prediction and spectrum monitoring. The spectrum prediction is a technique which allows the CU to fore-cast the future states (idle or active) of the channel on the basis of the historical information of channel states and at a particular time the emergence of PU can be predicted [66]. On the other hand, the spectrum monitoring is a technique which allows the CU to detect the emergence of PU during the CUs' data trans-mission by exploiting the received signal characteristics at the CU such as received error count (REC) [67].

The detailed description of spectrum sensing, spectrum analysis, and spectrum mobility techniques is reported in [55, 68–73], respectively. Moreover, the spec-trum prediction and spectrum monitoring techniques are discussed in detail in the later chapters. In the next section, we emphasize on the spectrum sharing/accessing techniques.

1.2.1 Spectrum Accessing/Sharing Techniques

The spectrum accessing/sharing is the third step of CR communication system after the spectrum sensing and spectrum analysis which is categorized into three types as discussed in [74] and shown in Figure 1.4. In the exclusive use model, the licensed user allocates the spectrum to the CU. The spectrum owner imposes some constraints and according to these constraints, the CU optimizes its parameters such as power, frequency/bandwidth to achieve the best performance. There are two types of exclusive use model named as (i) long-term exclusive use model and (ii) dynamic exclusive model. In the long-term exclusive use model, the licensed user allocates the spectrum to the CU for certain period of time, whereas in the dynamic exclusive model, the small chunks of spectrums are allocated to CU for a short duration of time.

Further, the spectrum common model is another kind of spectrum accessing technique in which all the CUs have same right to access the radio spectrum. It is also classified in the following major module, namely, uncontrolled, managed, and private commons sub-models. In the uncontrolled model, there is no owner of the spectrum such as Industrial Scientific Medical (ISM) (2.4 GHz) and U-NII (5 GHz) band which is being used by the entire CU. In addition to this, there is no control on the transmitted power of CU. In the managed commons sub-model, the radio spectrum is controlled jointly by a group of CUs; however, in the private commons sub-model, the spectrum owner specifies the technology and protocols to the CU. The spectrum owner provides a command to the CU and that command may contain the transmission parameters (e.g. time, frequency band, and transmit power). In the shared use model, the PU and CU share the spectrum opportunis-tically or simultaneously. However, the spectrum sharing models are classified on the basis of architecture, spectrum allocation behavior, spectrum access techni-ques, and scope as shown in Figure 1.2 [74].

Moreover, the spectrum sharing techniques on the basis of the architecture of network are classified as centralized spectrum sharing and distributed spectrum sharing. In the centralized approach, the spectrum allocation and access are controlled by the central entity/fusion center. The spectrum is allocated to all the CRs in the network via the central node which has the complete information about all the nodes in the network [75–78]. On the other hand, in the distributed approach, the spectrum allocation decision is taken by the individual node in the network [79–82]. However, the information about the spectrum is exchanged between the nodes in order to avoid the collision between the different CUs [83]. In addition to this, the spectrum sharing technique on the basis of spectrum allocation behavior is also categorized as cooperative and noncooperative spectrum sharing [84–91]. In the co-operative spectrum sharing, the interference at different CUs is noticed and the information of interference is exchanged between all the CUs and then the decision of spectrum allocation is taken [85, 87, 88, 92–94]. However, in the noncooperative spectrum sharing, the interference on the particular CU is measured and the message is sent to all the CUs. The spectrum sharing techniques can also be classified on the basis of scope like the spectrum sharing inside the CRN (intra-network) and the spectrum sharing within two or more networks (inter-network) as discussed in [55, 95]. In the intra-network spectrum sharing technique, the CUs share the spectrum within the network and try to access the spectrum without causing interference to the PU. On the other hand, in the inter-network technique, the CUs need to share the spectrum with the CUs of the other network. This technique is useful in the scenarios where multiple CRNs need to establish on the same location and same spectrum. The spectrum sharing techniques are also categorized on the basis of accessing of the spectrum which is described in Figure 1.5 [96] and tabulated in Table 1.2.

1.2.1.1 Interweave Spectrum Access

It is an opportunistic spectrum accessing (OSA) technique in which the CU is not allowed to share the spectrum until and unless it is used by the PU. This accessing technique can be applied in the temporal or spatial domain [97]. In the temporal domain, the CU has to exploit the temporal opportunities to access the spectrum from the burst traffic of PU. The spectral accessing technique in a temporal domain requires the joint design of the spectrum sensing and spectrum accessing [98]. However, in the spatial domain, the CU has to exploit the frequency band which is unused by the PU in a particular geographical region. The data rates achieved with the interweave spectrum accessing techniques are significantly high as compared to the underlay technique; however, this spectrum accessing technique has following limitations such as: (i) CU has to wait until the PU is using the spectrum, to get hold of spectrum holes and (ii) CU has to stop the communication if the spectrum is needed by the PU.

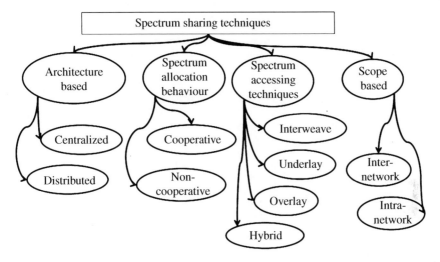

Figure 1.5 Classification of spectrum sharing technique in the CR communication system. *Source*: Based on Sharma et al. [96].

Table 1.2 Spectrum sharing techniques based on the spectrum accessing approaches (SAAs).

SAA metrics	Interweave spectrum access	Underlay spectrum access	Overlay spectrum access	Hybrid spectrum access
Accessing technique	Opportunistic spectrum accessing (OSA) technique	Spectrum sharing technique	Interference avoidance technique using advanced coding and transmission strategies at the CU	Has features of both interweave and underlay approaches
Time division (TD) requirement	CU uses the spectrum when PU is idle. (TD is required)	Both the PU and CU use the spectrum simultaneously. (TD is not required)	Both the PU and CU use the spectrum simultaneously. (TD is not required)	Can use the spectrum opportunistically or simultaneously. (TD is required)

(Continued)

Table 1.2 (Continued)

SAA metrics	Interweave spectrum access	Underlay spectrum access	Overlay spectrum access	Hybrid spectrum access
Domain	Can exploit the spectrum in temporal domain and/or spatial domain	Can access the spectrum at the same time, at same frequency, and at same location under some interference power constraints	Interference of the CU transmitter to the PU receiver is compensated by using a part of the CUs' power to relay the primary message, i.e. time, space, and frequency domain	Overcome the limitations of both the approaches, i.e. interweave and underlay
Data rate	Data rates are significantly high as compared to underlay technique	Data rates are significantly low as compared to interweave technique due to power constraints	Complex nature due to high cognition demand as compared to other techniques	More complex nature as compared to interweave and underlay approach; however, with reference to overlay technique, it is less complex
Pros and/or cons Critical performance metrics	CU has to wait until the PU is using the spectrum and has to stop the communication if PU needs spectrum	CU has no need to stop the communication in the presence of PU	CU has no need to stop the communication in the presence of PU	CU has no need to stop the communication in the presence of PU as it reduces its power levels
Critical performance metrics	Probability of false-alarm and probability of misdetection, interference efficiency, throughput, data-loss, energy efficiency	Throughput, transmission power level of CU, interference temperature at PU	Complexity of approach, maximum power that can be transmitted by PU as well as CU, throughput	Probability of false-alarm and probability of misdetection, interference efficiency, throughput, data-loss, energy efficiency, interference temperature at PU

1.2.1.2 Underlay Spectrum Access

"According to this technique, both the PU and CU can access the spectrum at the same time, at the same frequency, and at the same location under some interference power constraints. However, the main focus is on the power control of CU to avoid the interference to PU. The interference threshold [99] value at the primary receiver is modeled by the interference temperature, which is defined by Federal Communication Commission (FCC) special policy task force as discussed in [100]. The main advantage of the underlay spectrum accessing over the interweave spectrum accessing technique is that the CUs are able to communicate without waiting for the spectrum hole/white space. However, the limitation of underlay spectrum accessing is the low data rates due to the limited power allocation. Therefore, the optimal power allocation to CU should be in such a way that maximizes the capacity and throughput of the cognitive link and avoid the interference to the PU. In addition to this, the optimal power allocation to CU is a crucial issue.

1.2.1.3 Overlay Spectrum Access

It is an interference avoidance technique to share the spectrum with the help of advanced coding and transmission strategies at the secondary transmitters. In this technique, the CU can transmit simultaneously with the PU. The interference of secondary transmitter to the primary receiver is compensated by using a part of the CUs' power to relay the PUs' message as discussed in [101]. However, the CU should be aware of the PUs' channel gains and codebooks. Various types of pre-coding techniques enable the overlay networks to exist such as "dirty paper coding" [102] and "Gelfand-Pinsker binning" [103]. Even though this approach appears very effective, however, due to high cognition demand between the PU and CUs, it becomes difficult to implement this technique. However, the possible solution to achieve waveform characteristic is to use the estimation theory. The recent overlay spectrum access technique is proposed which is based on the cooperation among CU and PU. The interference between the PUs and the CUs is be offset by relying on some of the CUs to act as relay node [104, 105].

1.2.1.4 Hybrid Spectrum Access

"The spectrum underlay approach disables the CU to transmit the signal with maximum power. Due to unawareness of CU from PUs' activities, the CU is unable to use the idle band efficiently. However, the spectrum interweave approach restricts the CU to wait until the PU is present which is not suitable for real-time applications. To overcome the limitations of both the approaches, the hybrid approach is an emerging approach as reported in the literature [106–109]. Therefore, in the hybrid approach, the CU has the freedom to access the channel with maximum power in the absence of PU and with constrained power in the presence of PU. Sharma et al. [106] have proposed a hybrid cognitive transceiver architecture and the performance of proposed approach has been evaluated in terms of the achievable throughput considering the periodic sensing and simultaneous sensing/transmission schemes. Furthermore, the proposed hybrid scheme achieves significantly higher throughput than that of the conventional spectrum sharing approaches. Jiang et al. [107] have proposed a full–partial access strategy for DSA. To identify the location of PU, a double threshold energy detection method is proposed. To analyze the proposed strategy, a Markov chain model is developed and it is reported that the performance (in terms of interfering probability) of the hybrid system is improved by 46.86% over the interweave systems.

In [108], the authors have studied a hybrid interweave-underlay spectrum access system with relaying transmission to improve the system performance. In addition to this, the authors have accessed the system performance in terms of outage probability, symbol error rate (SER), and outage capacity over the Nakagami-m fading environment by deriving the respective analytical expressions and demonstrated the hybrid approach outperforms over the conventional underlay scheme. A comparison of interweave and underlay CU transmission is presented in [110] and it is depicted that due to the overhearing nature of cognitive transmitters, i.e. learning phase, the performance of interweave approach degrades. In addition to this, the use of interference decoding methods in the underlay system can approach the performance of the interweave system. Thakur et al. [111, 112] have exploited the advanced frame structures in hybrid spectrum access technique in order to improve the throughput and data loss of the CRN. From the above discussion, it is revealed that the interweave, underlay, and overlay are the fundamental approaches of spectrum accessing whereas the hybrid approach is considered as a combination of interweave and underlay approaches. Therefore, we have illustrated a flow diagram of these accessing techniques in Figure 1.6a and b. Figure 1.6a contains the fundamental approaches whereas in Figure 1.6b, we have considered the hybrid approach. As the CU has to access the spectrum, it has choices of interweave, underlay, overlay, or hybrid accessing approaches. However, the CU chooses the approach according to its requirements and applications. If the CR user is interested to continue by using the interweave approach, it should have sensing ability, which confirms the presence or absence of the PU in the environment and if the PU is absent, i.e. spectrum hole is present, then spectrum analysis needs to be performed to take out the best available spectrum. Furthermore, the decisive spectrum needs to access and communication is established on that spectrum. The process of sensing by one unit continues and as PU resumes its transmission, the CU needs to switch the communication on the new spectrum and this process is called as spectrum mobility.

Furthermore, if the CU is interested to continue to implement the overlay approach, the CU needs to use the advanced coding and transmission strategies, PUs' channel gain and codebooks, etc., and the communication is established on

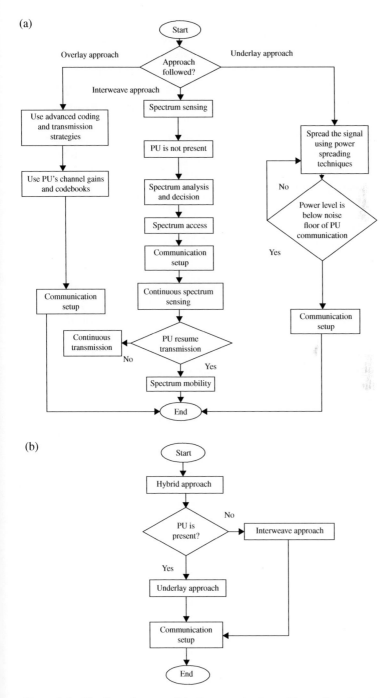

Figure 1.6 The flow diagram of the framework of cognitive radio network using (a) different spectrum accessing techniques and (b) hybrid spectrum accessing technique.

the same power, same spectrum, same time, and same space without interfering the PUs' communication. If CU follows the underlay approach, then there is no need to sense the environment and the information is spread using the power spreading techniques where the power level of the signal after spreading is checked; if it is below the threshold (defined by interference temperature), then communication is set up in the presence of PU, otherwise the signal need to spread again.

The flow of hybrid approach is shown in Figure 1.6b where the sensing plays an important role, i.e. if PU is present, CU follows the underlay approach, otherwise it goes with interweave approach. However, the communication is switched between the underlay and interweave approach.

1.3 Implementation of CR Networks

Various reported literatures emphasizing on the practical CRNs are presented in this section; however, some of the important articles concerned to the power allocation and imperfect channel state information (CSI) are discussed as follows. In [113], the authors have presented a technical review on the implementation issues of the practical CRNs. The prime focused areas of discussion are licensing models and applications, systems for spectrum management, cognitive architectures, and cognitive engines. In the licensing models, reports on the licensing strategies by various agencies such as United States Presidents' Council of Advisors on Science and Technology (US PCAST) [114], Authorized Shared Access (ASA) [115], US FCC [116], Notice of Proposed Rule Making NPRM [117], etc., are illustrated to reach a feasible and implementable licensing model. Further, several possible applications of the CR with their practical issues are discussed. In the systems and spectrum management part, the different issues such as legacy protection and commercial liability, selection of right architecture, network latency and CR complexity, etc., are explored for feasibility point of view. In the further sections, the spectrum search, sensing constraints and trade-offs, decision reliability, and security perspectives have been illustrated. Wang and Liu [45] have presented the fundamental concept about CR characteristics, functions, network architecture, and applications. Initially, the prerequisite requirement on deploying CR, i.e., spectrum sensing, and review on different types of detection techniques and cooperative spectrum sensing protocols is illustrated. In addition, recently proposed dynamic spectrum management and sharing schemes are reviewed, such as MAC, spectrum handoff, power control, routing, and cooperation enforcement. Moreover, some security issues in the implementation of the CRN are also discussed. In [118], the authors have presented a survey on the robust power control policies and beamforming techniques in the CRNs. Initially, the authors have illustrated the CR technology with different spectrum sharing techniques and network models. Further, the robust and stochastic optimization for the CRNs is illustrated and has presented the different robust design methods. In addition to this, the robust design for beamforming and power control techniques in the CRNs is presented and also presented the future research directions for feasibility point of view such as robust design of CRNs, orthogonal frequency division multiple access (OFDMA)-based CRNs, relay networks, and multi-antenna CRNs. Further, the future research highlights regarding the communication scenarios, algorithm types, robust design methods, energy efficiency, etc. Naeem et al. [119] have presented a survey on the resource allocation techniques in cooperative CRNs (CCRNs). The authors have explored that the potential techniques to improve the spectral utilization are cooperative communication, CR, and multi-antenna. Further, the authors have exploited cooperative communication and CR technologies simultaneously for more improvement in the spectral utilization and combined technology known as CCRNs. Initially, some light is focused on the cooperative communication, CR, spectrum sensing, and DSA techniques. Further, the resources for the cooperative CR technology such as power allocation, relay selection, user scheduling, routing, subcarrier allocation, etc., are illustrated. In addition to this, the taxonomy of the protocols and network configurations for CCRNs is presented. The authors have suggested the future research challenges and point out the issues namely, (i) multihop relaying, (ii) green cooperative CR, (iii) cooperative CR with imperfect CSI, and (iv) cooperation for machine to machine communication. The channel assignment algorithms, taxonomy, potential issues, and challenges in CR communication systems are well explored by Ahmed et al. [120]. The authors have started with channel assignment algorithms for the wireless networks and CRNs. Further, the comparison of the various channel assignment problems in CRNs is presented and taxonomy as well as the state-of-the-art of different channel assignment algorithms are illustrated. Moreover, the channel assignment problems for the CR ad-hoc networks (CRAHNs), CR cellular networks (CRCNs), cognitive wireless local area networks (CWLANs), cognitive wireless mesh networks (CWMNs), and CR sensor networks (CRSNs) are also explored and finally the critical challenges as well as potential solutions are proposed. Tenab and Hamouda [121] have presented a survey on the resource allocation techniques in underlay CRNs. The resources exploited in the underlay CRNs are the user assignment, band assignment, beam-forming matrix, power allocation, rate allocation, and network specific variables. Initially, the taxonomy of underlay resource allocation on the basis of approaches, criteria, techniques, and architecture is illustrated. Further, a comprehensive survey of the resource allocation algorithms for centralized and distributed architectures is presented. In addition to this, the open research challenges for underlay resource allocation such as network mobility, hybrid users, channel models, CR for the advanced LTE networks, and experimental test beds are explored.

1.4 Motivation

The study presented on the CR network reveals that the ultimate theme of the CR is to use the spectrum efficiently by exploiting its different functioning cycles. Therefore, the key concern of this book is to propose and explore the frameworks which can exploit the spectrum efficiently. It is perceived from the above-discussed spectrum sharing techniques in Section 1.2 that the hybrid spectrum accessing technique is a potential way for efficient spectrum utilization in the presence and absence of PU. Therefore, the exploitation of hybrid spectrum accessing technique on different frameworks may lead to improved throughput. The spectrum prediction is an important technique which predicts the future state of the channel/spectrum on the basis of historical information about the spectrum. The system impairments and spectrum prediction technique may cause the error in the process of prediction which is presented as the probability of spectrum prediction error and this error causes the wrong prediction about the state of the channel. Therefore, the effect of spectrum prediction and error in this process on the CRN performance needs to be analyzed. In addition to this, the detection of the emergence of PU during data transmission period is a much desired phenomenon which is known as spectrum monitoring. The error in the spectrum monitoring process causes the delay in the detection of the emergence of PU. This delay in the detection of emergence of PU degrades the performance of CU which needs to be analyzed and effective ways to diminish these effects are desired. The spectrum prediction and spectrum monitoring are the two prominent techniques to detect the emergence of PU during CUs' data transmission. Therefore, it is worth examining the combined effect of both the techniques for the detection of emergence of PU. The implementation of hybrid spectrum accessing strategy in multichannel CRNs with distributed environment is for efficient spectrum utilization is a research problem which needs to be resolved.

The potential techniques that provide the facility to exploit the spectrum efficiently are the non-orthogonal multiple access (NOMA) as well as MIMO and exploiting the simultaneous frameworks with CR-NOMA and MIMO-CR-NOMA can lead to more spectral efficiency with increased system complexity due to need of addressing the challenges of all the techniques. The interference is a prime factor that always comes into picture while exploring the wireless communication and deteriorates its performance; therefore, considering the interference, addressing and managing the challenges is of prime importance. Further, the applications of CR in the prominent application areas in the form of case study will be a leading edge for effective understanding to the reader; therefore, we have emphasized over the role of CR in IoTs and vehicular networks.

1.5 Organization of Book

This book comprises 13 chapters. Chapter 1 starts with the briefs about the connected environments and evolution of the wireless communication generations as well as technical perspectives by using 3 GPP. Further, it emphasizes on the significance of DSA over static spectrum access (SSA) and role of CR to achieve the DSA mechanism. The integral parts of CR namely, the spectrum sensing, spectrum analysis and decision, spectrum sharing/accessing, and spectrum mobility are briefly discussed. Further, the spectrum sharing and accessing techniques are illustrated in detail. The comparison of various spectrum accessing techniques such as interweave spectrum access, underlay spectrum access, overlay spectrum access, and hybrid spectrum access is presented.

In Chapter 2, a novel hybrid-cum-improved spectrum access (HISA) technique is exploited to improve the throughput and data-loss of the CRN. The HISA technique consists of two advanced frame structures that explore the hybrid spectrum accessing strategy to utilize the channel in the presence of PU. The closed-form expressions of the throughput and data-loss for the proposed CRN are derived numerically and further the simulation results are the witness of superior performance with reference to the throughput and data-loss. However, the selection of frame structures is the function of requirements and application of the CU such as the first frame structure performs well in terms of throughput; however, for improved data-loss, the second frame structure outperforms.

In Chapter 3, the CRN in the high traffic environments is investigated where the spectrum prediction plays an important role to select a particular channel for spectrum sensing from the pool of channels on the basis of idle prediction probability. The considered frame structure has spectrum prediction phase before the spectrum sensing and data transmission phase. We have exploited the concept of hybrid spectrum access strategy to improve the throughput of the considered frame structure. Moreover, the probability of the PUs' emergence in the data transmission period is very significant that needs to be detected; however, there is no mechanism to perform this function. Therefore, we have used the spectrum monitoring simultaneously to the data transmission period which is an advanced technique that exploits the received signal characteristics to detect the emergence of

PU. The use of spectrum monitoring improves the performance of high-traffic CRN (HTCRN) in terms of data-loss, power-loss, and interference-at-PU.

Chapter 4 exploits the potential issues concerning the random selection of spectrum sensing channel, after the prediction phase in the CRN. Therefore, a novel approach (Approach-1) of improved channel selection is proposed which relies on the probabilities of channels by which they are predicted idle. Further, the closed-form expressions for the throughput of CU in the conventional and proposed approaches are derived. In addition to this, one fundamental approach to compute the prediction probabilities is also proposed. Moreover, a new challenging issue, named, "sense and stuck" in the conventional approach is observed. The proposed approach is validated by comparing the results achieved with the conventional approach. However, to achieve the prediction probabilities, the pre-channel-state-information is prerequisite, which may be unavailable for particular scenarios; therefore, one modified selection method is introduced to avoid the sense and stuck problem. An algorithm to evaluate throughput using the random, improved, and modified selection method is presented with its space and time complexities. Furthermore, for additional improvement in the throughput of CU, a new frame structure is introduced, in which the spectrum prediction and spectrum sensing periods are exploited for simultaneous transmission of data via the underlay spectrum access technique (Approach-2). The simulated results of the Approach-2 are compared with our pre-obtained results of Approach-1, which confirm significant improvement in the throughput.

In Chapter 5, we have extended the work presented in Chapter 3 where the spectrum monitoring is assumed to be a perfect phenomenon that means there is no spectrum monitoring error. However, in practice, the hardware impairments and channel random nature result in the imperfections in the spectrum monitoring process which are presented through the probability of spectrum monitoring error. Therefore, in Chapter 5, we have introduced the concept of imperfect spectrum monitoring error and have analyzed the effect of imperfections on the data-loss, power-wastage, interference-efficiency, and energy-efficiency. The imperfection in spectrum monitoring degrades the performance of CRN when analyzing for the different scenarios of the traffic intensity and probability of spectrum monitoring error.

The key purpose of Chapter 6 is to diminish the effects of imperfection in spectrum monitoring error as discussed in Chapter 5. In order to achieve this, the cooperation among CUs is proposed for the homogeneous and heterogeneous CRN environments. A scenario where the cooperation among CUs is possible for spectrum monitoring is presented. The homogeneous CRN is defined when all the CUs have same probability of spectrum monitoring error; however, the CRN where the CUs have different probability of spectrum monitoring error is defined as the heterogeneous CRN. The Binomial and Poisson-Binomial distribution functions are used to compute the probability of spectrum monitoring error after cooperation in the homogeneous and heterogeneous CRN, respectively. The cooperation among CUs for spectrum monitoring improves the performance of CRN in terms of the data-loss, interference efficiency, and energy efficiency.

Chapter 7 explores the concept of spectrum mobility by using the spectrum prediction and spectrum monitoring techniques simultaneously, to detect the emergence of PU. In the proposed strategy, the decision results of the spectrum prediction and monitoring techniques are fused using AND and OR fusion rules, for the detection of emergence of PU during the data transmission. Further, the closed-form expressions of the resource wastage, achieved throughput, interference power at PU, and data-loss for the proposed approaches as well as for the prediction and monitoring approaches are derived. Moreover, the simulation results for the proposed approaches are presented and validation is performed by comparing the results with prediction and monitoring approach. In a special case, when the prediction error is zero, the graphical characteristics of all metric values overlies the spectrum monitoring approach, which further support the proposed approach.

In Chapter 6, we have analyzed that the CRN relies on the centralized architecture; however, in the centralized architecture, all the CUs are controlled by a centralized/controlling CU (CCU) and if CCU fails, the entire network gets down. In addition to this, the huge information needs to exchange between the CUs and CCU. The distributed architecture is suitable option to overcome the mentioned limitations of the centralized architecture. In the distributed architecture, all the CUs perform all the functions of spectrum accessing, individually. Therefore, the self-schedule multichannel medium access control (SMC-MAC) protocol plays a key role. In Chapter 8, we have proposed a hybrid framework in the distributed CRNs in which a novel frame structure is proposed and the SMC-MAC protocol is developed for the proposed frame structure. The hybrid spectrum accessing technique is used to exploit the active channels with constrained power transmission. The proposed framework is analyzed for the perfect and imperfect spectrum sensing scenarios. Through the simulation results, it is perceived that the proposed framework outperforms the conventional SMC-MAC protocols present in the literature with reference to the spectral efficiency/utilization as well as interference efficiency.

Chapter 9 exploits the other spectral efficient technique, that is, the NOMA technique. The potential frameworks of NOMA implementation over CR as well as the feasibility of proposed frameworks are presented which is named as CR-NOMA framework. Further, the differences between proposed CR-NOMA and conventional CR frameworks are discussed. Finally, the potential issues regarding the implementation of CR-NOMA frameworks are explored.

Chapter 10 emphasizes on one more spectral efficient technique, that is, the MIMO in the spatial multiplexing mode, and have collectively exploited CR, NOMA, and MIMO communication systems in order to propose a spectral-and-interference-efficient framework which is named as MIMO-based CR–NOMA communication system. The proposed framework is analyzed for the uplink and downlink scenarios. Further, the closed-form expressions for throughput at each user due to the number of transmitting- and receiving-antennas are derived numerically for the downlink- and uplink-scenarios. In addition to this, the total/sum throughput for different frameworks such as CR–NOMA, CR–MIMO, and MIMO-based CR–NOMA systems is derived for both the downlink and uplink scenarios. Moreover, in order to satisfy the interference constraints at the PU due to cell-edge/far-user transmission in the uplink scenario, a new metric known as interference-efficiency is derived. Furthermore, the proposed frameworks are simulated for downlink and uplink scenarios and the relationship between throughput of cell-centered/near-user and cell-edge/far-user are presented. Moreover, the total throughput for different frameworks and interference efficiency is also illustrated with the simulation results. The simulation results reveal that the proposed MIMO-based CR–NOMA system outperforms the existing MIMO–NOMA, CR–NOMA, and CR–OMA systems in terms of the CUs' individual throughput, total throughput, and interference efficiency of the system.

In Chapter 11, the interference perspectives and mathematical formulation of interference in the wireless communications are studied in detail. The CRNs are classified into interfering and non-interfering CRNs and interference scenarios are investigated in the CU-to-CU, CU-to-PU, and PU-to-CU cases. Further, the interference cancellation techniques in the CRN are explored and have also proposed the cross-layer interference mitigation in the CRNs. Moreover, the interference avoidance techniques by advancing the different constituents of cognitive cycle that are the spectrum sensing, spectrum accessing, spectrum monitoring, and spectrum mobility are illustrated.

Chapters 12 and 13 emphasize the potential applications of the CR that are the IoTs and internet-of-vehicles or vehicular networks. In Chapter 12, we have started with the digitization techniques of the fourth industrial revolution (4IR), where IoTs is a very popular technique that is subdivided into the industrial IoTs (IIoTs) and consumer (CIoTs). Further, in addition to other potential applications, the connected vehicles/internet of vehicles is a prominent part whose potential constituents that are vehicle-to-vehicle, vehicle-to-infrastructure, infrastructure-to-vehicle, vehicle-to-pedestrians are illustrated. Further, the potential simulation frameworks in order to simulate the vehicular networks are discussed and a comparison of those simulation techniques is illustrated.

Chapter 13 explores the radio resource management perspectives in the internet-of-vehicles (IoVs) networks. The authors have illustrated the different wireless access technologies for the IoVs networks that are vehicular communication, cellular communication, and short range static communication. The vehicular communication techniques comprise the DSRC, wireless access for vehicular communication (WAVE), communication architecture, and land mobile (CALM). The cellular communication techniques are 3G, 4G, and 5G, wireless access for microwave (WiMax), and satellite communication where LTE, LTE-Advanced, and NR are popular techniques. The short-range static communication techniques are Bluetooth, WiFi, and Zigbee. Further, the authors have presented the spectrum sharing perspectives in the IoVs networks and have proposed the probable CR-IoVs frameworks. Furthermore, the potential research challenges for particular constituents of the cognitive cycle are illustrated.

1.6 Summary

This chapter presents the briefs about the connected environments with recent developments, evolution of the wireless communication generations as well as technical perspectives by using 3GPP. Further, the role of CR in order to achieve the DSS is illustrated with various constituents of cognitive engine cycle that are spectrum sensing, spectrum analysis and decision, spectrum sharing and accessing. Further, the spectrum sharing and accessing techniques are illustrated in detail. The comparison of various spectrum accessing techniques such as interweave spectrum access, underlay spectrum access, overlay spectrum access, and hybrid spectrum access is presented. Moreover, for the implementation of CR, various efforts across the globe in terms of test beds or other hardware structures are discussed. Further, the organization of the book is presented.

References

1 Heslop, B. By 2030, each person will own 15 connected devices – here's what that means for your business and content. https://www.martechadvisor.com/articles/iot/by-2030-each-person-will-own-15-connected-devices-heres-what-that-means-for-your-business-and-content/ (accessed 23 September 2020).
2 Hatton, M. IoT News – The IoT in 2030: 24 billion connected things generating $1.5 trillion. *IoT Business News* (20 May 2020). https://iotbusinessnews.com/2020/05/20/03177-the-iot-in-2030-24-billion-connected-things-generating-1-5-trillion/ (accessed 23 September 2020).

3 Number of connected devices worldwide 2030. *Statista*. https://www.statista.com/statistics/802690/worldwide-connected-devices-by-access-technology/ (accessed 23 September 2020).

4 FCC-03-322A1.pdf [Online]. http://web.cs.ucdavis.edu/~liu/289I/Material/FCC-03-322A1.pdf (accessed 18 February 2018).

5 Mitola, J. and Maguire, G.Q. (1999). Cognitive radio: making software radios more personal. *IEEE Personal Communications* 6 (4): 13–18. https://doi.org/10.1109/98.788210.

6 Zhao, Q. and Sadler, B.M. (2007). A survey of dynamic spectrum access. *IEEE Signal Processing Magazine* 24 (3): 79–89. https://doi.org/10.1109/MSP.2007.361604.

7 Akyildiz, I.F., Lee, W.-Y., Vuran, M.C., and Mohanty, S. (2006). NeXt generation/dynamic spectrum access/cognitive radio wireless networks: a survey. *Computer Networks* 50 (13): 2127–2159. https://doi.org/10.1016/j.comnet.2006.05.001.

8 Thakur, P., Singh, G., and Satasia, S.N. (2016). Spectrum sharing in cognitive radio communication system using power constraints: a technical review. *Perspectives on Science* 8: 651–653. https://doi.org/10.1016/j.pisc.2016.06.048.

9 Lu, L., Zhou, X., Onunkwo, U., and Li, G.Y. (2012). Ten years of research in spectrum sensing and sharing in cognitive radio. *EURASIP Journal on Wireless Communications and Networking* 2012: 28. https://doi.org/10.1186/1687-1499-2012-28.

10 He, A., Bae, K.K., Newman, T.R. et al. (2010). A survey of artificial intelligence for cognitive radios. *IEEE Transactions on Vehicular Technology* 59 (4): 1578–1592. https://doi.org/10.1109/TVT.2010.2043968.

11 Alias, D.M. and Ragesh G.K. (2016). Cognitive radio networks: a survey. *Proceedings of International Conference on Wireless Communications, Signal Processing and Networking* (WiSPNET) (23–25 March 2016), 1981–1986, doi: https://doi.org/10.1109/WiSPNET.2016.7566489.

12 Liang, Y.C., Chen, K.C., Li, G.Y., and Mahonen, P. (2011). Cognitive radio networking and communications: an overview. *IEEE Transactions on Vehicular Technology* 60 (7): 3386–3407. https://doi.org/10.1109/TVT.2011.2158673.

13 Sasipriya, S. and Vigneshram, R. (2016). An overview of cognitive radio in 5G wireless communications. *Proceedings of IEEE International Conference on Computational Intelligence and Computing Research* (ICCIC) (15–17 December 2016), 1–5, doi: https://doi.org/10.1109/ICCIC.2016.7919725.

14 Hu, F., Chen, B., and Zhu, K. (2018). Full spectrum sharing in cognitive radio networks toward 5G: a survey. *IEEE Access* 6 (2): 15754–15776. https://doi.org/10.1109/ACCESS.2018.2802450.

15 Amjad, M., Rehmani, M.H., and Mao, S. (2018). Wireless multimedia cognitive radio networks: a comprehensive survey. *IEEE Communication Surveys and Tutorials* 20 (2): 1056–1103. https://doi.org/10.1109/COMST.2018.2794358.

16 Paul, C., Ganesh, A., and Sunitha, C. (2018). An overview of IoT based smart homes. *Proceedings of 2nd International Conference on Inventive Systems and Control* (ICISC), Coimbatore, India (19–20 January 2018), 43–46, doi: https://doi.org/10.1109/ICISC.2018.8398858.

17 Khan, M., Silva, B.N., and Han, K. (2016). Internet of things based energy aware smart home control system. *IEEE Access* 4: 7556–7566. https://doi.org/10.1109/ACCESS.2016.2621752.

18 Santoso, F.K. and Vun, N.C.H. (2015). Securing IoT for smart home system. *Proceedings of International Symposium on Consumer Electronics* (ISCE), Madrid, Spain (24–26 June 2015), 1–2, doi: https://doi.org/10.1109/ISCE.2015.7177843.

19 Haverkort, B.R. and Zimmermann, A. (2017). Smart industry: how ICT will change the game! *IEEE Internet Computing* 21 (1): 8–10. https://doi.org/10.1109/MIC.2017.22.

20 Perumal, T., Ramli, A.R., and Leong, C.Y. (2011). Interoperability framework for smart home systems. *IEEE Transactions on Consumer Electronics* 57 (4): 1607–1611. https://doi.org/10.1109/TCE.2011.6131132.

21 Fernández-Caramés, T.M. and Fraga-Lamas, P. (2018). A review on human-centered IoT-connected smart labels for the industry 4.0. *IEEE Access* 6: 25939–25957. https://doi.org/10.1109/ACCESS.2018.2833501.

22 Harmon, R.R., Castro-Leon, E.G., and Bhide, S. (2015). Smart cities and the internet of things. *Proceedings of Portland International Conference on Management of Engineering and Technology* (PICMET), Portland, OR (2–6 August 2015), 485–494, doi: https://doi.org/10.1109/PICMET.2015.7273174.

23 Su, K., Li, J., and Fu, H. (2011). Smart city and the applications. *Proceedings of International Conference on Electronics, Communications and Control* (ICECC), Ningbo, China (9–11 September 2011), 1028–1031, doi: https://doi.org/10.1109/ICECC.2011.6066743.

24 Akyildiz, I.F. and Kak, A. (2019). The internet of space things/CubeSats. *IEEE Network* 33 (5): 212–218. https://doi.org/10.1109/MNET.2019.1800445.

25 Akyildiz, I.F. and Kak, A. (2019). The internet of space things/CubeSats: a ubiquitous cyber-physical system for the connected world. *Computer Networks* 150: 134–149. https://doi.org/10.1016/j.comnet.2018.12.017.

26 Hofmann, F., Mahdi, A.H., Brahmi, N. et al. (2019). 5G NetMobil: pathways towards tactile connected driving. *Proceedings of IEEE 2nd 5G World Forum* (5GWF), Dresden, Germany (30 September to 2 October 2019), 114–119, doi: https://doi.org/10.1109/5GWF.2019.8911621.

27 Nakamoto, Y. (2019). Some technology issues in a connected world. *Proceedings of IEEE Intl Conf on Dependable, Autonomic and Secure Computing, Intl Conf on Pervasive Intelligence and Computing, Intl Conf on Cloud and Big Data Computing, Intl Conf on Cyber Science and Technology Congress (DASC/PiCom/CBDCom/CyberSciTech)*, Fukuoka, Japan (5–8 August 2019), 663–666, doi: https://doi.org/10.1109/DASC/PiCom/CBDCom/CyberSciTech.2019.00125.

28 Chen, Y., Zhu, P., He, G., Yan, X., Baligh, H., and Wu, J. (2020). From connected people, connected things, to connected intelligence. *Proceedings of 2nd 6G Wireless Summit* (6G SUMMIT), Levi, Finland (17–20 March 2020), 1–7, doi: https://doi.org/10.1109/6GSUMMIT49458.2020.9083770.

29 Parkvall, S., Dahlman, E., Furuskar, A., and Frenne, M. (2017). NR: the new 5G radio access technology. *IEEE Communications Standards Magazine* 1 (4): 24–30. https://doi.org/10.1109/MCOMSTD.2017.1700042.

30 Rappaport, T.S., Xing, Y., Kanhere, O. et al. (2019). Wireless communications and applications above 100 GHz: opportunities and challenges for 6G and beyond. *IEEE Access* 7: 78729–78757. https://doi.org/10.1109/ACCESS.2019.2921522.

31 Akyildiz, I.F., Kak, A., and Nie, S. (2020). 6G and beyond: the future of wireless communications systems. *IEEE Access* 8: 133995–134030. https://doi.org/10.1109/ACCESS.2020.3010896.

32 Chowdhury, M.Z., Shahjalal, M., Ahmed, S., and Jang, Y.M. (2020). 6G wireless communication systems: applications, requirements, technologies, challenges, and research directions. *IEEE Open Journal of the Communications Society* 1: 957–975. https://doi.org/10.1109/OJCOMS.2020.3010270.

33 Zhu, J., Zhao, M., Zhang, S. et al. (2020). *China Communications* 17 (6): 51–67. https://doi.org/10.23919/JCC.2020.06.005.

34 Ziegler, V. and Yrjola, S. (2020). 6G indicators of value and performance. *Proceedings of 2nd 6G Wireless Summit (6G SUMMIT)*, Levi, Finland (17–20 March 2020), 1–5, doi: https://doi.org/10.1109/6GSUMMIT49458.2020.9083885.

35 Brito, J.M.C., Mendes, L.L., and Gontijo, J.G.S. (2020). Brazil 6G project – an approach to build a national-wise framework for 6G networks. *Proceedings of 2nd 6G Wireless Summit* (6G SUMMIT), Levi, Finland (17–20 March 2020), 1–5, doi: 10.1109/6GSUMMIT49458.2020.9083775.

36 Elmeadawy, S. and Shubair, R.M. (2019). 6G Wireless Communications: Future Technologies and Research Challenges. *Proceedings of IEEE International Conference on Electrical and Computing Technologies and Applications (ICECTA)*, Ras Al Khaimah, United Arab Emirates (19–21 November 2019), 1–5. IEEE, 2019.

37 Yang, P., Xiao, Y., Xiao, M., and Li, S. (2019). 6G wireless communications: Vision and potential techniques. *IEEE Network* 33 (4): 70–75.

38 Alsharif, M.H., Kelechi, A.H., Albreem, M.A. et al. (2020). Sixth generation (6G) wireless networks: vision, research activities, challenges and potential solutions. *Symmetry* 12 (4): 4. https://doi.org/10.3390/sym12040676.

39 Zhang, S., Xiang, C., and Xu, S. (2020). 6G: connecting everything by 1000 times price reduction. *IEEE Open Journal of Vehicular Technology* 1: 107–115.

40 Matinmikko-Blue, M., Yrjölä, S., and Ahokangas, P. (2020). Spectrum management in the 6G Era: the role of regulation and spectrum sharing. *Proceedings of 2nd 6G Wireless Summit* (6G SUMMIT), Levi, Finland (17–20 March 2020), 1–5.

41 Zhang, L., Liang, Y.-C., and Niyato, D. (2019). 6G Visions: mobile ultra-broadband, super internet-of-things, and artificial intelligence. *China Communications* 16 (8): 1–14. https://doi.org/10.23919/JCC.2019.08.001.

42 Mitola, J. (1995). The software radio architecture. *IEEE Communications Magazine* 33 (5): 26–38. https://doi.org/10.1109/35.393001.

43 Seshukumar, K., Saravanan, R., and Suraj, M.S. (2013). Spectrum sensing review in cognitive radio. *Proceedings of International Conference on Emerging Trends in VLSI, Embedded System, Nano Electronics and Telecommunication System* (ICEVENT), Tiruvannamalai, India (7–9 January 2013), 1–4, doi: https://doi.org/10.1109/ICEVENT.2013.6496549.

44 Yucek, T. and Arslan, H. (2009). A survey of spectrum sensing algorithms for cognitive radio applications. *IEEE Communication Surveys and Tutorials* 11 (1): 116–130. https://doi.org/10.1109/SURV.2009.090109.

45 Wang, B. and Liu, K.J.R. (2011). Advances in cognitive radio networks: a survey. *IEEE Journal on Selected Topics in Signal Processing* 5 (1): 5–23. https://doi.org/10.1109/JSTSP.2010.2093210.

46 Muchandi, N. and Khanai, R. (2016). Cognitive radio spectrum sensing: a survey. *Proceedings of International Conference on Electrical, Electronics, and Optimization Techniques* (ICEEOT), Tiruvannamalai, India (7–9 January 2013), 3233–3237, doi: https://doi.org/10.1109/ICEEOT.2016.7755301.

47 Sun, H., Nallanathan, A., Wang, C.X., and Chen, Y. (2013). Wideband spectrum sensing for cognitive radio networks: a survey. *IEEE Wireless Communications* 20 (2): 74–81. https://doi.org/10.1109/MWC.2013.6507397.

48 Malhotra, M., Aulakh, I.K., and Vig, R. (2015). A review on energy based spectrum sensing in Cognitive Radio Networks. *Proceedings of International Conference on Futuristic Trends on Computational Analysis and Knowledge Management* (ABLAZE), Noida, India (25–27 February 2015), 561–565, doi: https://doi.org/10.1109/ABLAZE.2015.7154925.

49 Perera, L.N.T. and Herath, H.M.V.R. (2011). Review of spectrum sensing in cognitive radio. *Proceedings of 6th International Conference on Industrial and Information Systems,* Kandy, Sri Lanka (16–19 August 2011), 7–12, doi: https://doi.org/10.1109/ICIINFS.2011.6038031.

50 Omer, A.E. (2015). Review of spectrum sensing techniques in cognitive radio networks. *Proceedings of International Conference on Computing, Control, Networking, Electronics and Embedded Systems Engineering* (ICCNEEE), Khartoum, Sudan (7–9 September 2015), 439–446, doi: https://doi.org/10.1109/ICCNEEE.2015.7381409.

51 Zeng, Y., Liang, Y.-C., Hoang, A.T., and Zhang, R. (2010). A review on spectrum sensing for cognitive radio: challenges and solutions. *EURASIP Journal on Advances in Signal Processing* 2010 (1): 381465/1–381465/15. https://doi.org/10.1155/2010/381465.

52 Wang, X., Ekin, S., and Serpedin, E. (2018). Joint spectrum sensing and resource allocation in multi-band-multi-user cognitive radio networks. *IEEE Transactions on Communications* 66 (8): 3281–3293. https://doi.org/10.1109/TCOMM.2018.2807432.

53 Lu, Y. and Duel-Hallen, A. (2018). A sensing contribution-based two-layer game for channel selection and spectrum access in cognitive radio ad-hoc networks. *IEEE Transactions on Wireless Communications* 17 (6): 3631–3640. https://doi.org/10.1109/TWC.2018.2810869.

54 Tong, J., Jin, M., Guo, Q., and Li, Y. (2018). Cooperative spectrum sensing: a blind and soft fusion detector. *IEEE Transactions on Wireless Communications* 17 (4): 2726–2737. https://doi.org/10.1109/TWC.2018.2801833.

55 Akyildiz, I.F., Lee, W.Y., Vuran, M.C., and Mohanty, S. (2008). A survey on spectrum management in cognitive radio networks. *IEEE Communications Magazine* 46 (4): 40–48. https://doi.org/10.1109/MCOM.2008.4481339.

56 Sharma, S.K., Bogale, T.E., Le, L.B. et al. (2018). Dynamic spectrum sharing in 5G wireless networks with full-duplex technology: recent advances and research challenges. *IEEE Communication Surveys and Tutorials* 20 (1): 674–707. https://doi.org/10.1109/COMST.2017.2773628.

57 Matinmikko, M., Mustonen, M., Roberson, D. et al. (2014). Overview and comparison of recent spectrum sharing approaches in regulation and research: from opportunistic unlicensed access towards licensed shared access. *Proceedings of IEEE International Symposium on Dynamic Spectrum Access Networks* (DYSPAN), McLean, VA (1–4 April 2014), 92–102, doi: https://doi.org/10.1109/DySPAN.2014.6817783.

58 Pandit, S. and Singh, G. (2017). An overview of spectrum sharing techniques in cognitive radio communication system. *Wireless Networks* 23 (2): 497–518. https://doi.org/10.1007/s11276-015-1171-1.

59 Zhang, L., Xiao, M., Wu, G. et al. (2017). A survey of advanced techniques for spectrum sharing in 5G networks. *IEEE Wireless Communications* 24 (5): 44–51. https://doi.org/10.1109/MWC.2017.1700069.

60 Baby, S.M. and James, M. (2016). A comparative study on various spectrum sharing techniques. *Procedia Technology* 25: 613–620. https://doi.org/10.1016/j.protcy.2016.08.152.

61 Xu, T., Li, Z., Ge, J., and Ding, H. (2014). A survey on spectrum sharing in cognitive radio networks. *KSII Transactions on Internet and Information Systems (TIIS)* 8 (11): 3751–3774. https://doi.org/10.3837/tiis.2014.11.006.

62 Ye, Y., Wu, D., Shu, Z., and Qian, Y. (2016). Overview of LTE spectrum sharing technologies. *IEEE Access* 4: 8105–8115. https://doi.org/10.1109/ACCESS.2016.2626719.

63 Salgado, C., Hernandez, C., Molina, V., and Beltran-Molina, F.A. (2016). Intelligent algorithm for spectrum mobility in cognitive wireless networks. *Procedia Computer Science* 83: 278–283. https://doi.org/10.1016/j.procs.2016.04.126.

64 Hernández, C., Pedraza, L., Páez, I., and Rodriguez-Colina, E. (2015). Algorithms for spectrum allocation in cognitive radio networks. *Information Tecnology* 26 (6): 169–186. https://doi.org/10.4067/S0718-07642015000600018.

65 Nguyen, K.-H. and Hwang, W.-J. (2012). An efficient power control scheme for Spectrum mobility management in cognitive radio sensor networks. In: *Embedded and Multimedia Computing Technology and Service, Lecture Notes in Electrical Engineering*, vol. 181 (eds. J. Park, Y.S. Jeong, S. Park and H.C. Chen), 667–676. Dordrecht: Springer https://doi.org/10.1007/978-94-007-5076-0_81.

66 Xing, X., Jing, T., Cheng, W. et al. (2013). Spectrum prediction in cognitive radio networks. *IEEE Wireless Communications* 20 (2): 90–96. https://doi.org/10.1109/MWC.2013.6507399.

67 Boyd, S.W., Frye, J.M., Pursley, M.B., and Royster, T.C. IV (2012). Spectrum monitoring during reception in dynamic spectrum access cognitive radio networks. *IEEE Transactions on Communications* 60 (2): 547–558. https://doi.org/10.1109/TCOMM.2011.122111.100603.

68 Oyewobi, S.S. and Hancke, G.P. (2017). A survey of cognitive radio handoff schemes, challenges and issues for industrial wireless sensor networks (CR-IWSN). *Journal of Network and Computer Applications* 97: 140–156. https://doi.org/10.1016/j.jnca.2017.08.016.

69 Christian, I., Moh, S., Chung, I., and Lee, J. (2012). Spectrum mobility in cognitive radio networks. *IEEE Communications Magazine* 50 (6): 114–121. https://doi.org/10.1109/MCOM.2012.6211495.

70 Akyildiz, I.F., Lo, B.F., and Balakrishnan, R. (2011). Cooperative spectrum sensing in cognitive radio networks: A survey. *Physics Communications* 4 (1): 40–62. https://doi.org/10.1016/j.phycom.2010.12.003.

71 Thakur, P., Kumar, A., Pandit, S. et al. (2017). Spectrum mobility in cognitive radio network using spectrum prediction and monitoring techniques. *Physics Communications* 24: 1–8. https://doi.org/10.1016/j.phycom.2017.04.005.

72 Kumar, K., Prakash, A., and Tripathi, R. (2016). Spectrum handoff in cognitive radio networks: a classification and comprehensive survey. *Journal of*

Network and Computer Applications 61: 161–188. https://doi.org/10.1016/j. jnca.2015.10.008.

73 Soto, J. and Nogueira, M. (2017). A framework for resilient and secure spectrum sensing on cognitive radio networks. *Computer Networks* 115: 130–138. https://doi. org/10.1016/j.comnet.2017.01.012.

74 Hossain, E., Niyato, D., and Han, Z. (2009). *Dynamic Spectrum Access and Management in Cognitive Radio Networks*, 1ste. New York, NY, USA: Cambridge University Press.

75 Ileri, O., Samardzija, D., Sizer, T., and Mandayam, N.B. (2005). Demand responsive pricing and competitive spectrum allocation via a spectrum server. *Proceedings of 1st IEEE International Symposium on New Frontiers in Dynamic Spectrum Access Networks*, (DySPAN 2005), Baltimore, MD (8–11 November 2005), 194–202, doi: https://doi.org/10.1109/DYSPAN.2005.1542635.

76 Kwon, B. and Copeland, J.A. (2009). A centralized spectrum sharing scheme for access points in IEEE 802.11 networks. *Proceedings of 6th IEEE Consumer Communications and Networking Conference*, Las Vegas, NV (10–13 January 2009), 1–5, doi: https://doi.org/10.1109/CCNC.2009.4784697.

77 Salami, G., Durowoju, O., Attar, A. et al. (2011). A comparison between the centralized and distributed approaches for spectrum management. *IEEE Communication Surveys and Tutorials* 13 (2): 274–290. https://doi.org/10.1109/ SURV.2011.041110.00018.

78 Liang, Q., Sun, G., and Wang, X. (2012). Random leader: a distributed-centralized spectrum sharing scheme in Cognitive Radios. *Proceedings. IEEE International Conference on Communications* (ICC), Ottawa, ON (10–15 June 2012), 1806–1810, doi: https://doi.org/10.1109/ICC.2012.6363808.

79 Lin, Y.Y. and Chen, K.C. (2010). Distributed spectrum sharing in cognitive radio networks – game theoretical view. *Proceedings of 7th IEEE Consumer Communications and Networking Conference*, Las Vegas, NV (9–12 January 2010), 1–5, doi: https://doi.org/10.1109/CCNC.2010.5421750.

80 Ding, L., Pudlewski, S., Melodia, T. et al. (2010). Distributed spectrum sharing for video streaming in cognitive radio Ad Hoc networks. In: *Ad Hoc Networks*, Lecture Notes of the Institute for Computer Sciences, Social Informatics and Telecommunications Engineering, vol. 28 (eds. J. Zheng, S. Mao, S.F. Midkiff and H. Zhu), 855–867. Berlin, Heidelberg: Springer https://doi.org/10.1007/978-3-642-11723-7_59.

81 Cai, M. and Laneman, J.N. (2017). Wideband distributed spectrum sharing with multichannel immediate multiple access. *Analog Integrated Circuits and Signal Processing* 91 (2): 239–255. https://doi.org/10.1007/s10470-017-0934-2.

82 Chen, X. and Huang, J. (2013). Database-assisted distributed spectrum sharing. *IEEE Journal on Selected Areas in Communications* 31 (11): 2349–2361. https:// doi.org/10.1109/JSAC.2013.131110.

83 Zhao, Q., Tong, L., Swami, A., and Chen, Y. (2007). Decentralized cognitive MAC for opportunistic spectrum access in ad hoc networks: a POMDP framework. *IEEE Journal on Selected Areas in Communications* 25 (3): 589–600. https://doi.org/10.1109/JSAC.2007.070409.

84 Zheng, H. and Cao, L.(2005). Device-centric spectrum management. *Proceedings of First IEEE International Symposium on New Frontiers in Dynamic Spectrum Access Networks* (DySPAN 2005), Baltimore, MD (8–11 November 2005), 56–65, doi: https://doi.org/10.1109/DYSPAN.2005.1542617.

85 Feng, X., Sun, G., Gan, X. et al. (2014). Cooperative spectrum sharing in cognitive radio networks: a distributed matching approach. *IEEE Transactions on Communications* 62 (8): 2651–2664. https://doi.org/10.1109/TCOMM.2014. 2322352.

86 Wang, H., Gao, L., Gan, X., Wang, X., and Hossain, E. (2010). Cooperative spectrum sharing in cognitive radio networks: a game-theoretic approach. *Proceedings of IEEE International Conference on Communications*, Cape Town, South Africa (23–27 May 2010), 1–5, doi: https://doi.org/10.1109/ICC.2010. 5502052.

87 Manna, R., Louie, R.H.Y., Li, Y., and Vucetic, B. (2011). Cooperative spectrum sharing in cognitive radio networks with multiple antennas. *IEEE Transactions on Signal Processing* 59 (11): 5509–5522. https://doi.org/10.1109/TSP.2011. 2163068.

88 Kader, M.F., Irfan, M., Shin, S.Y., and Chae, S. (2017). Cooperative spectrum sharing in cognitive radio networks: an interference free approach. *Physics Communications* 25: 66–74. https://doi.org/10.1016/j.phycom.2017.09.001.

89 Liu, H., Hua, S., Zhuo, X. et al. (2016). Cooperative spectrum sharing of multiple primary users and multiple secondary users. *Digital Communications and Networks* 2 (4): 191–195. https://doi.org/10.1016/j.dcan.2016.10.005.

90 Manna, R.F., Al-Qahtani, F.S., and Zummo, S.A. (2017). A full diversity cooperative spectrum sharing scheme for cognitive radio networks. *IEEE Access* 5: 17722–17732. https://doi.org/10.1109/ACCESS.2017.2732221.

91 Amjad, M.F., Chatterjee, M., Nakhila, O., and Zou, C.C. (2016). Evolutionary non-cooperative spectrum sharing game: long-term coexistence for collocated cognitive radio networks. *Wireless Communications and Mobile Computing* 16 (15): 2166–2178. https://doi.org/10.1002/wcm.2674.

92 Feng, X., Wang, H., and Wang, X. (2015). A game approach for cooperative spectrum sharing in cognitive radio networks. *Wireless Communications and Mobile Computing* 15 (3): 538–551. https://doi.org/10.1002/wcm.2364.

93 Bhattarai, S., Park, J., Gao, B. et al. (2016). An overview of dynamic spectrum sharing: ongoing initiatives, challenges, and a roadmap for future research. *IEEE Transactions on Cognitive Communications and Networking* 2 (2): 110–128.

94 Tang, Y. (2017). Cooperative spectrum sharing in cognitive radio networking. PhD Thesis, University of Waterloo, Waterloo, Ontario, Canada (April 2017).

95 Yao, Y., Ngoga, S.R., Erman, D., and Popescu, A. (2012). Performance of cognitive radio spectrum access with intra- and inter-handoff. *Proceedings of IEEE International Conference on Communications* (ICC'2012), Ottawa, ON (10–15 June 2012), 1539–1544, doi: https://doi.org/10.1109/ICC.2012.6364230.

96 Sharma, S.K., Bogale, T.E., Chatzinotas, S. et al. (2015). Cognitive radio techniques under practical imperfections: a survey. *IEEE Communication Surveys and Tutorials* 17 (4): 1858–1884. https://doi.org/10.1109/COMST.2015.2452414.

97 Fujii, T. and Suzuki, Y. (2005). Ad-hoc cognitive radio – development to frequency sharing system by using multi-hop network. *Proceedings of First IEEE International Symposium on New Frontiers in Dynamic Spectrum Access Networks,* (DySPAN 2005), Baltimore, MD (8–11 November 2005), 589–592, doi: https://doi.org/10.1109/DYSPAN.2005.1542675.

98 Zhao, Q., Tong, L., and Swami, A. (2005). Decentralized cognitive mac for dynamic spectrum access. *Proceedings of First IEEE International Symposium on New Frontiers in Dynamic Spectrum Access Networks* (DySPAN 2005), Baltimore, MD (8–11 November 2005), 224–232, doi: https://doi.org/10.1109/DYSPAN.2005.1542638.

99 Ban, T.W., Choi, W., Jung, B.C., and Sung, D.K. (2009). Multi-user diversity in a spectrum sharing system. *IEEE Transactions on Wireless Communications* 8 (1): 102–106. https://doi.org/10.1109/T-WC.2009.080326.

100 Establishment of an Interference Temperature Metric to Quantify and Manage Interference and to Expand Available Unlicensed Operation in Certain Fixed, Mobile and Satellite Frequency Bands. *Federal Communications Commission* (25 December 2015). https://www.fcc.gov/document/establishment-interference-temperature-metric-quantify-and-manage-interference-and-expand-0 (accessed 18 February 2018).

101 Goldsmith, A., Jafar, S.A., Maric, I., and Srinivasa, S. (2009). Breaking spectrum gridlock with cognitive radios: an information theoretic perspective. *Proceedings of the IEEE* 97 (5): 894–914. https://doi.org/10.1109/JPROC.2009.2015717.

102 Costa, M. (1983). Writing on dirty paper (Corresp.). *IEEE Transactions on Information Theory* 29 (3): 439–441. https://doi.org/10.1109/TIT.1983.1056659.

103 Gel'f, S.I. and Pinsker, M.S. Coding for channel with random parameters. *ResearchGate.* https://www.researchgate.net/publication/266509306_Coding_for_channel_with_random_parameters (accessed 18 February 2018).

104 Liang, W., Ng, S.X., and Hanzo, L. (2017). Cooperative overlay spectrum access in cognitive radio networks. *IEEE Communication Surveys and Tutorials* 19 (3): 1924–1944. https://doi.org/10.1109/COMST.2017.2690866.

105 Ghane, A.H. and Harsini, J.S. (2018). A network steganographic approach to overlay cognitive radio systems utilizing systematic coding. *Physics Communications* 27: 63–73. https://doi.org/10.1016/j.phycom.2018.01.008.

106 Sharma, S.K., Chatzinotas, S., and Ottersten, B. (2014). A hybrid cognitive transceiver architecture: sensing-throughput tradeoff. *Proceedings of 9th International Conference on Cognitive Radio Oriented Wireless Networks and Communications* (CROWNCOM), Oulu, Finland (2–4 June 2014), 143–149, doi: https://doi.org/10.4108/icst.crowncom.2014.255366.

107 Jiang, X., Wong, K.K., Zhang, Y., and Edwards, D. (2013). On hybrid overlay-underlay dynamic spectrum access: double-threshold energy detection and Markov model. *IEEE Transactions on Vehicular Technology* 62 (8): 4078–4083. https://doi.org/10.1109/TVT.2013.2258360.

108 Chu, T.M.C., Phan, H., and Zepernick, H.-J. (2014). Hybrid interweave-underlay spectrum access for cognitive cooperative radio networks. *IEEE Transactions on Communications* 62 (7): 2183–2197. https://doi.org/10.1109/TCOMM.2014.2325041.

109 Gmira, S., Kobbane, A., and Sabir, E. (2015). A new optimal hybrid spectrum access in cognitive radio: overlay-underlay mode. *Proceedings of International Conference on Wireless Networks and Mobile Communications* (WINCOM), Marrakech, Morocco (20–23 October 2015), 1–7, doi: https://doi.org/10.1109/WINCOM.2015.7381314.

110 Blasco-Serrano, R., Lv, J., Thobaben, R., Jorswieck, E., Kliks, A., and Skoglund, M. (2012). Comparison of underlay and overlay spectrum sharing strategies in MISO cognitive channels. *Proceedings of 7th International ICST Conference on Cognitive Radio Oriented Wireless Networks and Communications* (CROWNCOM), Stockholm, Sweden (18–20 June 2012), 224–229, doi: https://doi.org/10.4108/icst.crowncom.2012.248283.

111 Thakur, P., Kumar, A., Pandit, S., Singh, G., and Satasia, S.N. (2016). Frame structures for hybrid spectrum accessing strategy in cognitive radio communication system. *Proceedings of Ninth International Conference on Contemporary Computing* (IC3), Noida, India (11–13 August 2016), 1–6, doi: https://doi.org/10.1109/IC3.2016.7880206.

112 Thakur, P., Kumar, A., Pandit, S. et al. (2017). Advanced frame structures for hybrid spectrum access strategy in cognitive radio communication systems. *IEEE Communications Letters* 21 (2): 410–413. https://doi.org/10.1109/LCOMM.2016.2622260.

113 Dudley, S.M., Headley, W.C., Lichtman, M. et al. (2014). Practical issues for spectrum management with cognitive radios. *Proceedings of the IEEE* 102 (3): 242–264. https://doi.org/10.1109/JPROC.2014.2298437.

114 President's Council of Advisors on Science and Technology (U.S.), Report to the President: realizing the full potential of government-held spectrum to spur economic growth. https://science.osti.gov/About/PCAST (accessed 22 December 2020).

115 EUR-Lex - 32012D0243 - EN - EUR-Lex. http://eur-lex.europa.eu/legal-content/EN/ALL/?uri=CELEX%3A32012D0243 (accessed 18 February 2018).

116 Buddhikot, M.M. (2007). Understanding dynamic pectrum access: models, taxonomy and challenges. *Proceedings of 2nd IEEE International Symposium on New Frontiers in Dynamic Spectrum Access Networks* (April 2007), 649–663, doi: https://doi.org/10.1109/DYSPAN.2007.88.

117 Enabling Innovative Small Cell Use In 3.5 GHZ Band NPRM & Order. *Federal Communications Commission* (12 December 2015). https://www.fcc.gov/document/enabling-innovative-small-cell-use-35-ghz-band-nprm-order (accessed 18 February 2018).

118 Xu, Y., Zhao, X., and Liang, Y.C. (2015). Robust power control and beamforming in cognitive radio networks: A survey. *IEEE Communication Surveys and Tutorials* 17 (4): 1834–1857. https://doi.org/10.1109/COMST.2015.2425040.

119 Naeem, M., Anpalagan, A., Jaseemuddin, M., and Lee, D.C. (2014). Resource allocation techniques in cooperative cognitive radio networks. *IEEE Communication Surveys and Tutorials* 16 (2): 729–744. https://doi.org/10.1109/SURV.2013.102313.00272.

120 Ahmed, E., Gani, A., Abolfazli, S. et al. (2016). Channel assignment algorithms in cognitive radio networks: taxonomy, open issues, and challenges. *IEEE Communication Surveys and Tutorials* 18 (1): 795–823. https://doi.org/10.1109/COMST.2014.2363082.

121 Tanab, M.E. and Hamouda, W. (2017). Resource allocation for underlay cognitive radio networks: a survey. *IEEE Communication Surveys and Tutorials* 19 (2): 1249–1276. https://doi.org/10.1109/COMST.2016.2631079.

2

Advanced Frame Structures in Cognitive Radio Networks

2.1 Introduction

We have perceived the potentials of spectrum accessing techniques in the cognitive radio networks where the hybrid spectrum access (HSA) is an advanced technique which exploits the interweave and underlay spectrum access strategies simultaneously to enhance the spectral efficiency and throughput [1–5]. In HSA, the cognitive user (CU) senses its environment and transmits the data with full and constrained power on the idle and active sensed channels, respectively [3, 5]. As the spectrum sensing is a very important phenomenon, therefore its performance should be significantly high. The performance metrics of the spectrum sensing technique are the probability of detection (P_d) and probability of false alarm (P_f) [6–12] and values of these metrics must be high and low, respectively. To achieve this criterion, the constraint is on the sensing time that it needs to be high, which imposes the lower-bound on the value of sensing time [13, 14]. It is worth mentioning that the spectrum sensing and data transmission time shows inversely proportional relation with each other which decreases the achievable throughput of the cognitive radio network with increase in sensing time. This inverse relation between sensing reliability and throughput is known as sensing-throughput trade-off [15–20]. To improve the throughput, we need to improve the sensing-throughput trade-off and modify the relation between sensing and data transmission time. Various researchers have modified this relation for the conventional (interweave approach) CRNs by introducing improved frame structures [21–23]. The throughput maximization for CU is the prime objective of the CRNs; therefore, numerous researchers have exploited the HSA and frame structures individually for throughput maximization [1–23]. However, to the best of the author's knowledge, the improved frame structures are not incorporated in the HSA techniques to analyze the combined effects on the throughput of CU and data loss. Therefore, in this chapter, we have presented a novel hybrid-cum-improved

Spectrum Sharing in Cognitive Radio Networks: Towards Highly Connected Environments,
First Edition. Prabhat Thakur and Ghanshyam Singh.

spectrum access technique (HISA) for significant improvement in the throughput, data-loss rate, and interference at the PU.

Further, this chapter is structured as follows. Section 2.2 comprises the previous work related to spectrum accessing and potential frame structures. In Section 2.3, the proposed frame structures and their combination with HSA are illustrated in detail. Section 2.4 consists of the mathematical analysis of the throughput and data loss. Further, the proposed work is simulated in Section 2.5 and finally concluding remarks are presented in Section 2.6.

2.2 Related Work

The spectrum accessing strategies and frame structures are the key elements of the HISA techniques; therefore, the work related to these elements is presented as follows.

2.2.1 Frame Structures

The conventional frame structure for CRN comprises two phases namely, the sensing phase and data transmission phase as shown in Figure 2.1a [3]. Here, the CU senses the channel for time τ and rest of the time $(T - \tau)$ is used for data transmission. However, the major confront of this frame structure is the sensing-throughput trade-off [15]. Therefore, in order to conquer this limitation, a novel architecture has been proposed in [15], where the sensing and transmission are simultaneous phenomena. The sensing information achieved through Nth frame is used for data transmission in $(N + 1)$th frame as shown in Figure 2.1b. However, the limitation of this approach is the use of outdated sensing information for data transmission. Since the sensing information of the previous frame is used for transmission in the current frame, the CU is unable to regulate its power due to unawareness of the PU status in the current frame. Thus, the entire data of colliding frame on the emergence of PU gets lost and results the decrease in throughput. To improve the throughput, a new framework is proposed by Pandit and Singh [21], in which the frame is divided into two or more blocks and each block comprises two sub-blocks, i.e. the header overhead and data payload block as depicted in Figure 2.1c, where the header overhead consists of flag-bit. The spectrum sensing is a continuous phenomenon in the entire time frame; however, the results are taken out at the particular overhead block. The flag-bit uses the sensing result of same frame which is computed up to that time (starting time of the header overhead block) and the flag-bit is set if the sensing results are different from previous frames' sensing results. The frame structure presented in Figure 2.1c conquers the

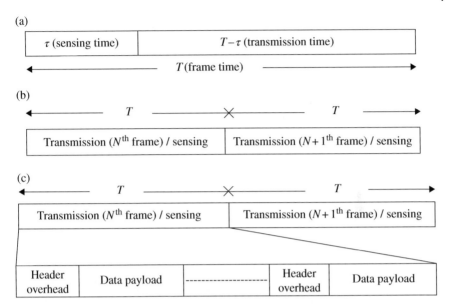

Figure 2.1 The frame structure for cognitive radio network: (a) conventional, (b) advanced, and (c) improved [4, 6]. *Source*: From Thakur et al. [5]. © 2017, IEEE.

limitations of previous frame structure shown in Figure 2.1b because the frames' sensing decision is used in the same frame to make spectrum access decision.

2.2.2 Spectrum Accessing Strategies

The spectrum accessing is an essential element of the CRNs and various researchers/scientists have illustrated the spectrum accessing techniques which are categorized as interweave, underlay, and hybrid [5]."

The interweave is the prime spectrum accessing strategy in which the CU senses its environment and sets up communication only on the idle sensed channels [24, 25], which means the spectrum sensing is a prerequisite for this approach as shown in Figure 2.2a. However, in the underlay technique [26], the sensing is not a compulsory phenomenon because the CU establishes communication to protect the PU from interference as shown in Figure 2.2b. The conventional strategies used to save the PU from the interference are power control, beamforming and spread spectrum, etc. [26]. The interweave and underlay techniques are incapable to use the spectrum shrewdly because, in the interweave approach, the CU needs to switch the communication on the emergence of PU; however, it has to stop the communication when all other channels are active/busy. On the contrary, in the underlay technique, the seamless communication is promising; however,

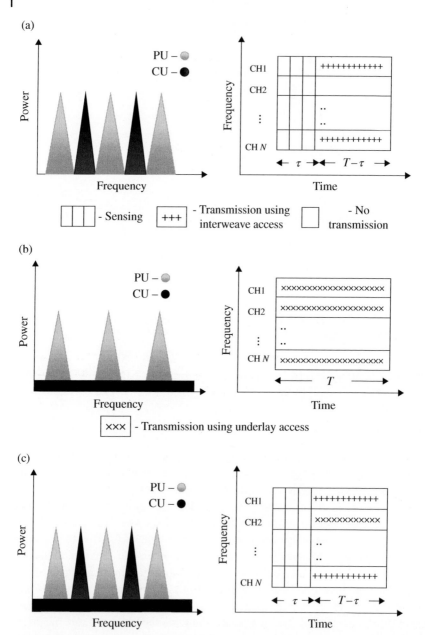

Figure 2.2 The spectrum access strategies in cognitive radio communication systems: (a) interweave, (b) underlay, and (c) hybrid access.

the achievable data rate is less due to low power transmission even in the absence of PU [26]. Therefore, to overcome the limitations of both the approaches, a hybrid approach which comprises these two approaches has been proposed [1–4], and the spectrum sensing is the prerequisite as shown in Figure 2.2c. In this technique, the active and idle channels are accessed via underlay and interweave approach, respectively. The authors' potential contribution in this chapter is summarized as follows."

- Two novel frame structures are proposed in which idle sensed channels are accessed via interweave strategy; however, active sensed channels are using underlay strategy.
- The closed-form expressions for the throughput of CU and data loss in the proposed frame structures are derived.
- The numerical simulations are illustrated to verify the effectiveness of the proposed frame structures.

2.3 Proposed Frame Structures for HSA Technique

The data communication using time frames is in fashion and is also exploited significantly in the HSA technique. However, the frame structure for this technique is still in its infantry stage as shown in Figure 2.2a, where the entire time frame is divided into two phases, namely, the sensing phase and data transmission phase and its limitation is similar to that of the conventional frame structure as shown in Figure 2.1a, i.e. the sensing-throughput trade-off. Therefore, in order to avoid this trade-off, we have proposed two novel frame structures for HSA technique as shown in Figures 2.3 and 2.4. The proposed frame structures outperform over the already illustrated frame structures and are shown in Figure 2.1b and 2.1c because the idle and active sensed channels are accessed using the interweave and underlay access strategies, respectively. However, in Figure 2.1b and 2.1c, only the idle sensed channels are accessed for communication via the interweave strategy and the active sensed channels remain unutilized. In the first proposed frame structure as shown in Figure 2.3, the spectrum sensing and data transmission are parallel phenomenon; however, the sensing decision at the previous frame is used for data transmission at the current frame as similar to Figure 2.1b. As the sensing information confirms the switching of channels from idle to busy/active or vice-versa, the CU needs to switch the transmission from interweave to underlay or vice-versa, respectively. However, the limitation of this model is similar to that of the frame structure in Figure 2.1b, i.e. as the channel switches from idle to active status, the data of colliding frame gets lost.

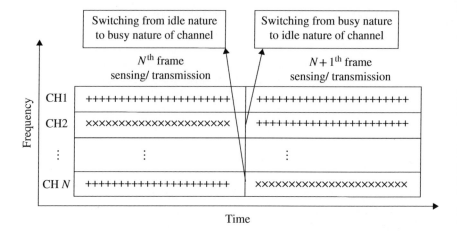

Figure 2.3 First proposed frame structure for hybrid-cum-improved-spectrum access technique.

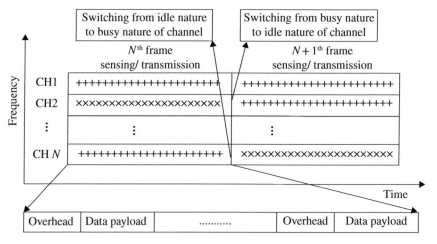

Figure 2.4 Second proposed frame structure for hybrid-cum-improved-spectrum access technique.

Therefore, in order to conquer this issue, one more frame structure for HSA technique is proposed as shown in Figure 2.4, where the frame is divided into two or more blocks and each block consists of two sub-blocks, namely, the overhead and data payload. The overhead block contains flag-bit and gets set if the sensing information (from the same frame up to the starting time of overhead block) decision is different from the previous frame. Thus, the problem of data loss in the colliding frame for the HSA strategy is conquered."

2.4   Analysis of Throughput and Data Loss

The throughput maximization for the conventional CRN has been exploited in detail in [3]. For the spectrum sensing, the binary hypothesis for the received signal $r(t)$ is as follows [6, 8]. The H_0 and H_1 confirms the absence and presence of the PU, respectively.

$$r(t) = \begin{cases} h.s(t) + w(t), & H_1 : (Presence\ of\ PU) \\ w(t), & H_0 : (Absence\ of\ PU) \end{cases}.$$

where h, $w(t)$, and $s(t)$ signifies the channel gain, additive white Gaussian noise (AWGN), and the transmitted signal of PU, respectively. For the simplicity of mathematical expressions, initially, we have derived the data rates for the four possible cases of the transmission as shown in Table 2.1, in which h_{ss} and h_{ps} are the channel gains of the CU-pair and PU transmitter to the CU receiver, respectively. The N_p and P_{pu} designates the noise power at CU receiver and PU transmitted power, respectively. The probability of correct detection of the PU (P_d) describes under the hypothesis H_1,whereas the probability of false alarm of the presence of PU(P_f) is defined under the hypothesis H_0 [27, 28]. For the better protection of PU, the value of P_d must be high; however, the CU can access the available band efficiently if the value of P_f is low. The expression for throughput of the conventional single channel cognitive radio is [3]:

$$TP = \frac{T-\tau}{T} \left[P(H_1)\left(1 - P_f\right) C_{00} + P(H_0)(1 - P_d) C_{01} \right], \tag{2.1}$$

where $P(H_0)$ and $P(H_1)$ are the probabilities of channel being idle and active, respectively. The throughput as given by Eq. (2.1) is with respect to the conventional frame structure for the single-channel CRN as shown in Figure 2.1a.

Table 2.1   The data rates of CU for various conditions.

Sensing state	Original state	Data rate
Idle	Idle	$C_{00} = log_2\left(1 + \frac{P_1 h_{ss}}{N_p}\right)$
Active	Idle	$C_{10} = log_2\left(1 + \frac{P_2 h_{ss}}{N_p}\right)$
Active	Active	$C_{11} = log_2\left(1 + \frac{P_2 h_{ss}}{N_p + P_{pu} h_{ps}}\right)$
Idle	Active	$C_{01} = log_2\left(1 + \frac{P_1 h_{ss}}{N_p + P_{pu} h_{ps}}\right)$

Now, we need to compute the throughput for HSA technique on the basis of the concept as discussed in Section 2.3. We assume that if the PU is accessing the channel with interweave approach, it transmits power P_1; otherwise, with underlay approach it transmits power P_2. The expression for throughput of the nth channel in Figure 2.2c is:

$$TP_c = \frac{T - \tau}{T} \left[P(H_0)(1 - P_f) C_{00} + P(H_1)(1 - P_d) C_{01} \right.$$

$$\left. + P(H_0)(P_f) C_{10} + P(H_1)(P_d) C_{11} \right]. \tag{2.2}$$

Now, the throughput of the first proposed scheme, i.e. TP_1 (as described in Eq. (2.3)), which is similar to that of Eq. (2.2) except the multiplying factor $(T - \tau)/T$, because the sensing and transmission are parallel phenomenon due to which zero-time is spent on the sensing only, and the entire time frame T is used for the data transmission. However, in the second proposed scheme, definite time is used for the control-overhead where it needs to stop data transmission. Therefore, to compute the achievable throughput, we have considered this overhead time and throughput in this is TP_o as given in Eq. (2.4), where x and k are the overhead time and number of overheads per frame, respectively, in Figure 2.4. The achievable throughput of the second proposed approach (TP_2) is the difference between the throughput of first proposed approach TP_1 and throughput in control-overhead time TP_o.

$$TP_1 = \left[P(H_0)(1 - P_f) C_{00} + P(H_1)(1 - P_d) C_{01} + P(H_0)(P_f) C_{10} \right.$$

$$\left. + P(H_1)(P_d) C_{11} \right], \tag{2.3}$$

$$TP_o = \frac{x \times k}{T} \left[P(H_0)(1 - P_f) C_{00} + P(H_1)(1 - P_d) C_{01} + P(H_0)(P_f) C_{10} \right.$$

$$\left. + P(H_1)(P_d) C_{11} \right], \tag{2.4}$$

$$TP_2 = TP_1 - TP_o. \tag{2.5}$$

The data loss occurs in the system when the data are transmitted with full-power P_1 (interweave transmission) and the PU resumes its transmission during data transmission period which means there is no data loss in the underlay transmission. As the hybrid transmission consists of interweave and underlay transmission, the data loss of interweave and hybrid approach will be the same. The data loss of conventional (interweave) is total transmitted data during the data transmission phase and is defined as:

$$DL_c = \frac{T - \tau}{T} \left[P(H_0)(1 - P_f) C_{00} + P(H_1)(1 - P_d) C_{01} \right]. \tag{2.6}$$

The first proposed frame structure transmits during the entire time frame; therefore, the data loss is defined as:

$$DL_1 = \left[P(H_0)(1-P_f)\,C_{00} + P(H_1)(1-P_d)\,C_{01}\right]. \tag{2.7}$$

The data loss in the first proposed approach is very high; however, in the second proposed model, only particular data payload block will be lost and the data loss is defined as:

$$DL_2 = \left[\frac{1}{k}\left\{P(H_1)(1-P_d)\,C_{01} + P(H_0)(1-P_f)\,C_{00}\right\}\right]$$
$$-\left[\frac{x}{T}\left\{P(H_0)(1-P_f)\,C_{00} + P(H_1)(1-P_d)\,C_{01}\right\}\right]. \tag{2.8}$$

Before moving to the next section on simulation results and discussion, it is worth naming all the considered cases in the simulation. We have considered six cases and shown the results for these cases which are as follows: (i) *Conv.* – When the data are transmitted with interweave access technique in the data transmission period of the frame structure depicted in Figure 2.1a. (ii) HSA with conventional frame structure (*Hybrid-Conv-F*) – When the data are transmitted with the HSA in the data transmission period of the frame structure depicted in Figure 2.1a. (iii) Conventional spectrum access with first advanced frame structure (*Conv-1st-F*) – When the data are transmitted with the interweave spectrum access in the data transmission period of the frame structure depicted in Figure 2.1a. (iv) Conventional spectrum access with second improved frame structure (*Conv-2nd-F*) – When the data are transmitted with the interweave spectrum access in the data transmission period of the frame structure depicted in Figure 2.1b. (v) HSA with first proposed frame structure (Hybrid-1st-prop-F) – When the data are transmitted with the HSA in the data transmission period of the frame structure depicted in Figure 2.3. (vi) HSA with second proposed frame structure (Hybrid-2nd-prop-F) – When the data are transmitted with the HSA in the data transmission period of the frame structure depicted in Figure 2.4.

2.5 Simulations and Results

The numerically simulated results of the proposed approach are illustrated and compared with the conventional approach. The values of simulation parameters are selected by inspiring from the IEEE 802.22 wireless regional area network (WRAN) standard [29–31]. The values of metrics considered for the numerical simulation are tabulated in Table 2.2. The simulation tool we have exploited for

Table 2.2 The numerical values of the simulation metrics.

Metric	Value	Metric	Value	Metric	Value
T	100 ms	P_p	4 W	fs	6 MHz
$P(H_1)$	0.1	h_{ss}	0.8	N_p	0.04 W
$P(H_0)$	0.9	h_{ps}	0.1	P_2	0.5 W
P_f	0.1	P_1	6 W	x	5 ms
k	4	τ	5 ms		

numerical simulations is MATLAB 2010 [32]. The value of P_d for the given values of τ, P_f, fs, and γ is computed using Eq. (2.5) as in [8]. The throughput of CU varies with sensing time for various spectrum access techniques as follows: (i) *Conv.*, (ii) Hybrid-Conv-F, (iii) Conv-1st-F, (iv) Conv-2nd-F, (v) Hybrid-1st-prop-F, and (vi) Hybrid-2nd-prop-F, and are illustrated in Figure 2.5.

It is perceived that the throughput in approaches with advanced frame structures is non-reducing as that of the conventional approaches. The proposed hybrid approaches, i.e. the Hybrid-1st-prop-F and Hybrid-2nd-prop-F, provide increased

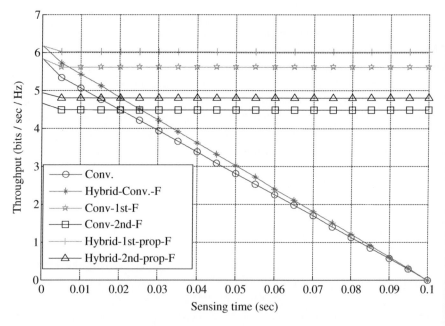

Figure 2.5 The throughput for conventional and proposed approaches.

throughput as compared to the conventional approaches with the similar frame structures, due to proficient use of active channels through the underlay transmission."

The variations of throughput of CU with the probability of channel being active $(P(H_1))$ are described in Figure 2.6. It is apparent that the Hybrid-1st-prop-F provides the highest throughput which is inversely proportional to the $P(H_1)$. The throughput of the Conv-2nd-F is minimum among all the considered scenarios; however, in the proposed Hybrid-2nd-prop-F, the throughput is low only till the $P(H_1)$ is 0.5 and above 0.5, it becomes the third highest throughput among all considered scenarios. Moreover, the throughput becomes zero in the conventional approaches; however, in the hybrid approaches, it never reaches zero due to underlay transmission in the presence of PU. In order to examine the data loss of proposed and conventional approaches, the relationship between the data loss and probability of channel being active is depicted in Figure 2.7. The data loss in the Hybrid-1st-prop-F and Conv-1st-F is maximum, as entire time frame gets lost on the emergence of PU. However, in the Hybrid-2nd-prop-F and Conv-2nd-F approaches, the data loss is relatively less because only the data of the particular data payload block gets lost.

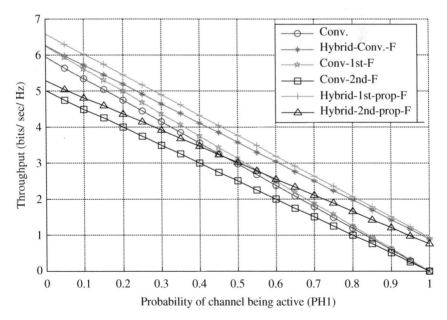

Figure 2.6 The throughput versus the probability of channel being active for various approaches.

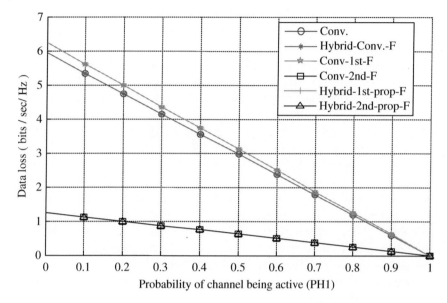

Figure 2.7 The data loss of the CU versus the probability of channel being active for various approaches.

Moreover, the data loss only occurs if the data are transmitted with full-power (P_1), i.e. interweave transmission and PU appear during the transmission; therefore, the data loss of interweave and hybrid transmission remains the same and obtained results also support this statement. Moreover, the data loss confirms the inverse relation with $P(H_1)$ because of the interweave approach in which the transmitted data decreases with increase in $P(H_1)$. "In addition to this, since in Hybrid-2nd-prop-F, the CU switches its interweave transmission to the underlay transmission on the emergence of PU, therefore the interference at the PU is also decreased. As a result, both the proposed approaches are suitable for different applications with the network desires, i.e. (i) for the CRNs with the high throughput demand, the first approach is appropriate, and (ii) to guarantee the reliable communication, i.e. minimum data loss and minimum interference to PU, the second approach outperforms."

2.6 Summary

In this chapter, a novel hybrid-cum-improved spectrum accessing strategy (HISA) is proposed according to which two frame structures for the HSA strategy are projected. Further, the mathematical expressions of the throughput and data loss for the HSA strategy with conventional and proposed frame structures are derived. The proposed approaches outperform the conventional approaches in terms of throughput and data loss; however, the selection of particular approach depends on the requirements and applications of the CRN. The first approach outperforms with reference to the throughput maximization; however, for the reduced data loss and reliable PU communication, the second approach performs well. The throughput and data loss optimization by constraining various metrics is an exigent task which is a potential issue for the future research."

References

1 Jiang, X., Wong, K.K., Zhang, Y., and Edwards, D. (2013). On hybrid overlay underlay dynamic spectrum access: double-threshold energy detection and Markov model. *IEEE Trans. Veh. Technol.* 62 (8): 4078–4083.

2 Thakur, P. and Singh, G. (2020). Power management for spectrum sharing in cognitive radio communication system: a comprehensive survey. *J. Electromagn. Wave Appl.* 34 (4): 407–461.

3 Yang, C., Lou, W., Fu, Y. et al. (2016). On throughput maximization in multichannel cognitive radio networks via generalized access strategy. *IEEE Trans. Commun.* 64 (4): 1384–1398.

4 Gmira, S., Kobbane, A., and Sabir, E. (2015). A new optimal hybrid spectrum access in cognitive radio: overlay-underlay mode. *2015 International Conference on Wireless Networks and Mobile Communications (WINCOM)*, Marrakech, Morocco (20–23 October 2015), 1–7.

5 Thakur, P., Kumar, A., Pandit, S. et al. (2017). Advanced frame structures for hybrid spectrum access strategy in cognitive radio communication systems. *IEEE Commun. Lett.* 17 (2): 410–413.

6 Yucek, T. and Arslan, H. (2009). A survey of spectrum sensing algorithms for cognitive radio applications. *IEEE Commun. Surv. Tutorials* 11 (1): 116–130.

7 Ali, A. and Hamouda, W. (2017). Advances on spectrum sensing for cognitive radio networks: theory and applications. *IEEE Commun. Surv. Tutorials* 19 (2): 1277–1304.

8 Gafoor, R.A., Kuriakose, R., Sibila, M. et al. (2018). Survey on traditional and advanced spectrum sensing techniques in cognitive radio networks. *2018 International Conference on Control, Power, Communication and Computing Technologies (ICCPCCT)*, Kanpur, India (23–24 March 2018), 65–72.

9 Deng, M., Hu, B.-J., and Li, X. (2017). Adaptive weighted sensing with simultaneous transmission for dynamic primary user traffic. *IEEE Trans. Commun.* 65 (3): 992–1004.

10 Wu, J., Wang, C., Yu, Y. et al. (2020). Performance optimization of cooperative spectrum sensing in mobile cognitive radio networks. *IET Commun.* 14 (6): 1028–1036.

11 Sharma, S.K., Bogale, T.E., Le, L.B. et al. (2017). Dynamic spectrum sharing in 5G wireless networks with full-duplex technology: recent advances and research challenges. *IEEE Commun. Surv. Tutorials* 20 (1): 674–707.

12 Karimi, M., Sadough, S.M.S., and Torabi, M. (2019). Improved joint spectrum sensing and power allocation for cognitive radio networks using probabilistic spectrum access. *IEEE Syst. J.* 13 (4): 3716–3723.

13 Owayed, A.A., Mohammed, Z.A., and Mosa, A.A.R. (2010). Probabilities of detection and false alarm in multitaper based spectrum sensing for cognitive radio systems in AWGN. *2010 IEEE International Conference on Communication Systems* (9–11 July 2015), 579–584.

14 Bhattacharjee, S., Das, P., Mandal, S., and Sardar, B. (2015). Optimization of probability of false alarm and probability of detection in cognitive radio networks using GA. *2015 IEEE 2nd International Conference on Recent Trends in Information Systems (ReTIS)*, Kolkata, India (9–11 July 2015) 53–57.

15 Liang, Y.-C., Zeng, Y., Peh, E.C.Y., and Hoang, A.T. (2008). Sensing-throughput tradeoff for cognitive radio networks. *IEEE Trans. Wireless Commun.* 7 (4).

16 Pei, Y., Hoang, A.T., and Liang, Y.-C. (2007). Sensing-throughput tradeoff in cognitive radio networks: how frequently should spectrum sensing be carried out? *2007 IEEE 18th International Symposium on Personal, Indoor and Mobile Radio Communications*, Athens, Greece (3–7 September 2007), 1–6.

17 Ahmad, A.W., Wang, H., and Lee, C. (2015). Maximizing throughput with wireless spectrum sensing network assisted cognitive radios. *Int. J. Dis. Sens. Netw.* 2015: 1–10.

18 Verma, G. and Sahu, O.P. (2017). Removal of sensing-throughput tradeoff barrier in cognitive radio networks. *Wireless Pers. Commun.* 94: 1477–1490.

19 Kan, C., Wu, Q., Ding, G., and Song, F. (2014). Sensing-throughput trade-off for interference-aware cognitive radio networks. *Frequenz* 68 (3–4): 97–108.

20 Bhogale, T.E., Vandendorpe, L., and Le, L.B. (2014). Sensing throughput tradeoff for cognitive radio networks with noise variance uncertainty. *2014 9th International Conference on Cognitive Radio Oriented Wireless Networks and Communications (CROWNCOM)*, Oulu, Finland (2–4 June 2014), 435–441.

21 Pandit, S. and Singh, G. (2014). Throughput maximization with reduced data loss rate in cognitive radio network. *Telecommun. Syst.* 57 (2): 209–215.

22 Stotas, S. and Nallanathan, A. (2010). Overcoming the sensing throughput tradeoff in cognitive radio networks. *IEEE International Conference on Communication (ICC)*, Cape Town (23–27 May 2010), 23–27.

23 Stotas, S. and Nallanathan, A. (2010). On the throughput maximization of spectrum sharing cognitive radio networks. *IEEE Global Telecommunications Conference (GLOBECOM 2010)*, Miami, FL (6–10 December 2010), 6–10.

24 Lee, Y.L., Saad, W.K., Abd El-Saleh, A., and Ismail, M. (2013). Improved detection performance of cognitive radio networks in AWGN and Rayleigh fading environments. *J. Appl. Res. Technol.* 11 (3): 437–446.

25 Liu, X., Jia, M., and Tan, X. (2013). Threshold optimization of cooperative spectrum sensing in cognitive radio networks. *Radio Sci.* 48 (1): 23–32.

26 Sharma, S.K., Bhogle, T.E., Chatzinotas, S. et al. (2015). Cognitive radio techniques under practical imperfections: a survey. *IEEE Commun. Surv. Tutorials* 17 (4): 1858–1854.

27 Suliman, I.M., Lehtomäki, J., and Umebayashi, K. (2015). On the effect of false alarm rate on the performance of cognitive radio networks. *EURASIP J. Wireless Commun. Netw.* 2015: 244.

28 Altrad, O. and Muhaidat, S. (2013). A new mathematical analysis of the probability of detection in cognitive radio over fading channels. *EURASIP J. Wireless Commun. Netw.* 2013 (1): 159.

29 Gupta, S. and Malagar, V. (2017). IEEE 802.22 standard for regional area networks. *2017 International Conference on Next Generation Computing and Information Systems (ICNGCIS)*, Jammu, India (11–12 December 2017), 126–130.

30 Ko, G., Franklin, A.A., You, S.-J. et al. (2010). Channel management in IEEE 802.22 WRAN systems. *IEEE Commun. Mag.* 48 (9): 88–94.

31 Stevenson, C.R., Chouinard, G., Lei, Z. et al. (2009). IEEE 802.22: the first cognitive radio wireless regional area network standard. *IEEE Commun. Mag.* 47 (1): 130–138.

32 MathWorks MathWorks introduceert Release 2015b van de MATLAB en Simulink productseries. https://nl.mathworks.com/company/newsroom/mathworks-announces-release-2015b-of-the-matlab-and-simulink-product-families.html (accessed 19 February 2018).

3

Cognitive Radio Network with Spectrum Prediction and Monitoring Techniques

3.1 Introduction

In Chapter 2, we have explored the hybrid-cum-improved spectrum accessing (HISA) strategy and have achieved the improved throughput and data loss. The key confront with the HISA technique is that the cognitive user (CU) is unable to detect the emergence of the primary user (PU) during CUs' data transmission, which is a desired phenomenon in the high-traffic wireless communication environments. The high-traffic wireless communication environment is defined as when the traffic intensity of the PUs' channel is greater than 0.5. The traffic intensity is defined later in this chapter in Section 3.3.1. It is perceivable from the discussion presented in Chapter 1 that the spectrum sensing has a significant role in the cognitive cycle to perceive the state of channels; therefore, its performance should be significantly accurate and advanced. The performance metrics of the spectrum sensing technique are the probability of detection (P_d) and probability of false-alarm (P_f) [1–7], where the CU requires high and low values of P_d and P_f, respectively. However, the spectrum sensing time needs to be significantly large to yield these requirements [8]. Since the spectrum sensing and data transmission time shows an inverse relation with each other, the researchers have modified this relation by introducing advanced frame structures to maximize the throughput and spectrum sensing reliability as discussed in Chapter 2 [9–13].

The PU channels are highly utilized in the high-traffic cognitive radio networks (HTCRNs) which results the decrease in the idle probability of channels. However, the CU picks the channel for spectrum sensing randomly which causes the high probability of active sensing of channels and decreases the throughput severely [14]. Thus, the significant data loss in HTCRNs affects the performance parameters such as an increase in the energy consumption and operating time, which are illustrated in [15–17]. Pei et al. [15] have presented an energy-efficient design of sequential channel sensing as well as optimized the power allocation, spectrum sensing time, and spectrum sensing order to improve the throughput of CU.

Spectrum Sharing in Cognitive Radio Networks: Towards Highly Connected Environments, First Edition. Prabhat Thakur and Ghanshyam Singh.
© 2021 John Wiley & Sons, Inc. Published 2021 by John Wiley & Sons, Inc.

The optimization problem where the throughput is maximized by constraining the spectrum sensing reliability is presented in [16], which is known as sensing-throughput trade-off [17–22].

The key objective of the spectrum prediction is to improve the spectrum sensing performance by sensing only the idle predicted channels and avoid the interference at PU. In order to consider the imperfect spectrum predictions in the HTCRN for throughput maximization, a novel frame structure has been presented in [14]. The presented time frame comprises three phases namely, the prediction phase, spectrum sensing phase, and data transmission phase. In the prediction phase, the CU predicts the idle channels on the bases of previous information whereas in the spectrum sensing phase, the CU only senses the idle predicted channels. Moreover, in the data transmission phase, the CU transmits data only on the idle sensed channels whereas it needs to wait on the active sensed channels which cause the problem of wait/switch/stop states.

- *Wait state*: If the PU is sensed as active during the spectrum sensing phase, the CU has to wait for the next frame to sense the other channels.
- *Switch state*: If the PU is detected as active during the data transmission period, the CU needs to switch its transmission on other idle channels.
- *Stop state*: If the PU is detected as active during the data transmission period, the CU needs to stop the communication if all other channels are active.

The key limitations of the proposed frame structure presented in [14] are as follows: (i) wait state and (ii) high data loss as well as the introduction of interference-at-PU, if the PU resumes communication during the data transmission period. The high data loss and interference occur because the CU is unable to detect the emergence of PU in the data transmission period and continue its data transmission with an interweave approach, even in the presence of PU [23–26]. However, if the CU is able to detect the emergence of PU in the data transmission period, the data transmission (using an interweave approach) can suspend or can use underlay approach for the data transmission. Thus, the data loss and interference-at-PU can be improved significantly. Therefore, the key contributions to resolve the aforementioned potential issues are summarized as follows:

- To conquer the issue of wait states, a novel approach is proposed in which the hybrid spectrum access technique named as Approach-1 is exploited in the same frame structure as reported in [14]. The proposed approach overcomes the challenges of wait state as well as improves the throughput of high-traffic CRN.
- Moreover, the concept of spectrum monitoring during the data transmission period is exploited to lighten the data loss of CU and interference-at-PU, and this is named as Approach-2.

The remainder of this chapter is structured as follows. In the next section, the related work is presented. Section 3.3 comprises the illustration of system models of the proposed approaches. In Section 3.4, the performance analysis of the proposed system models is illustrated. Further, the simulation results and analysis are presented in Section 3.5 and finally, Section 3.6 summarizes the work and discusses the future scope.

3.2 Related Work

3.2.1 Spectrum Prediction

The spectrum prediction is defined as a phenomenon of forecasting the future channel state on the basis of the pre-available information. Various researchers have proposed several approaches for spectrum prediction in the literature [14, 26–34]. In [26], Yarkan and Arslan have explored the concept of spectrum prediction in cognitive radio network by using a binary time series approach, which relies on the pre-available spectrum states and a linear prediction is performed to yield the next state of channel. However, the major limitation of this technique is the nondeterministic nature of the binary series. In [27], the two-channel state predictor based on the neural network using multilayer perceptron (MLP) and hidden Markov model (HMM) is presented and it is reported that both the schemes do not require the prior knowledge of the channel states, which is the more practical scenario. Moreover, it is illustrated that MLP is trained once whereas HMM is trained at particular intervals. However, the MLP provides facility to get train over the sequential interval in the dynamic environments. In [28], the authors have presented a survey in which various spectrum prediction techniques are described. Subsequently, Christian et al. [29] have presented an advanced low-interference channel status prediction algorithm (LICSPA) to improve the prediction errors and interferences at the PU. Further, it is reported that the proposed algorithm reduces the interference at PU up to 40% when compared with the conventional approach of prediction. In [30], the authors have explored the effects of cooperation among the CUs in order to improve the performance of spectrum prediction. In addition to this, the performance of various prediction approaches and fusion scenarios has been investigated by exploiting the different traffic conditions of PUs for cooperative prediction. Moreover, it is concluded that the prediction accuracy is also the function of PU traffic intensity and its rate of change. Furthermore, it is also illustrated that the cooperation among CUs enhances the accuracy of spectrum prediction at the cost of increased complexity and therefore, there must be a trade-off between these two metrics. The key intent of spectrum prediction is to enhance the spectrum sensing performance by sensing only the idle predicted channels and escaping the energy loss by avoiding interference at the PU. In order to consider the improved spectrum predictions in the HTCRN for throughput maximization, a new frame structure has been presented in [14]. In this frame structure, the time frame is divided into three phases: the spectrum prediction phase, spectrum sensing phase, and data transmission phase. In the prediction phase, the CU predicts the idle channels on the basis of previous available information whereas in the spectrum sensing phase, the CU senses only the randomly selected channel among idle predicted channels. Moreover, in the data transmission phase, the CU transmitted data only on the idle sensed channels. Recently, in [31], the authors have illustrated pattern-sequence-based forecasting (PSF) to predict the spectrum occupancy prediction in cognitive radio networks while [32] emphasized over the hybrid genetic and shuffled frog-leaping algorithm for neural network structure optimization.

3.2.2 Spectrum Monitoring

The classical frame structure for CRNs with the spectrum prediction ability is depicted in Figure 3.1b. The bottleneck criterion of this frame structure is the time-critical nature of the spectrum prediction, spectrum sensing, and data transmission, which needs to perform individually at specific time slots, which means it is almost impossible to execute two or more phenomena at the same time. This restricts the CU to detect the emergence of PU in the data transmission period which formulates the problems of data loss of CU and interference to PU. An effective method to trounce this issue is the spectrum monitoring which is a promising

(a)

(b)

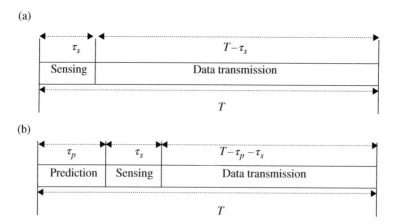

Figure 3.1 Frame structures in cognitive radio communication: (a) conventional and (b) using prediction slot [14]. *Source*: From Yang and Zhao [14].

technique that enables the CU to detect the emergence of PU even in the data transmission phase as reported in [35–39]. In [35], the authors have explored the concept of spectrum monitoring in detail and have presented the potential approaches for it on the basis of receiver statistics such as receiver error count (*REC*). In this approach, the CU receives the data from the transmitter and counts the error in the received packet; if PU resumes its data transmission, there will be deprivation in the performance of the received signal that results in a large number of errors in the received packet. Thus, the increased number of errors signifies the emergence of PU. In addition, the authors have illustrated the effect of cooperation on the spectrum monitoring using two CUs. Ali and Hamouda [36] have illustrated an energy ratio spectrum monitoring algorithm for OFDM and presented the mathematical expression for the detection probability and probability of false alarm for additive white Gaussian noise (AWGN) channels. Orooji et al. [37] have presented a novel decision statistic for spectrum monitoring by using REC and have analyzed the detection and false-alarm probabilities for the same.

3.3 System Models

3.3.1 System Model for Approach-1

A CRN consisting of the pair of transceivers in the high-traffic environment is considered, where the hybrid spectrum access technique is used to access the idle sensed channels. The PU network is assumed to be cellular and the PUs' activities on each channel are independent since these channels can belong to different networks. In case of the spectrum sensing, the binary hypotheses considered for the received signal $r(t)$ are H_0 and H_1, which confirms the absence and presence of PU, respectively, as discussed in Chapter 2, Section 2.4. To attain the reliable spectrum sensing results to protect the PU, the spectrum sensing duration must be greater than that of the predefined threshold value as defined in [40]:

$$\tau_{smin} = \frac{1}{\gamma^2 f_s} \left(Q^{-1}(P_f) - Q^{-1}(P_d)\sqrt{2\gamma + 1} \right), \tag{3.1}$$

where γ, f_s, P_d, and P_f denotes the signal-to-noise ratio (SNR) of CU signal (using an interweave approach) at cognitive receiver, sampling frequency, probability of correct detection, and probability of false-alarm, respectively. The traffic of PU is assumed to be a binary stochastic process, in which 0 and 1 represents the idle and active states of the channel, respectively. In addition, the average arrival and channel holding time of PU are modeled as the Poisson distribution (with parameter λ) and Binomial distribution (with parameter μ), respectively [41], because the cellular network follows the same nature. Thus, the active channel probability: $P(H_1) = \mu/\lambda$ and the idle channel probability: $P(H_0) = (\lambda - \mu)/\lambda$. In the proposed CRN, the probability of channel to be active is considered as traffic intensity ρ, that means $\rho = P(H_1) = \mu/\lambda$.

In the proposed frame structure for CRN, the frame time (T) is divided into three phases, i.e. (i) spectrum prediction phase (τ_p), (ii) spectrum sensing phase (τ_s), and (iii) data transmission phase ($T - \tau_p - \tau_s$) as shown in Figure 3.1b. Here, the CU performs spectrum prediction on the N number of channels on the basis of previous available information and furthermore, it senses randomly selected channel among the idle predicted channels. Moreover, the data are transmitted on the channel with full-power P_1 if it is sensed idle, otherwise with constrained power P_2, where $P_1 > P_2$.

The received SNR at CU receiver due to transmitted power P_1 and P_2 is denoted as SNR_{s1} and SNR_{s2}, respectively. However, the SNR at CU receiver due to the primary transmitted power P_p is denoted as SNR_p. The throughput of CU is the total transmitted data per unit time consumed. There are four possible conditions for data transmission in each frame as shown in Table 3.1.

3.3.2 System Model for Approach-2

The key concerns of conventional and proposed hybrid approaches are the data loss and interference-at-PU receiver, and the fundamental cause is the inability of CU to detect the emergence of PU during the data transmission period as discussed in Section 1.2.1.3. Therefore, to overcome the aforementioned issues, we have proposed a novel frame structure by exploiting the concept of spectrum monitoring in the data transmission phase as shown in Figure 3.2. The spectrum monitoring takes place at the CU receiver, simultaneous to the data reception via using the received signal characteristics such as REC. Further, the CU receiver and transmitter are synchronized with each other on the common control channel

Table 3.1 The throughput of CU for different conditions.

True channel state	Sensing state	Throughput
0	0	$c_0 = \dfrac{T - \tau_p - \tau_s}{T} \, log_2(1 + SNR_{s1})$
1	0	$c_1 = \dfrac{T - \tau_p - \tau_s}{T} \, log_2\left(1 + \dfrac{SNR_{s1}}{1 + SNR_p}\right)$
0	1	$c_2 = \dfrac{T - \tau_p - \tau_s}{T} \, log_2(1 + SNR_{s2})$
1	1	$c_3 = \dfrac{T - \tau_p - \tau_s}{T} \, log_2\left(1 + \dfrac{SNR_{s2}}{1 + SNR_p}\right)$

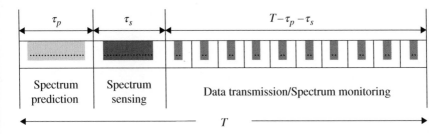

Figure 3.2 The proposed frame structure using spectrum monitoring.

(CCC) which is assumed to be always available [42]. In this scenario, we considered that a time frame consists of a number of data packets (*No*) in the data transmission period, where the packet duration and packet energy is denoted as *PT* and *PE*, respectively.

The proposed frame structure is very similar to the structure shown in Figure 3.1b in addition to the CU simultaneous functioning of data transmission and spectrum monitoring. Moreover, the flow diagram of proposed work in this chapter is illustrated in Figure 3.3. The entire process of the proposed work is as follows. Initially, the CU predicts the states of channels whether active or idle and performs spectrum sensing only on the idle predicted channels. Further, the CU starts data transmission on the idle sensed channels and there is a possibility that the PU resumes its communication during the data transmission. If it happens, then CU stops its communication instantly and particular data packet at the emergence of PU loses its data in the proposed approach (with spectrum monitoring); however, in the conventional approach, the CU continues its transmission till the frame completion time which results in the full-data loss of the CU after the emergence of PU, and the PU gets interference from the CU transmitted data. Further, if the CU completed its data transmission, then it stops its transmission; otherwise, go back to the spectrum prediction phase in the next time frame.

3.4 Performance Analysis

3.4.1 Throughput Analysis Using Approach-1

In the considered CRN, initially the spectrum prediction technique is exploited that is a binary hypothesis and the probability of wrong prediction (P_{pe}) is used to consider the imperfect spectrum prediction. To consider the true (actual, real) channel states (TCS) as active/idle, the probability distribution of true channel and

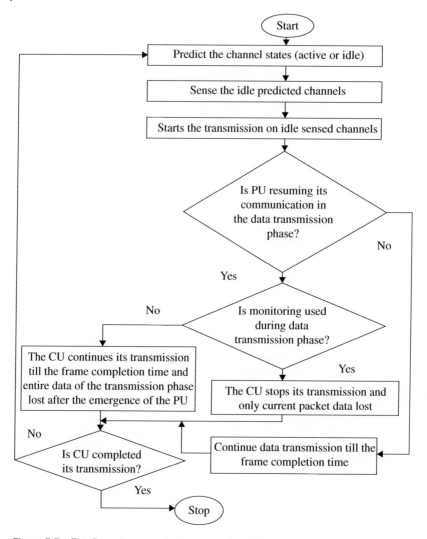

Figure 3.3 The flow diagram of the proposed work.

predicted state is presented in Table 3.2. Therefore, the probability of channel to be predicted idle is:

$$P_p^0 = \left(1 - P_{pe}\right) P(H_0) + P_{pe} P(H_1) \tag{3.2}$$

and the probability of channel to be predicted active is:

$$P_p^1 = \left(1 - P_{pe}\right) P(H_1) + P_{pe} P(H_0). \tag{3.3}$$

Table 3.2 The probability distribution of true and predicted channel states.

Prediction state (PS)	True channel state (TCS)	Probability
0	0	$(1 - P_{pe})\, P(H_0)$
0	1	$P_{pe}\, P(H_1)$
1	0	$P_{pe}\, P(H_0)$
1	1	$(1 - P_{pe})P(H_1)$

Further, the spectrum sensing is performed according to which the CU senses the idle predicted channels randomly. Here, we consider the k number of channels predicted to be idle among N number of channels; therefore, this event is similar to k repeated Bernoulli trials ($k < N$), and corresponding probability is defined as: $C_N^k \left(P_p^0\right)^k \left(P_p^1\right)^{(N-k)}$. Therefore, the probability of idle prediction for the entire CRN is given as:

$$P_N^0 = \sum_{k=1}^{N} C_N^k \left(P_p^0\right)^k \left(P_p^1\right)^{(N-k)} \tag{3.4}$$

and the probability of active prediction for whole CRN is provided by:

$$P_N^1 = C_N^0 \left(P_p^0\right)^0 \left(P_p^1\right)^{(N-0)} = \left(P_p^1\right)^N. \tag{3.5}$$

The P_N^1 is the probability that all channels in the CRN are active and therefore, the CU randomly selects the channel for spectrum sensing, and the probability distribution regarding true channel and spectrum sensing states is provided in Table 3.3. However, the spectrum prediction and spectrum sensing phenomena are independent of each other.

Table 3.3 The probability distribution of true and sensing channel states.

Sensing state (SS)	True channel state (TCS)	Probability
0	0	$(1 - P_f)$
0	1	$(1 - P_d)$
1	0	(P_f)
1	1	(P_d)

Table 3.4 The probability distribution of the combination of true channel, prediction, and sensing states.

T.C.S	P.S	S.S.	Probability
0	0	0	$P_1 = (1 - P_f)P(H_0)(1 - P_{pe})\dfrac{P_N^0}{P_p^0}$
0	0	1	$P_2 = (P_f)P(H_0)(1 - P_{pe})\dfrac{P_N^0}{P_p^0}$
0	1	0	$P_3 = (1 - P_f)P(H_0)(P_{pe})\dfrac{P_N^1}{P_p^1}$
0	1	1	$P_4 = (P_f)P(H_0)(P_{pe})\dfrac{P_N^1}{P_p^1}$
1	0	0	$P_5 = (1 - P_d)P(H_1)(P_{pe})\dfrac{P_N^0}{P_p^0}$
1	0	1	$P_6 = (P_d)P(H_1)(P_{pe})\dfrac{P_N^0}{P_p^0}$
1	1	0	$P_7 = (1 - P_d)P(H_1)(1 - P_{pe})\dfrac{P_N^1}{P_p^1}$
1	1	1	$P_8 = (P_d)P(H_1)(1 - P_{pe})\dfrac{P_N^1}{P_p^1}$

In addition, for the CU spectrum prediction, spectrum sensing and true channel states are also independent, therefore there will be eight possible combinations of probability denoted as $P_1, P_2, P_3, ..., P_8$ as shown in Table 3.4. Now, the throughput computation of CU network depends on the spectrum sensing state, which means the CU transmit with full-power at idle sensed state while with the constrained power on active sensed states. Therefore, by considering all possible combinations in Table 3.4, the throughput of the CU is given as:

$$R_{avg} = (P_1 + P_3)c_0 + (P_5 + P_7)c_1 + (P_2 + P_4)c_2 + (P_6 + P_8)c_3. \quad (3.6)$$

However, the throughput varies with a change in number of channels (N), therefore we need to normalize the throughput. It is evident that if the data are transmitted on the entire frame, the CU will achieve maximum throughput which is given as:

$$R_{up} = P(H_0)c_0 + P(H_1)c_1. \quad (3.7)$$

Now, using Eq. (3.7), the computed normalized throughput is defined as:

$$R_{norm} = \frac{R_{avg}}{R_{up}}. \tag{3.8}$$

Conventionally, before designing a receiver, the value of P_f, P_d, and τ_s are already defined and in the proposed CRN, these are considered as invariable quantities. Therefore, the throughput of the CRN is affected by three parameters, i.e. the traffic intensity (ρ), probability of wrong prediction (P_{pe}), and number of channels (N).

3.4.2 Analysis of Performance Metrics of the Approach-2

To analyze the performance of proposed Approach-2, we have considered three metrics, namely: (i) data-loss of the CU (DL), (ii) energy-loss of the CU (EL), and (iii) SNR at the PU receiver. The data loss is in the form of packet whereas the energy loss in the form of packet energy (PE). The data loss of the conventional approach (DL_{Conven}) is proportional to the traffic intensity, i.e. $\{DL_{Conven} = \rho \times No\}$ and the normalized value of the DL_{Conven} is denoted as $RL_{norm-Conven}$ and defined as $\{DL_{norm-Conven} = (\rho \times No)/\rho\}$. However, the data loss in the proposed approach (RL_{Prop}) is the particular packet on the emergence of PU. The normalized value of the DL_{Prop} is denoted as $DL_{norm-Prop}$, where $DL_{norm-Prop} = 1/No$.

Similarly, the energy loss for the conventional and proposed approach is defined as $EL_{Conven} = \rho \times PE \times No$ and $EL_{Prop} = PE$, respectively. The normalized values of both are $\{EL_{norm-Conven} = (\rho \times PE \times No)/(PE \times No) = \rho\}$ and $\{EL_{norm-Prop} = PE/(PE \times No) = 1/No\}$.

The SNR at the PU receiver due to primary and cognitive transmission is denoted as $SNRP_p$ and $SNRP_s$, respectively. Therefore, the total SNR at the PU receiver due to PU and CU transmission is $SNRP_{ps} = SNRP_p/(1 + SNRP_s)$. In the conventional approach, all the packets after the emergence of the PU get lost; however, the particular packet at the emergence of PU lost in the proposed approach. The starting time (PT_s) and ending time (PT_E) of that packet depend on the traffic intensity (ρ) of the PU and computed as:

$$PT_s = \{(1-\rho) \times (T - (\tau_s + \tau_p))\} + \{(\tau_s + \tau_p)\}, \tag{3.9}$$

$$PT_E = PT + PT_s, \tag{3.10}$$

where PT is the packet duration and defined as: $PT = (T - \tau_s - \tau_p)/No$. To compute the SNR at the PU receiver, an algorithm is presented below as Algorithm-3.1.

Algorithm 3.1   SNR Calculation at PU Receiver

Input $(T, \rho, \tau_s, \tau_p, PT)$
Output $(SNRP_{conven}, SNRP_{prop})$
BEGIN {
Step-1: Variable Declaration
    T: Frame Duration;
    $t = 0: 0.1: T;$
    ρ: Traffic intensity;
    τ_p: Prediction duration
    τ_s: Sensing duration
    No: Number of packets in data transmission phase.
    $SNRP_s$: SNR at PU receiver due to cognitive transmission;
    $SNRP_p$: SNR at PU receiver due to primary transmission;

Step-2: Computation of PT, PT_s, PT_E and $SNRP_{ps}$,

$$PT \leftarrow \left\{ \frac{T - \tau_s - \tau_p}{No} \right\};$$

$$PT_s \leftarrow \left\{ (1 - \rho) \times \left(T - \left(\tau_s + \tau_p \right) \right) \right\} + \left\{ \left(\tau_s + \tau_p \right) \right\};$$

$$PT_E \leftarrow \{ PT_s + PT \};$$

$$SNRP_{ps} \leftarrow \frac{SNRP_p}{1 + SNRP_s}$$

Step-3: SNR calculations at the PU receiver using conventional and proposed approaches
for $ii \leftarrow 1 : length(t)$
   If $(t\ (ii) \leq PT_s)$ **then**
           $SNRP_{conven}(ii) \leftarrow 0$
           $SNRP_{prop}(ii) \leftarrow 0$
      elseif $\{ (t\ (ii) \geq PT_s)\ \&\ \&\ (t\ (ii) \leq, PT_E) \}$
then
           $SNRP_{conven}(ii) \leftarrow SNRP_{ps}$
           $SNRP_{prop}(ii) \leftarrow SNRP_{ps}$
      elseif $(t\ (ii) \geq PT_E))$ **then**
           $SNRP_{conven}(ii) \leftarrow SNRP_{ps}$
           $SNRP_{prop}(ii) \leftarrow SNRP_p$
      end
   end
 end
} **END**

3.5 Results and Discussion

3.5.1 Proposed Approach-1

The numerically simulated throughput of the proposed Approach-1 in the CRN is presented and compared with the conventional approach reported in [10]. In [10], the authors have used an interweave spectrum access technique, which has introduced the problem of waiting probability. Due to this, the CU is unable to transmit on the active sensed channels and total throughput is given as:

$$R_{avgc} = (P_1 + P_3)c_0 + (P_5 + P_7)c_1 \tag{3.11}$$

and normalized throughput in this case is:

$$R_{normc} = (R_{avgc}/R_{up}). \tag{3.12}$$

In the numerically simulated results, the conventional approach reported in [11] is denoted as Conv whereas the proposed approach as Hybrid. Moreover, the values of simulation parameters are selected on the basis of IEEE 802.22 wireless regional area network (WRAN) standard and are presented in Table 3.5 [43–45].

The variations of normalized throughput with the probability of wrong prediction for several values of traffic intensity in the conventional and proposed hybrid approaches are described in Figure 3.4. It is apparent that the normalized throughput decays slowly for small values of wrong prediction whereas for large values it goes down hastily. Further, the proposed hybrid approach performs well as compared to the conventional approach, especially, for high values of the traffic intensity.

Figure 3.5 illustrates the relation between the normalized throughput and traffic intensity for various values of the probability of wrong prediction and it is

Table 3.5 The simulation parameters for the proposed CRN.

Parameter	Value	Parameter	Value
T	100 ms	f_s	1500 samples/s
τ_p	5 ms	SNR_{s1}	20 dB
τ_s	2.5 ms	SNR_{s2}	−5 dB
P_d	0.9	SNR_p	−15 dB
P_f	0.1	$SNRP_s$	5 dB
No	10	$SNRP_p$	15 dB
PE	1 mw		

Sources: From Ref. [43]; Ko et al. [44]; Stevenson et al. [45].

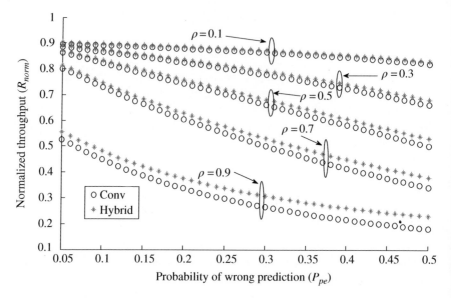

Figure 3.4 Normalized throughput versus the probability of wrong prediction for proposed hybrid and conventional approach ($N = 10$).

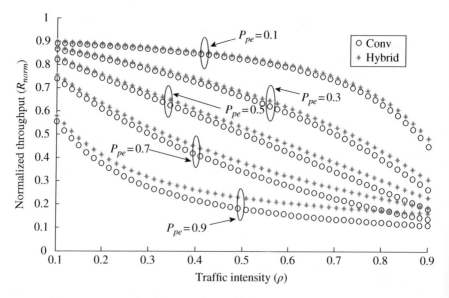

Figure 3.5 Normalized throughput versus traffic intensity for proposed hybrid and conventional approach ($N = 10$).

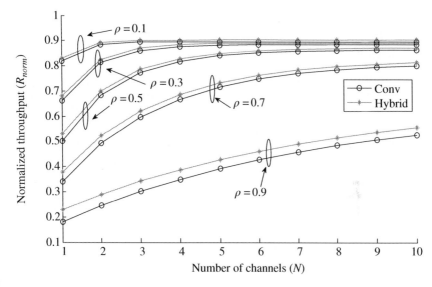

Figure 3.6 Normalized throughput versus number of channels for proposed hybrid and conventional approach.

perceived that the throughput decreases (hastily and slowly for large and small values of P_{pe}, respectively) with increase of the traffic intensity. The effect of number of channels on the normalized throughput for the different values of traffic intensity is depicted in Figure 3.6. For large values of the traffic intensity (0.7, 0.9), the normalized throughput is directly proportional to the number of channels; however, for small values of traffic intensity (0.1, 0.3), the normalized throughput improves for a particular value of the number of channels and then becomes constant. Thus, we conclude that there is a maximum limit for the number of channels needs to choose for maximum normalized throughput. From the above discussion, it is perceived that the proposed Approach-1 outperforms in the worst scenarios such as high traffic intensity as well as wrong predicted scenarios. Moreover, the proposed approach improves the throughput when compared with the conventional approach; however, the proposed Approach-1 performs well at large values of probability of wrong prediction."

3.5.2 Proposed Approach-2

In this scenario, the proposed Approach-1 and conventional approaches are assumed as a conventional approach (Conven), and proposed Approach-2 as a proposed approach (Prop). The three performance metrics, namely, the data, energy loss of the CU, and SNR at the PU receiver have been simulated and compared with the conventional approach. Here, it is considered that the PU definitely resumes its communication during the data transmission period. The simulation parameters are the same as in Table. 3.5.

In Figure 3.7, the variations of normalized data loss of the CU with the traffic intensity of PU in the conventional and proposed approaches are illustrated. It is clear that the normalized data loss shows directly proportional relation with traffic intensity for the conventional approach; however, for the proposed approach, it is constant and is not affected by the traffic intensity because the particular packet is lost on the emergence of PU. The effect of traffic intensity on the energy loss for the conventional and proposed approaches is presented in Figure 3.8. The normalized energy loss varies linearly with traffic intensity in the conventional approach whereas in the proposed approach, it remains constant. This happens since in the proposed approach, the particular data packet is lost when the CU stops the communication on the emergence of PU; however, in the conventional approach, the CU transmits continuously and gets collided with PU after the emergence of the PU. Moreover, the relation between the SNR of the PU with time frame at a particular value of traffic intensity ($\rho = 0.5$) is presented in Figure 3.9; however, to focus on the time of emergence of PU, we have shown only small part of time

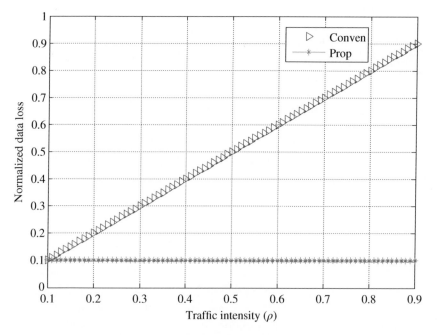

Figure 3.7 Normalized data loss of CU versus traffic intensity.

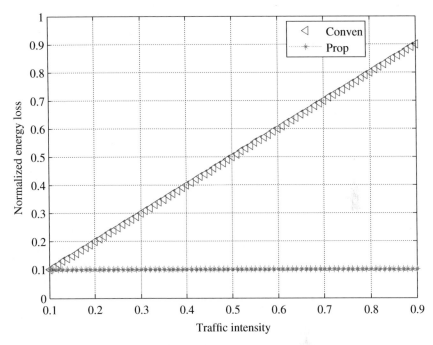

Figure 3.8 Normalized energy loss of CU versus traffic intensity.

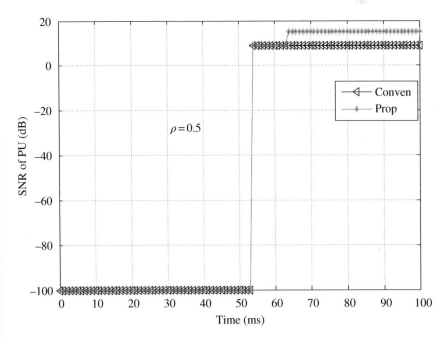

Figure 3.9 SNR at PU receiver versus time.

frame. It is apparent that SNR remains zero till the emergence of the PU and have suddenly jumped to the value of $SNRP_{ps}$ in the conventional approach and remains constant till the end of the time frame; however, in the proposed approach, SNR remains $SNRP_{ps}$ till the transmission of particular packet on the emergence of PU and after that it reaches to the full SNR (15 dB), i.e. $SNRP_p$ as the CU stops its transmission.

3.6 Summary

In this chapter, the concepts of spectrum prediction and hybrid spectrum access techniques are explored to improve the throughput by alleviating the spectrum sensing errors. Further, the problem of wait states is conquered via hybrid spectrum access technique. The closed-form mathematical expressions of the throughput for the proposed Approach-1 and conventional approaches are derived. The simulation results have witnessed that the proposed Approach-1 performs well when compared with the conventional approach in the worst scenarios such as very high traffic CRN and high prediction errors. Moreover, by exploiting the shared issues of conventional and proposed Approach-1, a new frame structure and flow diagram to resolve these issues are presented. The performance of proposed structures is analyzed using three metrics and have illustrated that proposed Approach-2 is better as compared to the Approach-1.

References

1 Pandit, S. and Singh, G. (2014). Throughput maximization with reduced data loss rate in cognitive radio network. *Telecommun. Syst.* 57 (2): 209–215.

2 Owayed, A.A., Mohammed, Z.A., and Mosa, A.A.R. (2010). Probabilities of detection and false alarm in Multitaper based spectrum sensing for cognitive radio systems in AWGN. *IEEE International Conference on Communication Systems* (17–19 November 2010) 579–584.

3 Bhattacharjee, S., Das, P., Mandal, S., and Sardar, B. (2015). Optimization of probability of false alarm and probability of detection in cognitive radio networks using GA. *IEEE 2nd International Conference on Recent Trends in Information Systems (ReTIS)* (9–11 July 2015), 53–57.

4 Suliman, I.M., Lehtomäki, J., and Umebayashi, K. (2015). On the effect of false alarm rate on the performance of cognitive radio networks. *EURASIP J. Wireless Commun. Netw.* 2015: 244.

5 Altrad, O. and Muhaidat, S. (2013). A new mathematical analysis of the probability of detection in cognitive radio over fading channels. *EURASIP J. Wireless Commun. Netw.* 2013 (1): 159.

6 Lee, Y.L., Saad, W.K., Abd El-Saleh, A., and Ismail, M. (2013). Improved detection performance of cognitive radio networks in AWGN and Rayleigh fading environments. *J. Appl. Res. Technol.* 11 (3): 437–446.

7 Liu, X., Jia, M., and Tan, X. (2013). Threshold optimization of cooperative spectrum sensing in cognitive radio networks. *Radio Sci.* 48 (1): 23–32.

8 Verma, G. and Sahu, O.P. (2016). Intelligent selection of threshold in cognitive radio system. *Telecommun. Syst.* 63 (4): 547–556.

9 Thakur, P., Kumar, A., Pandit, S. et al. (2016). Frame structures for hybrid spectrum accessing strategy in cognitive radio communication system. *9th International Conference on Contemporary Computing (IC3)* (11–13 August 2016), 1–6.

10 Thakur, P., Kumar, A., Pandit, S. et al. (2017). Advanced frame structures for hybrid spectrum access strategy in cognitive radio communication systems. *IEEE Commun. Lett.* 21 (2): 410–413.

11 Thakur, P. and Singh, G. (2020). Power management for spectrum sharing in cognitive radio communication system: a comprehensive survey. *J. Electromagn. Waves Appl.* 34 (4): 407–461.

12 Khan, A.U., Abbas, G., Abbas, Z.H. et al. (2020). Spectrum efficiency in CRNs using hybrid dynamic channel reservation and enhanced dynamic spectrum access. *Adhoc Netw.* 107: 1–16.

13 Vasquez-Toledo, L.A., Borja-Benitez, B., Marcelin-Jimenez, R. et al. (2020). Mathematical analysis of highly scalable cognitive radio systems using hybrid game and queuing theory. *AEU-Int. J. Electromagn. Commun.* 127: 153406.

14 Yang, J. and Zhao, H. (2015). Enhanced throughput of cognitive radio networks by imperfect spectrum prediction. *IEEE Commun. Lett.* 19 (10): 1738–1741.

15 Pei, Y., Liang, Y.C., Teh, K.C., and Li, K.H. (2011). Energy-efficient design of sequential channel sensing in cognitive radio networks: optimal sensing strategy, power allocation, and sensing order. *IEEE J. Sel. Areas Commun.* 29 (8): 1648–1659.

16 Chatterjee, S., Maity, S.P., and Acharya, T. (2014). Energy efficient cognitive radio system for joint spectrum sensing and data transmission. *IEEE J. Emerging Sel. Top. Circuits Syst.* 4 (3): 292–300.

17 Pei, Y., Hoang, A.T., and Liang, Y.C. (2007). Sensing-throughput tradeoff in cognitive radio networks: how frequently should spectrum sensing be carried out? *18th IEEE International Symposium on Personal, Indoor and Mobile Radio Communications* (2–4 June 2014), 1–5.

18 Bogale, T.E., Vandendorpe, L., and Le, L.B. (2014). Sensing throughput tradeoff for cognitive radio networks with noise variance uncertainty. *9th IEEE International Conference on Cognitive Radio Oriented Wireless Networks and Communications (CROWNCOM)* (3–7 September 2007), 435–441.

19 Kaushik, A., Sharma, S.K., Chatzinotas, S. et al. (2015). Sensing-throughput tradeoff for cognitive radio systems with unknown received power. *Cognitive Radio Oriented Wireless Networks* (October 2015), 308–320.

20 Kan, C., Wu, Q., Ding, G., and Song, F. (2014). Sensing-throughput trade-off for interference-aware cognitive radio networks. *Frequenz* 68 (3–4): 97–108.

21 Kumar, A., Pandit, S., and Singh, G. (2020). Optimization of censoring-based cooperative spectrum sensing approach with multiple antennas and imperfect reporting channel scenarios for cognitive radio network. *IET Commun.* 14 (16): 2666–2676.

22 Huang, S., Jiang, N., Gao, Y. et al. (2020). Radar sensing-throughput trade-off for radar assisted cognitive radio enabled vehicular ad-hoc networks. *IEEE Trans. Veh. Technol.* 67 (7): 7483–7492.

23 Hedhly, W., Amin, O., and Alouini, M.-S. (2020). Benefits of improper Gaussian signaling in interweave cognitive radio with full and partial CSI. *IEEE Trans. Cognit. Commun. Netw.* https://doi.org/10.1109/TCCN.2020.2971681.

24 Khan, A.U., Abbas, G., Abbas, Z.H. et al. (2020). HBLP: a hybrid underlay-interweave mode CRNs for the future 5G-based internet of things. *IEEE Access* 8: 63403–63420.

25 Jiang, X., Wong, K.K., Zhang, Y., and Edwards, D. (2013). On hybrid overlay underlay dynamic spectrum access: double-threshold energy detection and Markov model. *IEEE Trans. Veh. Technol.* 62 (8): 4078–4083.

26 Yarkan, S. and Arslan, H. (2007). Binary time series approach to spectrum prediction for cognitive radio. *66th IEEE Vehicular Technology Conference* (30 September to 3 October 2007), 1563–1567.

27 Tumuluru, V.K., Wang, P., and Niyato, D. (2012). Channel status prediction for cognitive radio networks. *Wireless Commun. Mobile Comput.* 12 (10): 862–874.

28 Xing, X., Jing, T., Cheng, W. et al. (2013). Spectrum prediction in cognitive radio networks. *IEEE Wireless Commun.* 20 (2): 90–96.

29 Christian, I., Moh, S., Chung, I., and Lee, J. (2012). Spectrum mobility in cognitive radio networks. *IEEE Commun. Mag.* 50 (6): 114–121.

30 Barnes, S.D., Maharaj, B.T., and Alfa, A.S. (2016). Cooperative prediction for cognitive radio networks. *Wireless Pers. Commun.* 89 (4): 1177–1202.

31 Patil, J., Bokde, N., Mishra, S.K., and Kulat, K. (2020). PSF-based spectrum occupancy prediction in cognitive radio. In: *Advanced Engineering Optimization Through Intelligent Techniques* (eds. R. Venkata Rao and J. Taler), 609–619. Singapore: Springer.

32 Supraja, P., Babu, S., Gayathri, V.M., and Divya, G. (2020). Hybrid genetic and shuffled frog-leaping algorithm for neural network structure optimization and learning model to predict free spectrum in cognitive radio. *Int. J. Commun. Syst.* https://doi.org/10.1002/dac.4532.

33 Ding, G., Jiao, Y., Wang, J. et al. (2018). Spectrum inference in cognitive radio networks: algorithms and applications. *IEEE Commun. Surv. Tutorials* 20 (1): 150–182.

34 Zhao, Y., Hong, Z., Luo, Y. et al. (2018). Prediction-based spectrum management in cognitive radio networks. *IEEE Syst. J.* 12 (4): 3303–3314.

35 Boyd, S.W., Frye, J.M., Pursley, M.B., and IV, T.C.R. (2012). Spectrum monitoring during reception in dynamic spectrum access cognitive radio networks. *IEEE Trans. Commun.* 60 (2): 547–558.

36 Ali, A. and Hamouda, W. (2015). Spectrum monitoring using energy ratio algorithm for OFDM-based cognitive radio networks. *IEEE Trans. Wireless Commun.* 14 (4): 2257–2268.

37 Orooji, M., Soltanmohammadi, E., and Naraghi-Pour, M. (2015). Improving detection delay in cognitive radios using secondary-user receiver statistics. *IEEE Trans. Veh. Technol.* 64 (9): 4041–4055.

38 Saifan, R., Kamal, A.E., and Guan, Y. (2011). Efficient spectrum searching and monitoring in cognitive radio network. *8th IEEE International Conference on Mobile Ad-Hoc and Sensor Systems* (17–22 October 2011), 520–529.

39 Thakur, P., Kumar, A., Pandit, S. et al. (2017). Effect of imperfect spectrum monitoring on cognitive radio network performance. *IEEE Fourth International Conference on Image Information Processing (ICIIP)*, Waknaghat, India (21–23 December 2017), 1–5.

40 Liang, Y.C., Zeng, Y., Peh, E.C.Y., and Hoang, A.T. (2008). Sensing-throughput tradeoff for cognitive radio networks. *IEEE Trans. Wireless Commun.* 7 (4): 1326–1337.

41 Masonta, M.T., Mzyece, M., and Ntlatlapa, N. (2013). Spectrum decision in cognitive radio networks: a survey. *IEEE Commun. Surv. Tutorials* 15 (3): 1088–1107.

42 Pandit, S. and Singh, G. (2015). Backoff algorithm in cognitive radio mac protocol for throughput enhancement. *IEEE Trans. Veh. Technol.* 64 (5): 1991–2000.

43 IEEE-SA – Contact Us. http://standards.ieee.org/contact/form.html (accessed 18 February 2018).

44 Ko, G., Franklin, A.A., You, S.-J. et al. (2010). Channel management in IEEE 802.22 WRAN systems. *IEEE Commun. Mag.* 48 (9): 88–94.

45 Stevenson, C.R., Chouinard, G., Lei, Z. et al. (2009). IEEE 802.22: the first cognitive radio wireless regional area network standard. *IEEE Commun. Mag.* 47 (1): 130–138.

4

Effect of Spectrum Prediction on Cognitive Radio Networks

4.1 Introduction

As discussed in Chapter 3, for the high-traffic cognitive radio networks (HTCRNs), the primary users (PUs) are active at most of the time in most of the channels. Due to the random spectrum sensing of channels in cognitive radio networks (CRNs), the probability of channel to be sensed active is significantly high, resulting in high data loss due to which the effective throughput of the network decreases significantly [1]. This reduction in throughput affects the performance metrics of the system such as transmission time and energy consumption as reported in [2, 3]. In [2], the authors have presented an energy-efficient design of the sequential channel sensing and optimized the sensing time, power allocation, and sensing order. Chatterjee et al. [3] have maximized the throughput of cognitive user (CU) using a constraint on the spectrum sensing reliability. Therefore, in order to avoid these challenging issues, Yang and Zhao [1] have introduced a new frame structure in which the CU predicts the channel state (active or idle) before spectrum sensing and senses only the idle predicted channels [1]. The selection of a channel for spectrum sensing among all idle predicted channels is a random phenomenon that can cause the "sense and stuck" problem.

Sense and stuck problem is defined in the CRNs in which the CU predicts the channel state before spectrum sensing; however, the selection of channels for sensing among all the idle predicted channels is a random phenomenon [4]. The CU senses channel and starts data transmission if the channel is sensed idle; otherwise, the CU selects the channel again randomly among all the idle predicted channels. Due to this random selection, the possibility of choosing the same channel again for spectrum sensing even it is already sensed as active is of significant importance. If it happens, the CU will get stuck in the same channel again and again. Therefore, in order to alleviate this problem and for the throughput enhancement, we have introduced a novel approach (named as *Approach-1* in this chapter) for improved selection of a channel for spectrum sensing on the basis of

Spectrum Sharing in Cognitive Radio Networks: Towards Highly Connected Environments,
First Edition. Prabhat Thakur and Ghanshyam Singh.

prediction probabilities. However, the value of prediction probabilities relies on the previous spectrum sensing information, i.e. the previous sensing states of the channel, which may be unavailable in a particular scenario. Therefore, a modified approach of channel selection is proposed, in which the CU senses the channel and if it sensed as active, a subsequent channel for spectrum sensing is selected from the directory of "all idle predicted channels excluding already sensed channels." Furthermore, for additional improvement in the throughput, a new frame structure is proposed, in which the communication is established in the underlay mode on spectrum prediction and sensing periods, parallel to the spectrum prediction and sensing phenomenon (named as *Approach-2* in this chapter)."

The potential contribution in this chapter is summarized as follows:

- A potential issue in the random selection method of the sensing channel among all idle predicted channels is observed which is named as *sense and stuck problem*.
- A very simple and fundamental method for computation of the probability of idle prediction of the channel is proposed which relies on previous spectrum sensing information.
- In order to conquer the sense and stuck problem and for the throughput enhancement, a method of improved selection of sensing channel is proposed and has been verified through numerical simulation.
- The unavailability of the prediction probabilities, due to pre-sensing-information requirement in particular scenarios led us to a modified approach of selection without prediction probabilities, in order to avoid the sense and stuck problem.
- An algorithm in order to evaluate the throughput using random, improved, and modified selection approaches is presented and has computed their space and time complexities.
- In order to get additional improvement in the throughput, a new approach is proposed to establish communication simultaneously with the spectrum sensing and prediction period.

Further, the work related to the spectrum accessing techniques is presented.

4.1.1 Spectrum Access Techniques

In addition to the spectrum accessing techniques presented in Chapter 2 [5–13], some recent works on the spectrum accessing techniques are presented in [14–17]. In [14], the authors have proposed an algorithm to not only select a channel for data transmission but also to predict how long the channel will remain unoccupied so that the time spent on channel sensing can be minimized. This algorithm learns in two stages – a reinforcement learning (RL) approach for channel selection and a Bayesian approach to determine the optimal duration for which sensing can be skipped. In [15], Sun et al. have proposed a multichannel spectrum access scheme based on RL in order to improve the spectrum access of the cognitive IoTs (CoIoTs), wherein the CoIoT can use multiple channels for transmissions in order to reduce the communication interruptions. The channels are ranked in the decreasing order of their predicted idle probabilities, which can make the CIoT find enough idle channels quickly via decreasing the number of sensing operations and spectrum handoffs. The superiority of the proposed scheme over the single-channel spectrum access scheme in terms of throughput, communication interruption, average collision probability, and average spectrum switching frequency is validated using the simulation results.

In [16], the authors have proposed a comprehensive solution to decrease the sensing latency and to make the CR networks (CRNs) aware of unlicensed user requirements by using a QoS-aware proactive spectrum sensing technique. A modified hybrid spectrum accessing technique is proposed by Bhowmick et al. [17]. In this technique, the CU performs spectrum sensing and data transmission using underlay approach simultaneously, during the sensing phase and results in significant improvement in the throughput. Moreover, the closed-form expressions for the throughput and outage probability of the CRN have been derived in [17].

Moreover, the advances over the spectrum prediction techniques are illustrated in detail [18–26]. In [18], the authors have presented a cooperative prediction-and-sensing-based spectrum sharing method for CRNs. In [19], Bhowmick et al. have illustrated a novel prediction-based cooperative spectrum sensing scheme on the performance of energy harvesting CRNs. The authors have investigated the network parameters such as number of detection frames, number of cooperative CR users, splitting parameters, and collision probability on throughput performance.

Zhang et al. [20] have exploited the latest applications of the machine learning and artificial intelligence (AI) for the spectrum accessing and discuss the problems during the incorporation of intelligence for CRNs. In [21], the authors have emphasized over the deep learning-based spectrum prediction technique where cooperative spectrum prediction with neural network predictors is investigated. Further, the soft cooperative fusion is implemented by exploiting the spatial dependency of spectrum measurement data and provides the spectrum status. In [22], the authors have proposed a convolution long short-term memory (LSTM)-based deep learning neural network for a long-term temporal prediction that is trained to learn joint spatial–spectral–temporal dependencies observed in spectrum usage.

In [23], the authors have well explored the analysis of CRNs using the channel prediction probabilities and improved frame structures (Figure 4.1). The authors

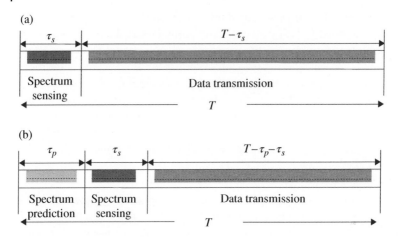

Figure 4.1 Frame structures in cognitive radio communication: (a) conventional and (b) using prediction.

have proposed a statistical prediction classification framework to categorize and assess current spectrum occupancy models in [24]. Further, an overview of statistical sequential prediction is presented. In [25], Fan et al. have emphasized over modeling of PU behavior, which is used for the recurrent neural network (RNN) for designing of the spectrum occupancy state predictor. Further, the authors in [26] have focused over the cooperative HMM-based spectrum prediction technique. The rest of the chapter is structured as follows. The system model for the proposed approach is illustrated in Section 4.2. In Section 4.3, the expressions of throughput are presented. The simulation results are discussed in Section 4.4 and finally, Section 4.5 summarizes the chapter and discusses the future scope.

4.2 System Model

In the proposed system (*Approach-1*) model, a CU pair of transceivers in the high-traffic environment with multiple PU channels (N) is considered in which the CU accesses the idle sensed channel with an interweave approach. As all the channels may belong to different networks, therefore, the activity of PU on each channel is assumed as an independent phenomenon. The spectrum sensing process relies on the binary hypothesis H_0 and H_1 considered for the received signal $r(t)$, which confirms the idle and active state of the channel, respectively, as discussed in

Chapter 2. However, for the reliable spectrum sensing results, the sensing duration must be above the predefined threshold (τ_{smin}) as illustrated in Chapter 3 and well explored in [17].

The PU traffic is considered as a binary stochastic process in which 0 and 1 denote the idle and active states of the channel, respectively. Furthermore, the traffic intensity(ρ) is defined as probability of the channel to be active as discussed in Section 3.3.1, which means $\rho = P(H_1) = \mu/\lambda$ [27]. Since we have considered the multichannel environment, therefore, it is worth mentioning that this traffic intensity is defined for the single channel selected for data transmission after the spectrum sensing. In addition to this, the traffic intensity is used to represent the original (true/actual) state of that selected channel.

In the proposed model (*Approach-1*), the time frame of duration T is divided into three phases, namely, the spectrum prediction phase (τ_p), sensing phase (τ_s), and data transmission phase ($T - \tau_p - \tau_s$). The frame structure for the conventional [1] and proposed system model, however, varies in the selection of the channel which means, in the proposed model, the sensing channel is selected on the basis of their probability whereas in the conventional model, the CU selects sensing channel randomly among all idle predicted channels. However, the formation of these probabilities is an exigent task; therefore, we have proposed an approach to form the probabilities of channel to be predicted idle on the basis of previous sensing results. A binary series of the channel states have been analyzed for n states which is used to yield the probability of the channel to be predicted idle for the $(n + 1)$ th state.

$$S_{n+1} \leftarrow S_1, S_2, S_3, S_4, ..., S_n. \tag{4.1}$$

A conventional theory of the probability is according to which the probability of success in the next event (P_{sn}) is defined as the ratio of the number of successful events to the total number of events occurred [28]. Therefore, in the binary series, the probability of the $(n + 1)$th state to be idle, $P_{pi}(S_{n+1})$, is defined as the ratio of the number of idle sensed channels (n_i) to the total number of channels (n).

$$P_{pi}(S_{n+1}) = \frac{n_i}{n}. \tag{4.2}$$

As the formation of probabilities relies on the pre-sensing-states of the channel, whose availability may be questionable in particular state of affairs. Therefore, to avoid the sense and stuck problem in such scenarios, a modified selection method of the channel is introduced as illustrated in the flow diagram as shown in Figure 4.2. Initially, the CU predicts the channel's state (active or idle) either with the probabilities by which the channel is predicted idle (P_{pi}) or without probabilities. If the probabilities are available, then selection of the channel is performed

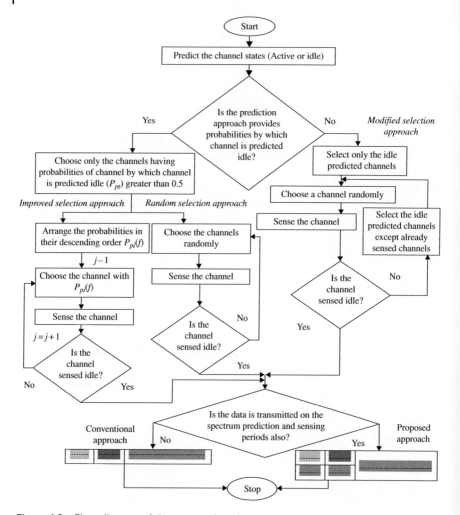

Figure 4.2 Flow diagram of the proposed work.

either by random selection or improved selection approach. However, the unavailability of prediction probabilities leads CU to the modified selection approach.

In the improved selection approach, the probabilities, P_{pi}, are sorted in the descending order, and then the channel with highest P_{pi} is sensed. If that channel is sensed idle, the data transmission starts; otherwise, the channel with second highest probability is sensed and so on, this process continues till all the idle predicted

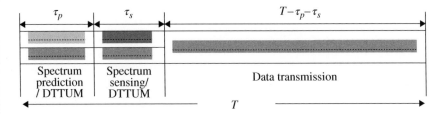

Figure 4.3 The proposed frame structure of cognitive radio networks.

channels are sensed. However, in the random selection approach, one channel is selected randomly among all predicted channels. If the spectrum sensing results confirm the idle nature of the channel, then the data transmission starts; otherwise, the CU again selects one channel among all the idle predicted channels, and so on. This process continues again and again due to which it suffers from the sense and stuck problem if zero channel is sensed idle. Therefore, in the modified selection approach, the CU senses one channel which is selected among all the idle predicted channels. If it is sensed as idle, the data transmission starts; otherwise, a new channel is selected for spectrum sensing from a set of all the idle predicted channels excluding already sensed channel, and so on. Furthermore, the data transmission is also possible in two ways, i.e. the conventional and proposed way.

In the conventional way, the data are transmitted only on the data transmission phase via an interweave approach when the channel is sensed idle. However, in the proposed way (*Approach-2*), the data are transmitted on the spectrum prediction and sensing phases using the underlay mode and on the data transmission phase using the interweave mode as presented in Figure 4.3. The complete approach of random, improved, and modified selection of channel for spectrum sensing (*Approach-1*) is presented in Algorithm 4.1.

The space and time complexity of all the approaches are also evaluated [29]. The time complexity of random selection approach is $O(G)$, where G is the maximum number of iterations. However, in the improved and modified selection approaches, it is $O(N.log(N))$ and $O(N)$, respectively. Thus, it is clear that the proposed approaches reduce the complexity of the algorithm significantly. The space complexity for all the approaches is computed as $O(N)$.

Further, in the conventional method of data transmission, after the spectrum sensing, if the CU finds an idle channel, then it starts data transmission with full power in the data transmission period. Now, the throughput in the data transmission period of the CU receiver becomes the function of original (true) state of the PU which may be either active or idle.

Algorithm 4.1   The Data Transmission Using Various Approaches

	Time complexity	Space complexity

Input $(T, P_p, N, G$)
Output (P_{pi})
BEGIN {

Step-1: *Variable Declaration*

N: Total number of PU's channels; ⟶	$O(1)$		$O(N)$	
P_p: The probability with which channel is predicted idle;⟶	$O(1)$		$O(N)$	
P_{pi}: Probabilities of the idle predicted channels; ⟶	$O(1)$		$O(N)$	
SR: Sensing result; ⟶	$O(1)$	$O(1)$	$O(1)$	$O(N)$
I: Idle sensed channels; ⟶	$O(1)$		$O(1)$	
$I \leftarrow 0$; ⟶	$O(1)$		$O(1)$	
$G \leftarrow$ Maximum number of iterations ⟶	$O(1)$		$O(1)$	

Step-2: Predict the channel states with their probabilities P_p of N number of channels; $O(1)$ $O(1)$ $O(1)$ $O(1)$

Step-3: Selection of the Sensing Channels

For $j \leftarrow 1: N$ ⟶	$O(N)$		$O(1)$	
$ii \leftarrow 1$; ⟶	$O(1)$		$O(1)$	
while $P_p(j) > 0.5$ ⟶	$O(N)$	$O(N^2)$	$O(1)$	$O(1)$
do $P_{pi}(ii) \leftarrow P_p(j)$;⟶	$O(1)$		$O(1)$	
$ii \leftarrow ii+1$; ⟶	$O(1)$		$O(1)$	
end				

End

Step 4. Data Transmission using Random Selection Approach

(i) Prepare a directory of all idle predicted channels with P_{pi}	$O(1)$		$O(1)$	
(ii) Selection of the idle sensed channel randomly ⟶	$O(1)$		$O(1)$	
SC: Randomly select channel from the directory. ⟶	$O(1)$		$O(1)$	
(iii) Sense the SC ⟶	$O(1)$		$O(1)$	
If (SR $\leftarrow 0$) ⟶	$O(1)$		$O(1)$	
$P_{pi} \leftarrow P_{pi}$ (*randomly selected*) ⟶	$O(1)$	$O(G)$	$O(1)$	$O(1)$
else				
go to **(ii)** ⟶	$O(G)$		$O(1)$	
End				
(iv) Start data transmission using interweave spectrum access approach ⟶	$O(1)$		$O(1)$	

	Time complexity	Space complexity

Step. 5. Data Transmission using Improved Selection Approach

(i) *Local variable declaration*

$jj \leq$ length (P_{pi}); $\longrightarrow$ $O(1)$ $\qquad$ $O(1)$

$jj \leftarrow 1$; $\longrightarrow$ $O(1)$ $\qquad$ $O(1)$

(ii) *Arrange these probabilities in their descending order (sorting (descend))*

$P_{pi} \leftarrow$ sort (P_{pi}): $\longrightarrow$ $O(N.log(N))$ $\qquad$ $O(1)$

(iii) *Selection of the idle sensed channel with highestprobability*

$\qquad$ Sense the channel with $P_{pi}(jj)$ $\longrightarrow$ $O(1)$ $\qquad$ $O(1)$

$\qquad$ **if** (SR $\leftarrow$ 0) $\longrightarrow$ $O(1)$ $\qquad$ $O(1)$

$\qquad\qquad$ $P_{pi} \leftarrow P_{pi}(jj)$ $\longrightarrow$ $O(1)$ $\qquad$ $O(1)$

$\qquad$ **else**

$\qquad\qquad$ $jj \leftarrow jj + 1,$ $\longrightarrow$ $O(1)$ $\qquad$ $O(1)$

$\qquad\qquad$ go to **(iii)** $\longrightarrow$ $O(N)$ $\qquad$ $O(1)$

$\qquad$ **End**

(iv) Start data transmission using interweave spectrum access approach $\longrightarrow$ $O(1)$ $\qquad$ $O(1)$

(Time complexity for Step 5: $O(N.log(N))$ overall; Space complexity: $O(1)$ overall)

Step 6. Data Transmission using Modified Selection Approach

(i) Prepare a directory of all idle predicted channels with P_{pi} $\longrightarrow$ $O(1)$ $\qquad$ $O(1)$

(ii) Selection of the idle sensed channel randomly $\longrightarrow$ $O(1)$ $\qquad$ $O(1)$

$\qquad$ SC: Randomly selected channel from the directory. $\longrightarrow$ $O(1)$ $\qquad$ $O(1)$

(iii) Sense the SC $\longrightarrow$ $O(1)$ $\qquad$ $O(1)$

$\qquad$ **if** (SR $\leftarrow$ 0) $\longrightarrow$ $O(1)$ $\qquad$ $O(1)$

$\qquad\qquad$ $P_{pi} \leftarrow P_{pi}$ (randomly selected) $\longrightarrow$ $O(1)$ $\qquad$ $O(1)$

$\qquad$ **else**

$\qquad\qquad$ Modify the directory by excluding already sensed channels $\longrightarrow$ $O(1)$ $\qquad$ $O(1)$

$\qquad\qquad$ go to **(ii)** $\longrightarrow$ $O(N)$ $\qquad$ $O(1)$

End

(Time complexity for Step 6: $O(N)$ overall; Space complexity: $O(1)$ overall)

If the original state of the PU is idle, then throughput of the CU is C_0 and defined as:

$$C_0 = \frac{T - \tau_p - \tau_s}{T} \, log_2(1 + SNR_{s1}) \qquad (4.3)$$

and if the original state of PU is active, then the throughput of the CU is C_1 and defined as:

$$C_1 = \frac{T - \tau_p - \tau_s}{T} \, log_2\left(1 + \frac{SNR_{s1}}{1 + SNR_p}\right), \qquad (4.4)$$

where SNR_{s1} and SNR_p are the signal-to-noise ratios (SNR) at the CU receiver due to CU (interweave mode) and PU transmission, respectively. Additionally, in order to enhance the throughput of network, a new frame structure is proposed as shown in Figure 4.3 and applied in Figure 4.2. This frame is very much similar to the time frame shown in Figure 4.1b; however, in the proposed frame, the CU transmits data in the spectrum prediction and sensing phase $(\tau_p + \tau_s)$ using the constrained power to avoid interference with the PU, i.e. the data transmission in underlay mode (DTIUM) of spectrum access technique [30]. The additional data transmission in the prediction and sensing phase contributes to significant enhancement in the throughput of the CRN and total throughput of the network in the proposed system will be the function of the presence or absence of the PU.

If the original state of PU is idle, then the throughput of CU in the prediction and sensing period is C_2 and defined as:

$$C_2 = \frac{\tau_p + \tau_s}{T} \, log_2(1 + SNR_{s2}). \tag{4.5}$$

Otherwise, if the original state of the PU is active, then the throughput of the CU in the prediction and sensing period is C_3 and defined as:

$$C_3 = \frac{\tau_p + \tau_s}{T} \, log_2\left(1 + \frac{SNR_{s2}}{1 + SNR_p}\right), \tag{4.6}$$

where SNR_{s2} denotes the SNR at the CU receiver due to CU transmission in underlay mode, i.e. constrained power."

4.3 Throughput Analysis

Initially, the CU predicts the probabilities of N number of channels by which they are predicted idle (P_{pi}) and selects only the channels having probability greater than 0.5. Further, sort these probabilities in the descending order $P_{pi}(j)$, where $j = 1, 2, ... N_i$ and N_i is the number of channels having probability greater than 0.5. Now, the selection of channel for spectrum sensing among these channels is performed using conventional (random) and proposed (improved, *Approach-1*) approaches. In the random selection approach, the CU selects the channel for spectrum sensing randomly and performs sensing on that channel. If the channel is sensed idle, then CU establishes communication on that particular channel; otherwise, again sense the channel which is randomly selected among all, and so on. The defined process may result in the sense and stuck problem. Therefore, in the proposed approach, the process of selection of the channel for spectrum sensing is improved, which means, on the basis of their probabilities to be predicted idle (P_{pi}). Firstly, the spectrum sensing is performed on the channel having highest probability to be predicted idle. If it is sensed idle, the CU starts data transmission; otherwise, sense the next channel with second highest probability to be predicted idle, and so on.

The spectrum prediction and sensing are considered as an independent phenomenon and the probability of channel to be idle and active/busy is denoted as P_i and P_b, which is computed as reported in [1]. The error in spectrum sensing and prediction processes also affects the selection of the channel for data transmission. Thus, the probability of channel selected as idle or active/busy for data transmission is the function of the traffic intensity of channel ($P(H_1)$), the prediction probability with which the channel is predicted as idle (P_{pi}) and the probability of sensing errors, i.e. probability of false-alarm (P_f) and probability of misdetection ($1 - P_d$). The computation of probability of the channel to be idle is as follows. Initially, the channel is predicted idle with the probability P_{pi} and further, in the spectrum sensing there are two possibilities: (i) the actual idle channel is sensed as idle, i.e. no false alarm when original state of the channel is H_0 and (ii) the actual active channel is sensed as idle, i.e. there is misdetection when the original state of the channel is H_1. Thus, the probability P_1 is the first case whereas P_2 is for the second case and the total probability of channel to be sensed as idle is $P_1 + P_2$, which is derived as follows:

$$P_i = P_1 + P_2 = P(H_0)P_{pi}(1 - P_f) + P(H_1)P_{pi}(1 - P_d), \qquad (4.7)$$

where $P_1 = P(H_0)P_{pi}(1 - P_f)$ is the probability when the channels' original state is idle and also sensed as idle. However, $P_2 = P(H_1)P_{pi}(1 - P_d)$ is the probability when the original state of the channel is active and sensed as idle. Similarly, the probability of the channel to be predicted active/busy is computed as follows:

$$P_b = P_3 + P_4 = P(H_0)\left(P_{pi}\right)\left(P_f\right) + P(H_1)\left(P_{pi}\right)\left(P_d\right), \qquad (4.8)$$

where $P_3 = P(H_0)(P_{pi})(P_f)$ is the probability when the spectrum sensing results show the false alarm even in the absence of PU. However, $P_4 = P(H_1)(P_{pi})(P_d)$ is the probability when the presence of the PU is sensed correctly. As the data are transmitted on the idle sensed channels only, therefore the average throughput (R_{avgp}) of the CU for *Approach-1* (improved selection) becomes the function of P_1, P_2, C_0, and C_1. Since P_1 is the probability of the channel to be detected idle truly/ actually and C_0 is the data rate when actually/truly the channel is idle, similarly P_2 is the probability of the channel to be detected idle falsely and C_1 is the data rate when actually the channel is active. Therefore, the average throughput (R_{avgp}) is defined using Eqs. (4.3), (4.4), (4.7), and (4.8) as:

$$R_{avgp} = P_1 C_0 + P_2 C_1. \qquad (4.9)$$

The process of throughput computation for the random selection approach is similar to that of the *Approach-1*, except the probability of selected channel to be predicted idle. Therefore, we assume that the probability of channel to be idle for random selection is P_{ir}, which is defined as:

$$P_{ir} = P_5 + P_6 = P(H_0)P_{pir}(1 - P_f) + P(H_1)P_{pir}(1 - P_d),\qquad(4.10)$$

where P_{pir} is the probability of channel to be predicted idle, which is randomly selected for spectrum sensing. Therefore, the average throughput in random selection approach is R_{avgr} and defined as:

$$R_{avgr} = P_5C_0 + P_6C_1.\qquad(4.11)$$

Now, the way of throughput computation for the proposed approach (*Approach-2*) is similar to the pre-discussed *Approach-1*; however, the throughput in the spectrum prediction and spectrum sensing phase also contributes to the total throughput. Therefore, the total throughput in the *Approach-2* using improved selection (RP_{avgp}) and random selection (RP_{avgr}) method is:"

$$RP_{avgp} = P_1C_0 + P_2C_1 + P(H_0)C_2 + P(H_1)C_3,\qquad(4.12)$$

$$RP_{avgr} = P_5C_0 + P_6C_1 + P(H_0)C_2 + P(H_1)C_3.\qquad(4.13)$$

4.4 Simulation Results and Discussion

The numerically simulated results for the proposed *Approach-1* are presented and have been compared with that of the conventional approach as presented in [31]. The numerical values of the simulation parameters are selected using the standard of IEEE 802.22 [32], which is the first presented standard based on cognitive radio to construct the wireless regional area network (WRAN) as shown in Table 4.1.

Table 4.1 Simulation parameters with their values.

Parameter	Value	Parameter	Value
T	100 ms	f_s	1500 samples/s
τ_p	5 ms	P_f	0.1
τ_s	2.5 ms	SNR_{s1}	20 dB
P_d	0.9	SNR_p	-15 dB

The throughput of CU varies with number of channels; therefore, to neutralize this effect, we have normalized the average throughput of the CU. The CU achieves maximum throughput R_{up} if CU transmits continuously during the data transmission period regardless of the PUs' activity.

$$R_{up} = P(H_0)C_0 + P(H_1)C_1. \qquad (4.14)$$

The normalized throughput of CU in the proposed (Conv-Pro-Sel – *conventional proper selection method*) and conventional (Conv-Rand-Sel – *conventional random selection method*) approach are denoted as R_{normp} and R_{normc} and defined as: $R_{normp} = R_{avgp}/R_{up}$ and $R_{normr} = R_{avgr}/R_{up}$, respectively. However, the maximum throughput achieved by the CU in the proposed *Approach-2* is R_{up} plus the throughput achieved at the spectrum prediction and spectrum sensing period which is denoted as RP_{up}. The RP_{up} is defined as:

$$RP_{up} = P(H_0)C_0 + P(H_1)C_1 + P(H_0)C_2 + P(H_1)C_3$$
$$= P(H_0)(C_0 + C_2) + P(H_1)(C_1 + C_3). \qquad (4.15)$$

Therefore, the normalized throughput in the proposed *Approach-2* for improved/proper (Prop-Pro-Sel – Proposed proper selection method) and random selection method (Prop-Rand-Sel – Proposed random selection method) is $RP_{normp} = RP_{avgp}/RP_{up}$ and $RP_{normr} = RP_{avgr}/RP_{up}$, respectively. The traffic intensity of the PU (ρ) and total number of channels in the network (N) affect the throughput of CU. Therefore, in the simulations, the throughput of CU is considered as the function of these two variables, i.e. N and ρ. Moreover, it is worth mentioning that since the spectrum prediction and spectrum sensing systems may have some errors, due to which a channel can be predicted and sensed as idle when its original state is active which is represented by traffic intensity. Therefore, the selection of channel for data transmission with high traffic intensity reduces the throughput drastically. For the generalization and in order to consider the random events such as P_{pi}, we have computed the throughput of CU for various scenario up to 10 000 times and have averaged these values to achieve the final results as shown in Figures 4.4–4.7."

It is palpable that the improved selection method allows the CU to select the channel having low traffic intensity (since the channel with the highest prediction probability to be idle is selected) as compared to that of the channel selected by the random selection approach. However, for the simulations, we have assumed the event when the selected channels by the improved and random selection methods have different prediction probabilities but same traffic intensities. Figures 4.4 and 4.5 refer to the Approach-1; however, Approach-2 is described in Figures 4.6 and 4.7. The relation between normalized throughput of CU and traffic intensity with

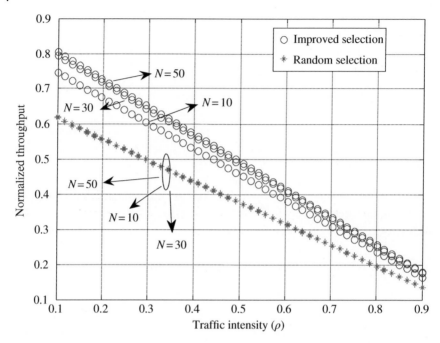

Figure 4.4 Normalized throughput of the CU versus traffic intensity for improved and random selection approach (SNR at CU due to interweave transmission = 20 dB) (Approach-1).

reference to different number of channels for the improved and random selection approach is presented in Figure 4.4. It is apparent that with an increase in the traffic intensity, the throughput of CU decreases linearly for random and improved selection approach. The throughput in the improved selection approach is more as compared to that of the random selection approach and the effect of different number of channels is also visualized in the improved selection approach (this effect appeared for different number of channels since we have computed the throughput for 10 000 runs for random values of P_{pi}); however, in the random selection approach, the throughput remains almost same for various values of number of channels. The important point which needs mentioning is that for $\rho = 0.1$, the difference between the throughput in random and improved selection approach is sufficient; however, it decreases with increases in traffic intensity. The variations of normalized throughput with number of channels for several values of the traffic intensity in the improved and random selection approach are illustrated in Figure 4.5. The throughput in the improved selection approach is more as compared to that of the random selection approach. The key point which needs to

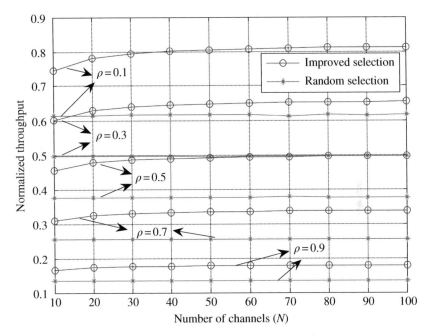

Figure 4.5 Normalized throughput of the CU versus number of channels for improved and random selection approach (SNR at CU due to interweave transmission = 20 dB) (Approach-1).

focus is that the throughput increases rapidly till the certain value of N such as 30, and after that the rate of increment slows down. Therefore, we conclude that the number of channels needs to be selected optimally by maximizing the throughput and minimizing the sensing load.

The relation between normalized throughput of the CU and SNR at CU due to an interweave (full-power) transmission is shown in Figure 4.6. It is clear that the throughput of the improved selection method is sufficiently high as compared to that of the random selection method. Moreover, the throughput characteristics are the witness of prominent improvement in the normalized throughput of CU in *Approach-2* as compared to that of the *Approach-1*. In Figure 4.7, the variations of the normalized throughput with the SNR at CU due to underlay transmission for different number of channels and at certain values of the traffic intensity and SNR at CU due to interweave transmission are presented. With an increase in the SNR at CU due to underlay transmission, the improvement in the proposed approach is significant."

The effect due to number of channels on the throughput is perceptible for the improved selection method; however, indistinguishable for the random selection

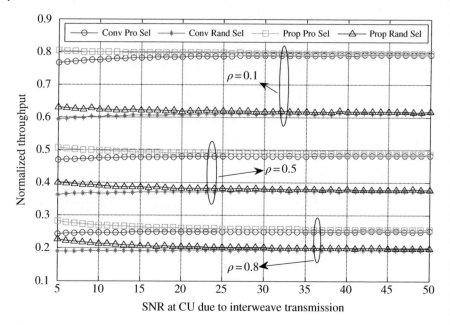

Figure 4.6 Normalized throughput versus SNR at CU due to interweave transmission (SNR at CU due to underlay transmission = 0 dB and N = 30) (*Approach-2*).

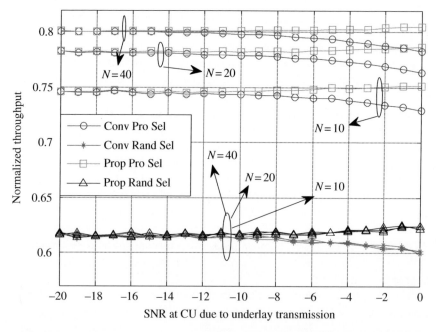

Figure 4.7 Normalized throughput versus SNR at CU due to underlay transmission (SNR at CU due to interweave transmission = 10 dB and ρ = 0.1) (*Approach-2*).

method. Therefore, we conclude that the proposed approach provides a significant improvement in the normalized throughput if the SNR difference at CU due to an interweave and underlay transmission is significantly less, and improvement decays with an increase in that SNR difference. This all happens because the increase in SNR due to an interweave transmission nullifies the effect of underlay transmission in the spectrum prediction and sensing phase and this is the key limitation of this proposed approach.

4.5 Summary

In this chapter, the challenging issues regarding the selection of channel for spectrum sensing after spectrum prediction are exploited to enhance the throughput of CU. However, in the pre-existing approaches, the spectrum sensing channel is selected randomly among all the idle predicted channels, nevertheless, in the proposed approach (*Approach-1*), the selection relies on the highest probability of the channel to be predicted idle. For the computation of prediction probabilities, we have presented an approach based on the conventional probability theory. Moreover, the simulation results are presented which are the witnesses of the significant improvement in the performance of the proposed *Approach-1*. Moreover, with the unavailability of prediction probabilities in meticulous scenarios, a modified selection approach is introduced to depart the sense and stuck problem. In addition to this, to evaluate the throughput of the CRN using random, improved, and modified selection approaches, an algorithm is illustrated, where the space and time complexities are also computed. It is concluded that the proposed approaches outperform the random selection approach in terms of time complexity; however, the space complexity is equal for all approaches. Furthermore, a new approach (*Approach-2*) for additional enhancement in the throughput of CU by exploiting the underlay communication in the spectrum sensing and prediction phases is presented and the results are compared with pre-obtained results of *Approach-1*. The results are evidence for the significant improvement in the throughput by using this approach.

References

1 Yang, J. and Zhao, H. (2015). Enhanced throughput of cognitive radio networks by imperfect spectrum prediction. *IEEE Commun. Lett.* 19 (10): 1738–1741.
2 Pei, Y., Liang, Y.C., Teh, K.C., and Li, K.H. (2011). Energy-efficient design of sequential channel sensing in cognitive radio networks: optimal sensing strategy, power allocation, and sensing order. *IEEE J. Sel. Areas Commun.* 29 (8): 1648–1659.

3 Chatterjee, S., Maity, S.P., and Acharya, T. (2014). Energy efficient cognitive radio system for joint spectrum sensing and data transmission. *IEEE J. Emerging Sel. Top. Circuits Syst.* 4 (3): 292–300.

4 Thakur, P., Kumar, A., Pandit, S. et al. (2017). Spectrum mobility in cognitive radio network using spectrum prediction and monitoring techniques. *Phys. Commun.* 24: 1–8.

5 Xing, X., Jing, T., Cheng, W. et al. (2013). Spectrum prediction in cognitive radio networks. *IEEE Wireless Commun.* 20 (2): 90–96.

6 Liang, W., Ng, S.X., and Hanzo, L. (2017). Cooperative overlay spectrum access in cognitive radio networks. *IEEE Commun. Surv. Tutorials* 19 (3): 1924–1944.

7 Ghane, A.H. and Harsini, J.S. (2018). A network steganographic approach to overlay cognitive radio systems utilizing systematic coding. *Phys. Commun.* 27: 63–73.

8 Sharma, S.K., Chatzinotas, S., and Ottersten, B. (2014). A hybrid cognitive transceiver architecture: sensing-throughput tradeoff. *9th IEEE International Conference on Cognitive Radio Oriented Wireless Networks and Communications (CROWNCOM)*, Oulu, Finland, 143–149.

9 Jiang, X., Wong, K.K., Zhang, Y., and Edwards, D. (2013). On hybrid overlay underlay dynamic spectrum access: double-threshold energy detection and Markov model. *IEEE Trans. Veh. Technol.* 62 (8): 4078–4083.

10 Chu, T.M.C., Phan, H., and Zepernick, H.-J. (2014). Hybrid interweave-underlay spectrum access for cognitive cooperative radio networks. *IEEE Trans. Commun.* 62 (7): 2183–2197.

11 Gmira, S., Kobbane, A., and Sabir, E. (2015). A new optimal hybrid spectrum access in cognitive radio: overlay-underlay mode. *International Conference on Wireless Networks and Mobile Communications (WINCOM)*, Marrakech, Morocco, 1–7.

12 Blasco-Serrano, R., Lv, J., Thobaben, R. et al. (2012). Comparison of underlay and overlay spectrum sharing strategies in MISO cognitive channels. *7th IEEE International ICST Conference on Cognitive Radio Oriented Wireless Networks and Communications (CROWNCOM)*, Stockholm, Sweden, 224–229.

13 Thakur, P., Kumar, A., Pandit, S. et al. (2016). Frame structures for hybrid spectrum accessing strategy in cognitive radio communication system. *9th IEEE International Conference on Contemporary Computing (IC3)*, Noida, India, 1–6.

14 Raj, V., Dias, I., Tholeti, T., and Kalyani, S. (2018). Spectrum access in cognitive radio using a two-stage reinforcement learning approach. *IEEE J. Sel. Top. Signal Process.* 12 (1): 20–34.

15 Sun, C., Ding, H., and Liu, X. (2020). Multichannel spectrum access based on reinforcement learning in cognitive internet of things. *Ad Hoc Netw.* 106: 1–26.

16 Ozturk, M., Akram, M., Hussain, S., and Imran, M.A. (2019). Novel QoS-aware proactive spectrum access techniques for cognitive radio using machine learning. *IEEE Access* 7: 70811–70827.

17 Bhowmick, A., Prasad, B., Roy, S.D., and Kundu, S. (2016). Performance of cognitive radio network with novel hybrid spectrum access schemes. *Wireless Pers. Commun.* 91 (2): 541–560.

18 Nguyen, V.D. and Shin, O.-S. (2018). Cooperative prediction-and-sensing-based spectrum sharing in cognitive radio networks. *IEEE. Trans. Cognitive Commun. Netw.* 4 (1): 108–120.

19 Bhowmick, A., Yadav, K., Roy, S.D., and Kundu, S. (2017). Throughput of an energy harvesting cognitive radio network based on prediction of primary user. *IEEE Trans. Veh. Tech.* 66 (9): 8119–8128.

20 Zhang, J., Chen, Y., Liu, Y., and Wu, H. (2020). Spectrum knowledge and real-time observing enabled smart spectrum management. *IEEE Access*: 44153–44162. https://doi.org/10.1109/ACCESS.2020.2978005.

21 Shawel, B.S., Woldegebreal, D.H., and Pollin, S. (2018). Deep-learning based cooperative spectrum prediction for cognitive networks. *2018 International Conference on Information and Communication Technology Convergence (ICTC)*, Jeju, South Korea, 133–137.

22 Shawel, B.S., Woldegebreal, D.H., and Pollin, S. (2019). Convolutional LSTM-based long-term spectrum prediction for dynamic spectrum access. *2019 27th European Signal Processing Conference (EUSIPCO)*, A Coruna, Spain, 1–5.

23 Thakur, P., Kumar, A., Pandit, S. et al. (2017). Performance analysis of cognitive radio networks using channel-prediction-probabilities and improved frame structure. *Digital Commun. Netw.* 4 (4): 287–295.

24 Eltom, H., Kandeepan, S., Evans, R.J. et al. (2018). Statistical spectrum occupancy prediction for dynamic spectrum access: a classification. *EURASIP J. Wireless Commun. Netw.* 2018: 1–17.

25 Fan, R., Guo, H., Di, L., and Ling, X. (2019). Spectrum occupancy state predictor based on recurrent neural network. *J. Phys.: Conf. Ser.* 1345: 042020.

26 Eltom, H., Kandeepan, S., Liang, Y.-C., and Evans, R.J. (2018). Cooperative soft fusion for HMM-based spectrum occupancy prediction. *IEEE Commun. Lett.* 22 (10): 2144–2147.

27 Pandit, S. and Singh, G. (2014). Throughput maximization with reduced data loss rate in cognitive radio network. *Telecommun. Syst.* 57 (2): 209–215.

28 Masonta, M.T., Mzyece, M., and Ntlatlapa, N. (2013). Spectrum decision in cognitive radio networks: a survey. *IEEE Commun. Surv. Tutorials* 15 (3): 1088–1107.

29 Papoulis, A. and Pillai, S.U. (2002). *Probability, Random Variables and Stochastic Processes*. McGraw-Hill Education. https://www.mheducation.co.in/

w9780070486584-india-probability-random-variables-and-stochastic-processes (accessed 18 February 2018).

30 Kleinberg, J. and Tardos, E. (2014). *Algorithm Design*, 1e. Edinburgh: Pearson Education Limited.

31 Thakur, P., Kumar, A., Pandit, S. et al. (2017). Advanced frame structures for hybrid spectrum access strategy in cognitive radio communication systems. *IEEE Commun. Lett.* 21 (2): 410–413.

32 IEEE-SA – Contact Us. http://standards.ieee.org/contact/form.html (accessed 18 February 2018).

5

Effect of Imperfect Spectrum Monitoring on Cognitive Radio Networks

5.1 Introduction

As discussed in the Chapter 3, the primary user (PU) exploits the spectrum very frequently in the high-traffic cognitive radio network (HTCRN) and random selection of the channel for spectrum sensing results in the high probability of active detection of channel. This degrades the performance of spectrum sensing which leads to the wastage of sensing-power and sensing-time [1–4]. Therefore, it is worth predicting the state of channel before the spectrum sensing and if it is predicted as idle, only then perform spectrum sensing [5–15]. The complete communication process using time frame which comprises the spectrum prediction, spectrum sensing, and data-transmission phases proceeds as follows and is shown in Figure 5.1a. The cognitive user (CU) predicts the state of channels (active or idle) and senses only the idle predicted channels in order to alleviate the spectrum sensing errors and establish communication (with full-power, i.e. interweave approach) on the idle sensed channels [16, 17]. Even though, the CU knew the channel state before to start the data transmission, however, the possibility of reappearance of PU is a very prominent phenomenon. Therefore, two approaches, namely, the proactive and reactive are proposed in the literature to detect the reappearance of PU during data transmission. In the former approach, the CU exploits the prediction technique in order to detect the emergence (reappearance) of PU and switches its communication before the actual emergence of PU as reported in [16]. In the spectrum prediction approaches, the pre-available information about the channel states and training of the device/CU is required to forecast the future behavior of the channel. This information is used to improve the spectrum sensing reliability of CU by selecting only the idle predicted channel for sensing and spectrum switching with the emergence of PU [7, 18, 19]. Due to the dynamic and random nature of the channels, the spectrum prediction is a

Spectrum Sharing in Cognitive Radio Networks: Towards Highly Connected Environments, First Edition. Prabhat Thakur and Ghanshyam Singh.

(a)

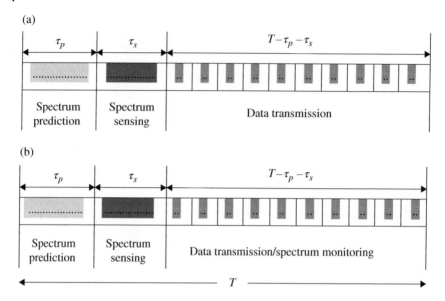

Figure 5.1 The time frame for HTCRN model: (a) conventional approach (*Source*: Based on Yang and Zhao [16]) and (b) proposed approach.

potential option to increase the spectrum sensing reliability because the channel access decision is obtained on the basis of sensing results rather than the prediction results. However, the spectrum sensing appears as a unfruitful approach for spectrum switching since the switching decision relies only on the prediction results; therefore, the reactive approach came into existence and is known as spectrum monitoring (SM) in which the CU has exploited the received signal statistics during data transmission in order to detect the emergence of PU [20–24]. Boyd et al. [20] have introduced this concept and have presented the techniques of SM using received signals' statistics such as receiver error count (REC). Moreover, the effect of cooperation between two CUs on the SM is also presented."

As per the authors' best knowledge, in the reported literature [20–23], the researchers have considered the perfect SM that is impractical scenario. The imperfections in SM disable the CU to detect the emergence of PU which results in the continuous transmission by CU with full-power, i.e. interweave approach. The continuous transmission even in the presence of PU results in the loss of energy and data as well as introduces prominent interference at the PU receiver. Therefore, to study the effect of imperfect SM, we have proposed a potential technique to analyze the achieved throughput, data loss, power wastage, interference at PU, interference efficiency, and energy efficiency in the HTCRN. The potential contribution in this chapter is summarized as follows:

- A modified frame structure is proposed for HTCRNs in order to detect the emergence of PU during data transmission with full power, in which the data transmission and SM are the parallel phenomena.
- A novel concept of imperfect SM is introduced and has proposed two algorithms to analyze the data loss and interference at PU for different scenarios of traffic intensity and monitoring error.
- The closed-form expressions for the ratio of achieved-throughput to data loss, interference efficiency, power wastage, and energy efficiency are derived, and numerically simulated results are presented to support the proposed system model. The randomness in the emergence of PU is explored and the Monte-Carlo simulations for 10 000 runs are exploited in order to further support the numerically simulated results.

In the next section, the work related to SM and spectrum prediction techniques are presented."

5.2 Related Work

5.2.1 Spectrum Sensing

The spectrum sensing is a popular approach in the CRN which allows the CU to sense the electromagnetic environment in order to perceive the state of channels (active or idle) [1–4, 9, 25–28]. If the channel is sensed as idle, only then the CU establishes communication, otherwise it senses another channel. The key technique used for spectrum sensing is the energy detection, where the CU receiver compares the received energy level with the predefined threshold and if it is greater than the threshold, the channel state is decided as active, otherwise idle [2]. In case of the spectrum sensing, the binary hypotheses considered for the received signal $r(t)$ are H_0 and H_1, which confirms the absence and presence of the PU, respectively, as discussed in Section 2.4.

$$r(t) = \begin{cases} h.s(t) + w(t) & H_1 \\ w(t) & H_0 \end{cases}, \tag{5.1}$$

where h is the magnitude of channel gain, $w(t)$ is the additive white Gaussian noise (AWGN), and $s(t)$ is the transmitted signal of the PU. The prominent errors in the spectrum sensing process are the false alarm and misdetection [2]. The false alarm is defined as the error of false detection of PU on the channel when originally/actually the PU is absent and represented by the probability of false alarm (P_f). The misdetection error occurs when the CU misses the detection of PU when originally PU is present on the channel which is represented by the probability of misdetection (P_m). The false alarm of PUs' presence on the channel restricts the CUs to access that channel which results the resource-wastage, particularly the spectrum wastage. On the other hand, the misdetection allows CUs to establish the communication simultaneous to the PU which results in the data-collision/ loss, power wastage, as well as introduces the interference at PU."

5.2.2 Spectrum Monitoring

The SM is an important phenomenon in which the CU monitors the spectrum during data transmission in order to detect the emergence of PU [20]. The prime difference between the spectrum sensing and SM is that the CU needs to detect the PU during the data transmission, which means the CU needs to perceive the presence of PU signal $s(t)$ with the already existing CU signal $C(t)$ and noise $w(t)$. Therefore, the binary hypothesis of SM considered for the received signal $r(t)$ is H_0 and H_1, which confirms the absence and presence of the PU, respectively.

$$r(t) = \begin{cases} h.s(t) + w(t) + C(t) & H_1 \\ w(t) + C(t) & H_0 \end{cases}. \tag{5.2}$$

The key intent of SM implementation is to detect the emergence of PU during data transmission. The error occurs in the SM process if the CU misses the detection of emergence of PU; therefore, the role of misdetection error is prominent however the false-alarm error does not have a significant role. Therefore, the probability of the SM error is inspired by the probability of misdetection. The probability of SM error is defined as the delay in the detection of the PU. Higher the value of probability of SM error, more will be the delay in the detection of the emergence of PU.

First time, this technique was proposed by Boyd et al. [20] in which the CU detects the emergence of PU by using the statistics of the received signal such as REC and further explored by various researchers [21–24]. The REC-based SM process works as follows, where the *packet* is defined as a data sequence comprising the binary symbols which are transmitted from the transmitter, travels through the communication channel, and received by the receiver. The communication channel affects the passing sequence due to multipath fading, shadowing, etc., and results in two possible cases: (i) the packet is decoded successfully at the receiver and the received sequence may not coincide with the transmitted data sequence at some places for the successfully decoded sequence. The number of places where the received data sequence does not coincide the transmitted data sequence is defined as the error count (EC) also known as receivers' EC. (ii) The packet does not decode successfully which results in the unavailability of received data sequence; therefore, the EC is undefined. The internal and complete framework of the receiver and transmitter is illustrated by Boyd et al. [20]. In the conventional session, which starts after the spectrum sensing, it is assumed that the CU receiver decodes the data sequence successfully and has a certain value of EC rely on the channel conditions. Moreover, the channel is of varying nature which results in the fluctuation in the value of REC; however, a maximum value of REC is defined so that the packet can be decoded successfully and that value is known as EC threshold. Furthermore, if the PU emerges and established communication during the data transmission of CU, packets received by CU get interference and EC increases. Therefore, if the EC increases and cross the threshold or the packet decoded successfully, the emergence of PU is confirmed. The recent works on the SM techniques is described in [20–25] as discussed in Section 3.2.1.

The remaining of this chapter is structured as follows. Section 5.3 depicts the system model of proposed technique. The performance analysis of proposed system using imperfect SM is exploited in Section 5.4. In Section 5.5, the simulation results are illustrated. Finally, Section 5.6 summarizes the chapter and presents the future perspectives.

5.3 System Model

In the proposed HTCRN, the CU is permitted to access the spectrum of PU if it is free/unutilized or underutilized from the PU communication. To detect the free/ idle available channel, the CU initially performs spectrum prediction on the number of channels and then senses a channel among the idle predicted channels and starts the data transmission if that channel is sensed as idle; otherwise, senses another channel. The entire time frame (T) of the proposed system consists of three phases, i.e. the spectrum prediction phase, spectrum sensing phase, and data transmission phase as shown in Figure 5.1b and discussed in Section 3.3.2. In the last phase, the SM is performed parallel to the data transmission, to perceive the emergence of PU. It is worth mentioning that the PUs' information via SM need to be obtained at the transmitter; therefore, the CU transmitter and CU receiver are assumed to be synchronized at the control channel which is considered to be always available. Thus, the monitoring information obtained at the CU receiver is also available at the CU transmitter. We have considered the imperfect SM in which the probability of SM error (P_{me}) is used to yield the imperfection of the system. However, the SM appears significant only if the PU resumes communication in the data transmission phase. Therefore, to consider this event, we have assumed that the probability of PUs' reappearance in the data transmission phase is provided by the traffic intensity (ρ) of the PU network. The traffic intensity of PU is assumed as a binary stochastic hypothesis, in which 0 and 1 represents the idle and active channels, respectively [21]. Further, the probability of channel being active is: $\rho = P(H_1) = \mu/\lambda$ and idle channel probability is: $P(H_0) = (\lambda - \mu)/\lambda$ as discussed in Section 3.3. In general, the traffic intensity represents the average channel holding time by the licensed user [29]; however, in the proposed model, we have reformed the concept of traffic intensity to verify the effect of SM.

According to the reformed concept, the high value of traffic intensity signifies the early emergence of PU in the data transmission period; however, low value of traffic intensity represents the emergence of PU in the later phase. The total number of packets transmitted in the data transmission phase is denoted as *No*. The SNR at the CU receiver due to CU transmission in interweave mode (SNR_s) and SNR at the PU receiver due to CU transmission interweave mode ($SNRP_s$) are related to P_1 as follows: $SNR_s = (P_1 \times h_{ss})/N_{PCU}$ and $SNRP_s = (P_1 \times h_{sp})/N_{PPU}$.

5.4 Performance Analysis of Proposed System Using Imperfect Spectrum Monitoring

In this section, we have examined the impact of imperfect SM on data loss of the proposed HTCRN and its flow diagram is presented in Figure 5.2. As the data loss take place in the network only, the PU emerges in the data transmission period and the probability of emergence is yielded by the traffic intensity of PU. The perfect SM system is very quick and ideal, even though a particular packet is required to compute decision statistics. Therefore, in the perfect SM, the emergence of PU is detected very quickly; that is, within a data packet time and particular data packet loss. However, in the case of imperfect SM, the data loss depends on the error probability of the system, i.e. the probability of SM error (P_{me}). In the proposed HTCRN, the certain number of packets will lost out of total number of packets, which relies on the SM error. In this scenario, the computation of number of packets lost among total number of packets approximately follows the Binomial distribution [30]. The Binomial distribution provides the discrete probability distribution, $P_p(x/X)$, to achieve exactly x successes out of X Bernoulli trials (where the result of every Bernoulli trial is true with probability p and false with probability $(1 - p)$). Similarly, in the proposed case, the discrete probability distribution $P(k/No)$ of obtaining exactly k number of packets lost among *No* (where the data loss result is true with probability P_{me} and false with probability $(1 - P_{me})$). Therefore, the probability of data loss due to SM error has been formulated as a Binomial distribution, which is defined as:

$$P\left(\frac{k}{No}\right) = \binom{N_0}{k} (P_{me})^k (1 - P_{me})^{No-k}. \tag{5.3}$$

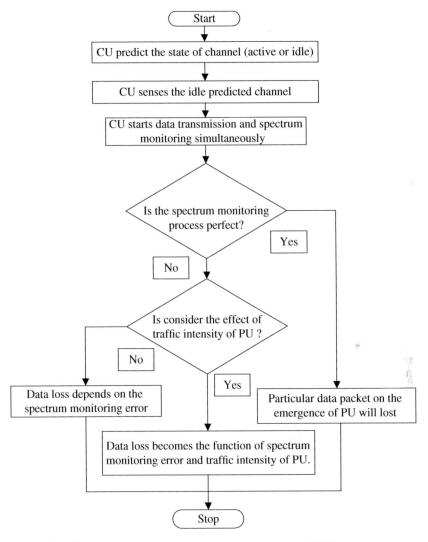

Figure 5.2 The flow diagram for the functioning of proposed HTCRN model.

The average value of the number of data packets lost is calculated by computing the expectation of variable and presented as:

$$k_{avg} = \sum_{k=1}^{No} k.P\left(\frac{k}{No}\right). \tag{5.4}$$

The average number of packets lost due to only the monitoring error is: k_{avg}; however, the particular one packet lost on the emergence of PU, even when $P_{me} = 0$. Therefore, the total number of packet loss in the network due to monitoring error is: $k_{an} = (1 + k_{avg})$. As in the proposed network, we have assumed the communication via frame structures that means after certain time interval, i.e. the frame time (T), the CU reiterates its steps of spectrum prediction, spectrum sensing, and data transmission.

Thus, it is a prominent phenomenon that the CU switches from data transmission mode to spectrum prediction mode of the upcoming frame, immediately on the emergence of PU due to the ending of data transmission period. Here, the number of packets lost relies on the time of emergence of PU, that is, the function of ρ in the considered CRN. Therefore, the full data loss in the proposed network (k_{comp}) is not only the function of the SM error but also of the traffic intensity of PU (ρ). Further, to achieve the full data loss in the proposed CRN, we have to compute the total number of packets to be transmitted after the emergence of PU (k_{TAEPU}), which depends on the traffic intensity and calculated as: $k_{TAEPU} = (1 - \rho) \times No$. The full data loss in the proposed CRN (k_{comp}) is:

$$k_{comp} = \begin{cases} k_{an} & \text{if } k_{TAEPU} \geq k_{an} \\ k_{TAEPU} & \text{if } k_{TAEPU} < k_{an} \end{cases}. \tag{5.5}$$

If the number of packets to be transmitted after the emergence of PU is more than the total number of packets lost due to SM error, then the full data loss in the proposed CRN is k_{an}; otherwise, k_{TAEPU}.

The full data loss of the proposed CRN is illustrated with Algorithm 5.1, in which three scenarios have been illustrated as follows. (i) The traffic intensity is zero which means the PU will not appear in the data transmission phase. Due to this, there will be no data loss even if the proposed CRN has SM error, which means, in this special case there is no prominent role of the SM. (ii) The value of traffic intensity is assumed any numerical value between 0 and 1 and the probability of SM error (P_{me}) is zero, which means the appearance of PU is confirmed and the monitoring system is perfect. In this case, the emergence of PU will be detected within a packet transmission time and that particular packet data will be lost. (iii) The numerical value of the traffic-intensity is in between 0 and 1, and the imperfect SM is considered with probability of SM error (P_{me}), which is very much a practical scenario. If we evade the effect of ρ, the data loss in this case due to SM error is the function of P_{me} as derived in Eq. (5.4). However, the full data loss in this case is the data loss due to P_{me} in addition to the one data packet, because the particular packet lost even if P_{me} is zero. On the other hand, the consideration of effect of ρ makes the full data loss in the proposed network as a function of both the P_{me} and ρ, which is calculated as given in Eq. (5.5)."

Algorithm 5.1   Computation of Data Loss in the HTCRN

Input (No, P_{me}, ρ)

Output $(k_{avg}, k_{an}, k_{TAEPU}, k_{comp})$

BEGIN

Step-1 Variable declaration

No : Total number of packets in the data transmission phase;

k : Number of packets lost due to spectrum monitoring error;

P_{me} : Probability of spectrum monitoring error;

k_{avg} : Average number of packets lost due to monitoring error;

ρ : Traffic intensity of the PU;

k_{an} : Average number of packets lost in the network without considering the effect of traffic intensity;

k_{TAEPU} : Total number of packets to be transmitted after the emergence of PU

k_{comp} : Complete data loss in the proposed network

Step-2 Computation of probability of k number of packet lost among No

$$P\left(\tfrac{k}{No}\right) \leftarrow {}_{k}^{No}Comb.(P_{me})^{k}(1 - P_{me})^{No-k};$$

Step-3 Computation of average number of packets lost due to monitoring error and of k_{an}

Let $k_{avg} \leftarrow 0$;

for $k \leftarrow 1 : No$

   $k_{avg} \leftarrow \left(k_{avg} + k.P\left(\tfrac{k}{No}\right)\right)$;

end

   $k_{an} = 1 + k_{avg}$;

Step-4 Computation of the total number of packets to be transmitted after the emergence of PU

$$k_{TAEPU} = (1 - \rho) \times No;$$

Step-4 Complete Data loss in the HTCRN

If $\{(\rho \leftarrow 0) \,\&\&\, (P_{me} \leftarrow \text{any value})\}$ **then**

   $k_{comp} \leftarrow 0$ packet;

elseif$\{(0 < \rho \leq 1) \&\& (P_{me} \leftarrow 0)\}$ **then**

   $k_{comp} \leftarrow 1$ packet;

elseif $\{(0 < \rho \leq 1) \,\&\&\, (0 < P_{me} \leq 1)\}$ **then**

   if $(k_{TAEPU} \geq k_{an})$ **then**

$$k_{comp} \leftarrow k_{an};$$

else   **then**

$$k_{comp} \leftarrow k_{TAEPU};$$

    end

   end

   end

 end

END

The imperfection in SM restricts the CU to detect the reappearance of PU hastily, and continue the data transmission, even on the emergence of PU that causes the interference at the PU receiver. The CU transmission starts interfering with the PU transmission when the PU emerges and continues till the detection of emergence of PU. Since the detection of emergence of PU during data transmission period is the function of P_{me}, the interference at PU also becomes the function of P_{me}. Here, the interference at PU is presented in the form of SNR at PU due to simultaneous transmission of the PU and CU as well as a new interference metric known as interference efficiency is also used. The SNR at the PU receiver due to PU and CU transmission is denoted as $SNRP_p$ and $SNRP_s$, respectively. Therefore, the total SNR at the PU receiver due to both the PU and CU transmission is: $SNRP_{ps} = SNRP_p/(1 + SNRP_s)$.

In the perfect SM, the particular packet at the emergence of PU is lost whereas in the imperfect monitoring, the number of packets lost after the emergence of PU relies on the P_{me}. The starting time (I_s) and ending time (I_E) of the interference at PU depends on the traffic intensity (ρ) of PU and on the P_{me}, which are computed as:

$$I_s = \left\lfloor \{(1-\rho) \times (T - (\tau_s + \tau_p))\} + \{(\tau_s + \tau_p)\} \right\rfloor, \tag{5.6}$$

$$I_E = I_s + (k_{comp}.PT), \tag{5.7}$$

where $\lfloor . \rfloor$ signifies the floor function. The PT is the packet duration and defined as: $PT = (T - \tau_s - \tau_p)/No$. The computation of SNR at PU receiver is presented with Algorithm 5.2.

Algorithm 5.2   SNR Computation at PU Receiver

Input $(T, \rho, \tau_s, \tau_p, PT)$
Output (*SNRP*)
BEGIN
   Step-1: Variable Declaration

T :	Frame Duration;
t =	0: 1: T;
τ_p:	Prediction duration;
τ_s:	Sensing duration;
$SNRP_s$:	SNR at PU receiver due to cognitive transmission;
$SNRP_p$:	SNR at PU receiver due to primary transmission;

   Step-2: Computation of I_s, I_E and $SNRP_{ps}$,
      Compute the value of I_s and I_E using Eqs. (5.6) and (5.7),

$$SNRP_{ps} \leftarrow \frac{SNRP_p}{1 + SNRP_s};$$

   Step-3: SNR calculations at the PU receiver using conventional and proposed approaches

for $ii \leftarrow 1{:}length(t)$
  If $(t\,(ii) \leq I_s)$ **then**
                $SNRP(ii) \leftarrow 0$
    elseif $\{(t(ii) \geq I_s)\ \&\&\ (t\,(ii) \leq I_E)\}$ **then**
                $SNRP(ii) \leftarrow SNRP_{ps};$
    elseif $(t(ii) \geq I_E))$ **then**
                $SNRP(ii) \leftarrow SNRP_{ps};$
    end
  end
end
END

Further, several performance metrics need to be computed by using the aforementioned analysis of the data loss and SNR at PU receiver in Algorithms 5.1 and 5.2. The performance metrics investigated further are the ratio of achieved throughput to the data loss, power wastage, interference efficiency, and energy efficiency which are evaluated as follows.

5.4.1 Computation of Ratio of the Achieved Throughput to Data Loss

The throughput obtained due to collision-free data transmission is defined as the achieved throughput (*RA*); however, the throughput obtained during collision is considered as data loss of the CRN. The achieved throughput of the network is computed as:

$$RA = \left(\left(I_s - \left(\tau_s + \tau_p \right) \right) / T \right) \times log_2(1 + SNR_s). \tag{5.8}$$

As in the perfect monitoring case, the reappearance of PU is detected within a packet time and loss occurs only for that time, therefore the data-loss in the CU network is computed as:

$$DL_{PM} = \left(\frac{PT}{T} \right) \times log_2(1 + SNR_s). \tag{5.9}$$

However, in case of the imperfect SM, the k_{comp} number of packets lost relies on the probability of the SM error. Thus, the total data loss time is: $k_{comp} \times PT$. Therefore, the data loss in this case is computed as:

$$DL_{IM} = \left(\frac{\left(k_{comp} \times PT \right)}{T} \right) \times log_2(1 + SNR_s). \tag{5.10}$$

Moreover, the data loss of PU will be same as of the CU if the packet length is same for both the users, i.e. PU and CU. On the other hand, if the packet length of PU is different, the PU data loss will be computed till the time of completion of that last packet which interferes with CU.

Now, the ratio of achieved throughput to data loss (*RARDL*) in the perfect and imperfect SM case is denoted as $RARDL_{PM}$ and $RARDL_{IM}$, respectively, which are defined as:

$$RARDL_{PM} = \frac{RA}{DL_{PM}}, \tag{5.11}$$

$$RARDL_{IM} = \frac{RA}{DL_{IM}}. \tag{5.12}$$

5.4.2 Computation of Power Wastage

Further, the data loss in the network results in the power wastage (*PW*) in the network; therefore, it is worth computing the power wastage in the proposed system. The power required by one packet for its complete processes such as transmission, channel passing and reception, etc., is denoted as P_P. Thus, the full-power wastage is the multiplication of P_P and total number of packets lost, which is evaluated for the perfect and imperfect SM cases as follows:

$$PW_{PM} = DL_{PM} \times P_P, \qquad (5.13)$$

$$PW_{IM} = DL_{IM} \times P_P. \qquad (5.14)$$

5.4.3 Computation of Interference Efficiency

The interference efficiency (IE) [31] is a key performance metric if the CU transmission causes interference at PU and is described as the number of bits transmitted per unit of energy imposed on the PU. In the proposed CRN, it is defined as the ratio of achieved throughput to the power received at the PU receiver when the true state of PU is active. Now, the power received at the PU receiver due to CU transmission is considered as the interference to the PU communication (IF) and for the perfect and imperfect SM symbolizes as IF_{PM} and IF_{IM}, respectively, which are evaluated as:

$$IF_{PM} = \left(\frac{PT}{T}\right) \times P_1 \times h_{sp}, \qquad (5.15)$$

$$IF_{IM} = \left(\frac{k_{comp} \times PT}{T}\right) \times P_1 \times h_{sp}. \qquad (5.16)$$

Moreover, the interference efficiency in the perfect and imperfect SM case is calculated as follows:

$$IE_{PM} = \frac{RA}{IF_{PM}}, \qquad (5.17)$$

$$IE_{IM} = \frac{RA}{IF_{IM}}. \qquad (5.18)$$

5.4.4 Computation of Energy Efficiency

The communication networks must be energy efficient in order to support the notion of green communication and to release the consumers from wasted power consumption. By keeping this in mind, we have analyzed the impact of perfect and imperfect SM on the energy efficiency (EE) of the proposed CRN. The EE is defined as the ratio of achieved throughput to the power consumed by the system and its units are bits/joule/Hz [32–37]. The power consumption in the perfect and imperfect SM case is derived as follows:

$$PC_{PM} = \left(\frac{(I_s + P_T) - (\tau_s + \tau_p)}{T} \times P_1\right) + P_s + P_{Pr}, \qquad (5.19)$$

$$PC_{IM} = \left(\frac{I_E - (\tau_s + \tau_p)}{T} \times P_1\right) + P_s + P_{Pr}, \qquad (5.20)$$

where P_{Pr} and P_S are the powers desired for the spectrum prediction and sensing processes. Thus, the *EE* in the perfect and imperfect SM case is evaluated as follows:

$$EE_{PM} = \frac{RA}{PC_{PM}}, \tag{5.21}$$

$$EE_{IM} = \frac{RA}{PC_{IM}}. \tag{5.22}$$

5.5 Results and Discussion

This section illustrates the results which are numerically simulated for the data loss, interference-at-PU, the ratio of achieved-throughput to the data loss, power wastage, interference efficiency, and energy efficiency of the proposed HTCRN model with imperfect SM. Moreover, the results are compared with the perfect SM of HTCRN. In addition to this, the simulation parameters considered for the proposed HTCRN are inspired by the WRAN standard and are presented in Table 5.1. Further, to validate the proposed system model, we have used the randomness in the emergence of PU and have computed all the performance metric for 10 000 random values of traffic intensity and further averaged those values (Monte-Carlo simulation for 10 000 runs) [38]. The imperfection of system is presented through the probability of SM error P_{me}.

For the perfect and imperfect SM systems, the value of P_{me} is 0 and $0 < P_{me} \leq 1$, respectively. The probability of packet loss versus number of packets is presented in Figure 5.3 for different values of P_{me}. It is apparent that for the small values of P_{me}, i.e. 0.1, the probability of losing less number of packets is large and continuously decreases with increase in the number of packets, whereas for large value of P_{me}, i.e. 0.9, the probability of losing less number of packets is small which increases regularly and achieved maximum value at 9 (number of packets) as

Table 5.1 The simulation parameters for the proposed HTCRN system.

Parameter	Value	Parameter	Value	Parameter	Value
T	100 ms	No	10	h_{ss}	0.8
τ_p	5 ms	P_1	6 W	h_{sp}	0.2
τ_s	2.5 ms	N_{PCU}	0.4 W	N_{PCU}	0.4 W
$SNRP_p$	15 dB	$P_P = P_{Pr}$	0.2 W	P_S	0.2 W

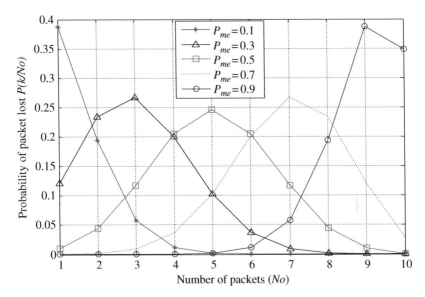

Figure 5.3 The probability of packet lost versus number of packets.

shown in Figure 5.3. However, for the medium values of P_{me}, i.e. 0.5, the probability of losing average number of packets (5 packets) is more; however, it decreases moving either right or left side on the horizontal axis.

The system with $P_{me} = 0.5$ shows more unpredicted behavior with reference to the data loss as almost all the number of data packets have certain probability of packet lost. However, the systems with $P_{me} = 0.1$ or 0.9 are more predicted as particular number of packets have higher probability to be lost and all others have very less. The relationship between the average number of packets lost due to monitoring error (k_{an}) and probability of SM error (P_{me}) for the perfect and imperfect monitoring system is illustrated in Figure 5.4.

For the perfect SM systems, the P_{me} is zero and only particular data packet will be lost. However, in the imperfect SM system, the data packet loss is increasing with probability of the SM error linearly. The variations of average power wastage with SM error are illustrated in Figure 5.5. The average power wastage shows the directly proportional relation with SM error. The interference at PU is measured in the form of SNR at PU, due to both the PU and CU transmission. The SNR at PU with respect to the time for various values of P_{me} is presented in Figure 5.6, where $P_{me} = 0$ shows the perfect monitoring and all other values of P_{me} show the imperfect SM. The SNR at PU receiver, on the emergence of the PU, becomes 8 dB. For the $P_{me} = 0$, the SNR remains 8 dB for time PT and then switches to 15 dB. However, for $P_{me} = 0.1$, the SNR switches after $2PT$ time and this time increases with

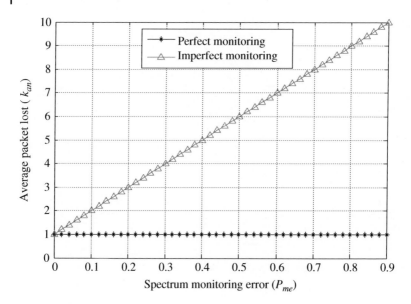

Figure 5.4 The average packet lost due to monitoring error versus probability of spectrum monitoring error.

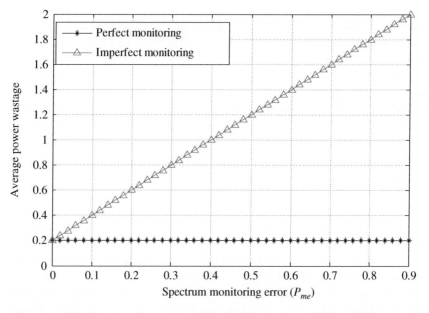

Figure 5.5 The variations of average power wastage (Watt) with the probability of spectrum monitoring error.

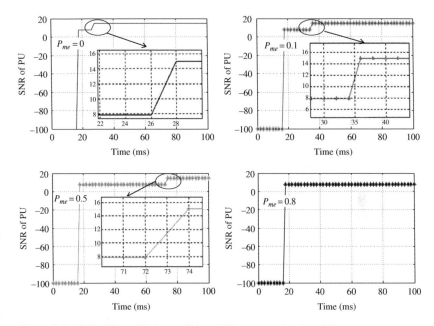

Figure 5.6 SNR (dB) at PU due to PU and CU transmission ($\rho = 0.9$).

increase in the P_{me}. Thus, it is clear that the increase in P_{me} indicates the increase in interfering time. The variations of interference efficiency with SM error for different values of the traffic intensity are shown in Figure 5.7. The interference efficiency is more for small values of traffic intensity and less for large values of the traffic intensity. Further, it is apparent that the increase in monitoring error worsens the interference efficiency which means higher the values of monitoring error probability, more will be the interference introduced at the PU. The variations of interference efficiency with increase in the "channel gain from CU transmitter to PU receiver" for different values of the traffic intensity are described in the Figure 5.8. The interference efficiency decays almost in exponential manner with increase in the "channel gain from CU transmitter to PU receiver." The relationship between the *RARDL* and probability of SM error for several values of traffic intensity is depicted in Figure 5.9. The small values of ρ results in the high value of *RARDL* and reduce with increase of ρ.

The *RARDL* shows the inverse relation with P_{me} for the imperfect SM; however, it remains constant for perfect SM case. Moreover, the values of *RARDL* using the Monte-Carlo simulations are obtained as shown in Figure 5.9 where the *RARDL* versus P_{me} follows the same nature as followed for the proposed approach which further validated the proposed approach. The relationship between the energy

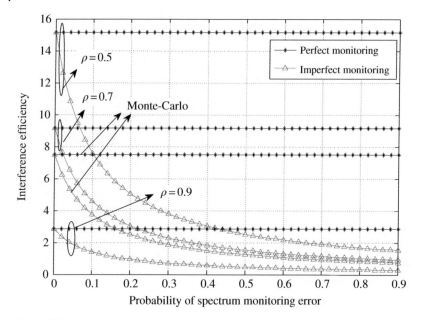

Figure 5.7 The variations of interference efficiency (bits/joule/Hz) with probability of spectrum monitoring error.

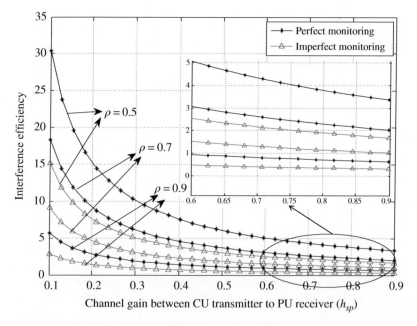

Figure 5.8 The relation between the interference efficiency (bits/joule/Hz) and channel gain from CU transmitter to PU receiver [$h_{ss} = 0.8$].

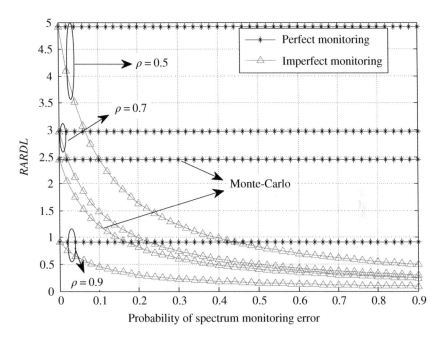

Figure 5.9 The ratio of the achieved throughput to the data loss (*RARDL*) versus probability of spectrum monitoring error.

efficiency and probability of SM error for different values of the traffic intensity are illustrated in Figure 5.10. The energy efficiency decays with increase in the probability of SM error almost in the exponential manner for the imperfect case; however, it remains constant for the perfect SM case. Moreover, the energy efficiency is inversely proportional to the traffic intensity, which means energy efficiency decreases with increase of the traffic intensity.

5.6 Summary

In this chapter, we have pioneered the notion of imperfect SM in the HTCRN and have investigated the data loss, interference at PU as well as the ratio of achieved-throughput to the data loss due to imperfect SM. It is concluded that the average value of probability of SM error shows more unpredicted behavior with context to the probability of data packet lost as compared to that of the higher or lower values of the probability of monitoring error (P_{me}). Moreover, the data loss (packet loss) is a linear function of the probability of SM error. Furthermore, the time of interference at PU increases with increase in the probability of the SM error. In addition,

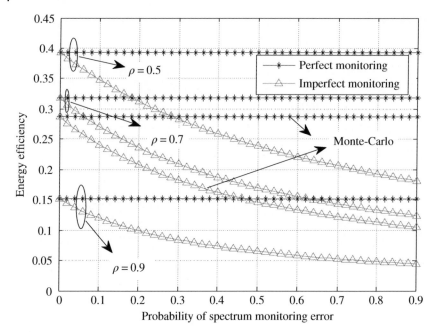

Figure 5.10 The variations of energy efficiency (bits/joule/Hz) with probability of spectrum monitoring error.

more performance metrics, namely, the ratio of achieved throughput to data loss, interference efficiency, and energy efficiency are also analyzed and it is concluded that the values of *RARDL*, interference, and energy efficiency decrease with increase in the probability of SM error. Further, to examine the impact of channel conditions and PU location, the relationship between interference efficiency and channel gain from CU transmitter to the PU receiver (h_{sp}) is demonstrated, which presents a fact that the worst nature of the channel between CU transmitter and PU receiver improves the interference efficiency. Furthermore, the Monte-Carlo simulations are explored to validate the numerically simulated results.

References

1 Ali, A. and Hamouda, W. (2017). Advances on spectrum sensing for cognitive radio networks: Theory and applications. *IEEE Commun. Surv. Tutor.* 19 (2): 1277–1304.

2 Yucek, T. and Arslan, H. (2009). A survey of spectrum sensing algorithms for cognitive radio applications. *IEEE Commun. Surv. Tutor.* 11 (1): 116–130.

3 Gupta, M.S. and Kumar, K. (2019). Progression on spectrum sensing for cognitive radio networks: A survey, classification, challenges and future research issues. *J. Netw. Comp. App.* 143: 47–76.

4 Kumar, A., Thakur, P., Pandit, S., and Singh, G. (2019). Analysis of optimal threshold selection for spectrum sensing in a cognitive radio network: an energy detection approach. *Wireless Netw.* 25 (7): 3917–3931.

5 Xing, X., Jing, T., Cheng, W. et al. (2013). Spectrum prediction in cognitive radio networks. *IEEE Wirel. Commun.* 20 (2): 90–96.

6 S. Yarkan and H. Arslan, "Binary time series approach to spectrum prediction for cognitive radio," *Proc. 66th IEEE Vehicular Technology Conference*, 2007, pp. 1563–1567. Baltimore, MD, USA, 30 Sept. 3 Oct. 2007.

7 Tumuluru, V.K., Wang, P., and Niyato, D. (2012). Channel status prediction for cognitive radio networks. *Wirel. Commun. Mob. Comput.* 12 (10): 862–874.

8 Barnes, S.D., Maharaj, B.T., and Alfa, A.S. (2016). Cooperative prediction for cognitive radio networks. *Wirel. Pers. Commun.* 89 (4): 1177–1202.

9 Zhao, Y., Hong, Z., Luo, Y. et al. (2018). Prediction-based spectrum management in cognitive radio networks. *IEEE Syst. J.* 12 (4): 3303–3314.

10 Ding, G., Jiao, Y., Wang, J. et al. (2017). Spectrum inference in cognitive radio networks: Algorithms and applications. *IEEE Communs. Sur. Tuts.* 20 (1): 150–182.

11 Yu, L., Chen, J., Ding, G. et al. (2018). Spectrum prediction based on taguchi method in deep learning with long short-term memory. *IEEE Access* 6: 45923–45933.

12 Sun, J., Shen, L., Ding, G. et al. (2017). Predictability analysis of spectrum state evolution: Performance bounds and real-world data analytics. *IEEE Access* 5: 22760–22774.

13 Sun, J., Wang, J., Ding, G. et al. (2018). Long-term spectrum state prediction: An image inference perspective. *IEEE Access* 6: 43489–43498.

14 Ge, C., Wang, Z., and Zhang, X. (2019). Robust long-term spectrum prediction with missing values and sparse anomalies. *IEEE Access* 7: 16655–16664.

15 Lin, F., Chen, J., Sun, J. et al. (2020). Cross-band spectrum prediction based on deep transfer learning. *China Commun.* 17 (2): 66–80.

16 Yang, J. and Zhao, H. (2015). Enhanced throughput of cognitive radio networks by imperfect spectrum prediction. *IEEE Commun. Lett.* 19 (10): 1738–1741.

17 Thakur, P., Kumar, A., Pandit, S. et al. (2017). Advanced frame structures for hybrid spectrum access strategy in cognitive radio communication systems. *IEEE Commun. Lett.* 21 (2): 410–413.

18 Anandkumar, A., Michael, N., Tang, A.K., and Swami, A. (2011). Distributed Algorithms for learning and cognitive medium access with logarithmic regret. *IEEE J. Sel. Areas Commun.* 29 (4): 731–745.

19 Macaluso, I., Ahmadi, H., and DaSilva, L.A. (2015). Fungible orthogonal channel sets for multi-user exploitation of spectrum. *IEEE Trans. Wirel. Commun.* 14 (4): 2281–2293.

20 Boyd, S.W., Frye, J.M., Pursley, M.B., and Royster, T.C. (2012). Spectrum monitoring during reception in dynamic spectrum access cognitive radio networks. *IEEE Trans. Commun.* 60 (2): 547–558.

21 Ali, A. and Hamouda, W. (2015). Spectrum monitoring using energy ratio algorithm for OFDM-based cognitive radio networks. *IEEE Trans. Wirel. Commun.* 14 (4): 2257–2268.

22 Orooji, M., Soltanmohammadi, E., and Naraghi-Pour, M. (2015). Improving detection delay in cognitive radios using secondary-user receiver statistics. *IEEE Trans. Veh. Technol.* 64 (9): 4041–4055.

23 R. Saifan, A. E. Kamal, and Y. Guan, "Efficient spectrum searching and monitoring in cognitive radio network," *Proc. 8th IEEE International Conference on Mobile Ad-Hoc and Sensor Systems*, 2011, pp. 520–529. Valencia, Spain, 17–22 Oct. 2011.

24 Thakur, P., Kumar, A., Pandit, S. et al. (2019). Performance analysis of cooperative spectrum monitoring in cognitive radio network. *Wirel. Netw.* 25 (3): 989–997.

25 Thakur, P., Kumar, A., Pandit, S. et al. (2017). Spectrum mobility in cognitive radio network using spectrum prediction and monitoring techniques. *Phys. Commun.* 24: 1–8.

26 Ding, G., Wang, J., Wu, Q. et al. (2013). Spectrum sensing in opportunity-heterogeneous cognitive sensor networks: How to cooperate? *IEEE Sens. J.* 13 (11): 4247–4255.

27 Soltanmohammadi, E., Orooji, M., and Naraghi-Pour, M. (2013). Improving the sensing throughput tradeoff for cognitive radios in Rayleigh fading channels. *IEEE Trans. Veh. Technol.* 62 (5): 2118–2130.

28 Thakur, P., Kumar, A., Pandit, S. et al. (2018). Performance analysis of high-traffic cognitive radio communication system using hybrid spectrum access, prediction and monitoring techniques. *Wirel. Netw.* 24 (6): 2005–2015.

29 Masonta, M.T., Mzyece, M., and Ntlatlapa, N. (2013). Spectrum decision in cognitive radio networks: A survey. *IEEE Commun. Surv. Tutor.* 15 (3): 1088–1107.

30 Morrison, M. and McKenna, J.F. (1965). Statistical estimates of traffic intensity in communications systems. *IEEE Trans. Mil. Electron.* 9 (2): 130–136.

31 Mili, M.R. and Musavian, L. (2017). Interference efficiency: A new metric to analyze the performance of cognitive radio networks. *IEEE Trans. Wirel. Commun.* 16 (4): 2123–2138.

32 Thakur, P. and Singh, G. (2019). Energy and spectral efficient SMC-MAC protocol in distributed cognitive radio networks. *IET Commun.* 13 (17): 2705–2713.

33 Maleki, S., Pandharipande, A., and Leus, G. (2011). Energy-efficient distributed spectrum sensing for cognitive sensor networks. *IEEE Sens. J.* 11 (3): 565–573.

34 Monemian, M. and Mahdavi, M. (2014). Analysis of a new energy-based sensor selection method for cooperative spectrum sensing in cognitive radio networks. *IEEE Sens. J.* 14 (9): 3021–3032.

35 Karmokar, A., Naeem, M., and Anpalagan, A. (2018). Green metric optimization in cooperative cognitive radio networks with statistical interference parameters. *IEEE Syst. J.* 1 (11): 1034–1037.

36 Tripathi, P.S.M. and Prasad, R. (2013). Energy efficiency in cognitive radio network. *Wireless VITAE 2013*, Atlantic City, NJ, 1–5.

37 Mili, M.R., Musavian, L., Hamdi, K.A., and Marvasti, F. (2016). How to increase energy efficiency in cognitive radio networks. *IEEE Trans. Commun.* 64 (5): 1829–1843.

38 S. Raychaudhuri "Introduction to Monte Carlo simulation," *Proc. Winter Simulation Conference*, Miami, FL, USA, USA, December 7–10, 2008.

6

Cooperative Spectrum Monitoring in Homogeneous and Heterogeneous Cognitive Radio Networks

6.1 Introduction

As discussed in Chapter 5, the spectrum mobility is an integral part of the cognitive cycle and spectrum monitoring (SM) is an important technique to detect the emergence of primary user (PU) simultaneous to the data transmission [1, 2]. In the SM, the cognitive user (CU) exploits the statistics of the received signal such as receiver error count (REC), energy level, etc., to detect the emergence of PU. However, the imperfections in SM cause the delay in the detection of the emergence of PU that results in the degradation in the performance of cognitive radio network (CRN) in terms of the data loss, interference efficiency, and energy efficiency [3].

A key implementation to manage the imperfections in SM is the cooperation among CUs [4–9]. Therefore, to improve the performance of CRN with imperfect SM, we have exploited the cooperation among CUs in this chapter for the homogeneous as well as heterogeneous CRN environments.

- *Homogeneous CRN*: A CRN is defined as a homogeneous CRN if all the CUs have same operating characteristics such as probability of error, channel conditions, hardware impairments, delay profiles, etc. [10, 11].
- *Heterogeneous CRN*: A CRN is defined as a heterogeneous CRN when all the CUs have different operating characteristics such as the probability of error, channel conditions, hardware impairments, delay profiles, etc. [12, 13].

In the proposed CRN, the homogeneous/heterogeneous nature is considered in terms of probability of SM error which is due to different channel conditions and hardware impairments.

Spectrum Sharing in Cognitive Radio Networks: Towards Highly Connected Environments,
First Edition. Prabhat Thakur and Ghanshyam Singh.
© 2021 John Wiley & Sons, Inc. Published 2021 by John Wiley & Sons, Inc.

The key contribution in this chapter is summarized as follows:

- We have proposed a potential and feasible framework of cooperative SM (CM) for CRN which is analyzed for different scenarios of the traffic intensity and SM error.
- The closed-form expressions of the achieved throughput, data loss, interference efficiency, and energy efficiency are derived.
- The Monte-Carlo simulations are used to consider the effect of random events, i.e. traffic intensity of PU.

The rest of the chapter is structured as follows. In Section 6.2, the background about the Binomial and Poisson-Binomial distributions is presented so that the models of homogeneous and heterogeneous CRN become easy to understand. The system model of proposed framework is described in Section 6.3. Section 6.4 comprises performance analysis of the proposed system model and simulation results with their analysis are presented in Section 6.5. Finally, Section 6.6 has summarized the work with future perspectives."

6.2   Background

In the cooperative CRN, all the CUs make a decision (yes/no) on the basis of received information from the electromagnetic environment and send this decision to the central node/fusion center (FC) in order to yield the final decision by combining all individual decisions. The type of cooperation where the individual CU sends the decision rather than the received information is known as hard-combining rules [5, 14, 15]. The hard-combining rule comprises OR, AND, and M-out-of-N-rules (MOON-rule). In the MOON rule, the final decision (yes/no) is confirmed if M out of N CUs support the same [16, 17]. The mathematical modeling of the distribution function of the M number of CUs' confirmation (yes/no) is achieved by the Binomial distribution if all the CUs decide the decision with same probability [18].

The Binomial distribution is a potential approach in order to yield the discrete probability distribution, $P_p(x/X)$, to achieve exactly x successes out of X Bernoulli trials (where the result of each Bernoulli trial is true with probability p and false with probability $(1 - p)$), which is well explored in the mathematics as well as implemented in various engineering applications such as communication engineering, computer engineering, signal and image processing, etc. One popular application of the Binomial distribution is in cooperative communication in order to know the probability of error after cooperation when the MOON rule is used to combine the results from various observation points (nodes/CUs). The key limitation of Binomial distribution is that it supports the homogeneous environment, i.e. when every Bernoulli trial is true with same probability p and false with $(1-p)$. However, in most of the feasible scenarios, every node has different probability of error either due to hardware or channel conditions. Therefore, it is worth analyzing the considered network for such feasible scenario and for this, we need to explore the Poisson-Binomial distribution [19]. The Poison-Binomial distribution is defined to yield the discrete probability distribution $P_p(x/X)$, to achieve exactly x successes out of X Bernoulli trials (where the result of ith Bernoulli trial is true with probability p_i and false with probability $(1-p_i)$). Various researchers have presented their investigation to compute the closed-form expression of the Poisson-Binomial distribution [19, 20]. Wang [19] has derived the probability of having x successful trials out of total X trials which is defined as:

$$P_p\left(\frac{x}{X}\right) = \sum_{A \in F_x} \prod_{i \in A} P_i \prod_{j \in A^c} (1 - P_j), \tag{6.1}$$

where F_x is the set of all subsets of x integers that can be selected from the set $\{1, 2, 3, 4 \dots X\}$. For instance, if $X = 4$ and $x = 2$, then $F_x = \{\{1, 2\}, \{1, 3\}, \{1, 4\}, \{2, 3\}, \{2, 4\}, \{3, 4\}\}$ and A^c denotes the complement of A. The number of elements in the set F_x are: $\dfrac{X!}{(X-x)!x!}$. The large value of X results in the large values of elements which is very difficult and complex to yield and results in the increase in computational complexity. Thus, the computational complexity is very large for this approach and by keeping this in mind, Fernandez and Williams [21] have exploited the discrete Fourier transform (DFT) to achieve the closed-form expression for the Poisson-Binomial distribution.

$$P_p\left(\frac{x}{X}\right) = \frac{1}{X+1} \sum_{l=0}^{X} C^{-lx} \prod_{m=1}^{X} 1 - P_m + C^l P_m, \tag{6.2}$$

where $C = e^{\left(\frac{2\pi i}{X+1}\right)}$ and $i = \sqrt{-1}$. In this chapter, $P_p\left(\frac{x}{X}\right)$ yields the complex number of probability which is against the norms/axioms of probability theory. Therefore, in order to overcome this issue, we have followed the fundamental steps of formation of the x successes out of X number of events [21] and have exploited the MATLAB 2010a [22]. We conclude that the expression in Eq. (6.2) has a deficiency which is avoided by taking the absolute value of the computed probability. Therefore, the improved form of Eq. (6.2) is:

$$P_p\left(\frac{x}{X}\right) = \left| \frac{1}{X+1} \sum_{l=0}^{X} C^{-lx} \prod_{m=1}^{X} 1 - P_m + C^l P_m \right|. \tag{6.3}$$

Further, in the MOON rule, the decision yes/no is confirmed when M number of nodes says yes/no. Therefore, the probability of decision yes/no is confirmed when

M or greater than M number of users will say yes/no. Thus, the probability of this event is defined as:"

$$P_p(yes/no) = \sum_{x=M}^{X} P_p\left(\frac{x}{X}\right). \tag{6.4}$$

6.3 System Model

In the proposed system model of homogeneous/heterogeneous CRN (HCRN), a transceiver pair of PU network communicates using a channel having bandwidth B as shown in Figure 6.1. The CRN consists of N number of CU pairs which are allowed to exploit the spectrum of PU in such a way that all the CUs get equal bandwidth and thus the bandwidth allocated to each CU is B/N. The CU communication is established via time frame (T) where time τ_s is devoted for spectrum sensing and $(T - \tau_s)$ for the data transmission. The number of packets transmitted during the data transmission period without the emergence of PU is N_0. In the proposed system model, a high-traffic environment with traffic intensity of PU greater than 0.5 is considered. Therefore, the CUs' perform spectrum prediction in the previous time frame during the data transmission to select the channel with highest idle probability for spectrum sensing in the current frame. Consequently, there is improvement in the spectrum sensing process since it saves the sensing time and energy spent on the sensing of active channels [23]. In addition to the potential discussion over spectrum sensing in Chapter 5, some recent advances about cooperative spectrum sensing techniques are illustrated in [24–26]. Since the spectrum prediction is performed in the background, the time devoted is null; however, the

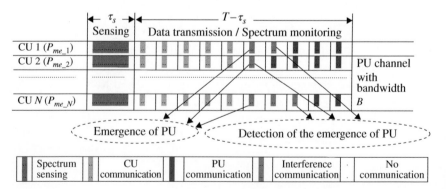

Figure 6.1 The scenario of the proposed cooperative spectrum monitoring in cognitive radio networks.

certain power (P_{Pr}) is required to execute this process. The spectrum prediction is well-matured technique presented in the literature [27–29] and discussed in Section 3.2.1; therefore, this chapter mainly emphasizes on the cooperative SM.

In the current time frame, the CU performs spectrum sensing for time τ_s and starts data transmission for time ($T - \tau_s$) with power P_1 on the idle sensed channel. In the data transmission period, the process of SM is performed, simultaneous to the data transmission to know the emergence of PU. The role of SM seems significant only when the PU recommences its communication during the data transmission; therefore, for the consideration of this event, the probability of PUs' emergence in the data transmission period is provided by the traffic intensity of PU (ρ). The traffic intensity of PU is presumed to be a binary stochastic hypothesis, in which 0 and 1 represents the idle and active states of channels, respectively [16]. Further, the average arrival and channel holding time of PU is modeled as the Poisson distribution (parameter λ) and Binomial distribution (parameter μ), respectively [30]. The higher value of traffic intensity signifies that the PU emerges in the starting phase of data transmission period; however, the lower value represents the emergence of PU in the later part. Moreover, every CU senses the state of channel (active or idle) during the sensing period exclusively and cooperates with each other to formulate the final decision, i.e. either idle or active. If the final decision is idle, all the CUs start data transmission in the form of packets. Due to prospect of the PUs' emergence during the CUs' data transmission, it is the responsibility of the CU to detect the emergence PU and stop its communication. The detection of emergence of PU is a potential phenomenon which is achieved by applying SM. In the proposed system model, the SM is assumed as the feasible imperfect phenomenon and imperfections in this process are presented by the probability of SM error (P_{me}). In the proposed CRN, the P_{me_i} denotes the probability of SM error for the ith CU where $i = 1, 2, 3, \dots N$. The imperfection in SM means the emergence of PU is not detected immediately; however, detected after certain detection delay.

This imperfection causes due to channel uncertainties and/or the hardware system impairments. The higher value of P_{me} indicates that there is more delay in the detection of PU. The delay in detection of the emergence of PU results in the continuation of CU communication and causes the data collision/loss as well as introduces the significant interference at PU. In addition to this, the interference power at PU depends on the channel gain between CU transmitter and PU receiver (h_{sp}) which is also known as interference link. Therefore, to manage the effect of imperfections, the cooperation between CUs plays a key role. The CUs decide the state of emergence of PU either 0 or 1 and reports to the FC where the final decision is achieved by combining all the reported state. If the final decision confirms the emergence of PU, all the CUs stop their communication, otherwise continues. Further, the cooperation in the SM improves the probability of SM error. When the data are fused using the MOON rule, the probability of SM error (Q_{me}) for the homogeneous CRN is defined as presented in Eq. (6.4).

$$Q_{me} = \sum_{j=l}^{N} \binom{N}{j} (P_{me})^j (1 - P_{me})^{N-j}. \tag{6.5}$$

Further, the probability of SM error (Q_{me}) for the heterogeneous CRN is defined and presented in Eq. (6.4).

$$Q_{me} = \sum_{j=M}^{N} \left| \frac{1}{N+1} \sum_{l=0}^{N} C^{-lM} \prod_{m=1}^{N} 1 - P_{me_m} + C^l P_{me_m} \right|. \tag{6.6}$$

Moreover, N_{PPU}, N_{PCU}, h_{ss}, and h_{sp} denote the noise power at PU receiver, noise power at CU receiver, channel gain from CU transmitter to CU receiver, channel gain from CU transmitter to PU receiver, respectively."

6.4 Performance Analysis of Proposed CRN

In this section, the effect of imperfect SM on the data loss of proposed homogeneous/heterogeneous CRN is illustrated and derived the closed-form expressions of the achieved throughput, data loss, interference efficiency, and energy efficiency. The data loss occurs in the CRN when the PU reappears during the data transmission of CU and the probability of reappearance is yielded by the traffic intensity of PU. The perfect SM is an impractical scenario in which the reappearance of PU is detected very quickly which means within a single data packet time (loss of that single packet). On the other hand, the imperfect SM is a feasible scenario in which the data loss is the function of the probability of SM error (P_{me}). In the considered HCRN, the fixed number of packets gets lost among the total number of packets (No), which depends on the SM error. In the considered CRN, the computation of number of packets lost among total number of packets follows the Binomial distribution [18]. Therefore, in the proposed CRN, the discrete probability distribution $P(k/No)$ of obtaining exactly k number of packets lost out of No (where the data loss result is true for the ith CU with probability P_{me_i} and false with probability $(1 - P_{me\_i})$). Thus, the probability of data loss due to probability of SM error for the ith CU in the noncooperative SM (NCM) is defined in Eq. (6.7). The subscript NCM, CM, and i denotes the noncooperative SM, cooperative SM, and ith CU, respectively.

$$P_{NCM_i} \left(\frac{k}{No} \right) = \binom{No}{k} (P_{me_i})^k (1 - P_{me_i})^{No-k}. \tag{6.7}$$

Similarly, the probability of data loss due to probability of SM error in the cooperative monitoring (CM) for all the CUs is presented as:

$$P_{CM}\left(\frac{k}{No}\right) = \binom{No}{k}(Q_{me})^k(1-Q_{me})^{No-k}. \tag{6.8}$$

Further, the average number of data packets lost (k_{avg}) for the ith CU in case of cooperative and noncooperative monitoring are evaluated by computing the expectation of variables as follows:

$$k_{NCM\_avg\_i} = \sum_{k=1}^{No} k.P_{NCM\_i}\left(\frac{k}{No}\right), \tag{6.9}$$

$$k_{CM\_avg} = \sum_{k=1}^{No} k.P_{CM}\left(\frac{k}{No}\right). \tag{6.10}$$

There is loss of one single packet even for $P_{me} = 0$. Therefore, to compute the total number of packet lost in the CRN (k_{an}) for both the cases, one packet need to add to k_{avg} which is presented as: $k_{an\_NCM\_i} = (1 + k_{avg\_NCM\_i})$ and $k_{an\_CM} = (1 + k_{avg\_CM})$.

In the proposed CRN, the communication is established using frame structures which means after fixed time interval, i.e. the frame time (T), the CU periodically repeats the process of spectrum sensing and data transmission. Therefore, it is possible that the CU switches from data transmission mode to the spectrum sensing mode of the next frame, immediately after the emergence of PU because of ending of the data transmission interval.

In this case, the number of packets lost relies on the time of emergence of PU which is the function of traffic intensity (ρ) in the proposed CRN. Thus, the complete data loss in the proposed HCRN (k_{comp}) is not only the function of the SM error but also of the traffic intensity of PU. Therefore, to find the complete data loss in the proposed HCRN, there is a need to compute the total number of packets to be transmitted after the emergence of PU (k_{TAEPU}) which relies on the traffic intensity and computed as: $k_{TAEPU} = (1 - \rho) \times No$. The complete data loss in the noncooperative ($k_{comp\_NCM\_i}$) CRN for the ith CU and in cooperative (k_{comp_CM}) for all CUs is:

$$k_{comp\_NCM\_i} = \begin{cases} k_{an\_NCM\_i} & if \ k_{TAEPU} \geq k_{an\_NCM\_i} \\ k_{TAEPU} & if \ k_{TAEPU} < k_{an\_NCM\_i} \end{cases}, \tag{6.11}$$

$$k_{comp\_CM} = \begin{cases} k_{an\_CM} & if \ k_{TAEPU} \geq k_{an\_CM} \\ k_{TAEPU} & if \ k_{TAEPU} < k_{an\_CM} \end{cases}. \tag{6.12}$$

If the number of packets to be transmitted after the emergence of PU is greater than the total number of packets lost due to monitoring error for the noncooperative (for ith CUs) and cooperative case (for all CUs), then the complete data loss in the proposed CRN is $k_{an\_NCM\_i}$ and k_{an_CM}, respectively, otherwise k_{TAEPU}. The average data loss of the complete HCRN in the noncooperative case is: $k_{comp\_NCM} = \frac{1}{N}\sum_{i=1}^{N} k_{comp\_NCM\_i}$; however, remains same in the cooperative case as in Eq. (6.12). The complete data loss of the ith CU in proposed CRN for the non-cooperative is illustrated in Algorithm 6.1, where three scenarios are discussed as follows. (i) $\rho = 0$, which means the PU will not appear in the data transmission phase which results in no data loss even if the proposed system has SM error which means, in this special case there is no significant role of the SM. (ii) $(0 < \rho < 1)$ && $(P_{me} = 0)$, which means the emergence of PU is confirmed and the SM system is perfect. In this case, the emergence of PU is detected within a packet transmission time and that particular packet data get lost. (iii) $(0 < \rho < 1)$ && $(P_{me} = 0)$, which is a practical state of affairs. If we avoid the effect of ρ, the data loss in this case due to monitoring error is the function of P_{me} as derived in Eqs. (6.9) and (6.10). However, the total data loss is the number of packets lost due to P_{me} plus one data packet, because the particular packet lost even when P_{me} is zero. On the other hand, on the consideration of the effect of ρ, the complete data loss in the proposed HCRN becomes the function of both the P_{me} and ρ, which is computed as given in Eq. (6.11). Similarly, in the cooperative case, the data loss becomes the function of Q_{me} and ρ, which is derived in Eq. (6.12)."

Algorithm 6.1 Calculation of Data Loss in the HCRN for ith CU in the Noncooperative Spectrum Monitoring

---

Input (No, P_{me}, ρ)

Output $(k_{avg}, k_{an}, k_{TAEPU}, k_{comp})$

BEGIN{

Step-1 Variable declaration

No:	Total number of packets in the data transmission phase;
k:	Number of packets lost due to spectrum monitoring error;
P_{me_i} :	Probability of SM error of the ith CU;
k_{avg} :	Average number of packets lost due to SM error;
ρ :	Traffic intensity of PU;
k_{an} :	Average number of packets lost in the network without considering the effect of traffic intensity;
k_{TAEPU} :	Total number of packets to be transmitted after the emergence of PU.
k_{comp} :	Complete data-loss in the proposed CRN.

Step-2 Computation of probability of k number of packet lost among No

$$P_{NCM\_i}\left(\frac{k}{No}\right) \leftarrow {}_k^{No}Comb.(P_{me\_i})^k(1 - P_{me\_i})^{No-k};$$

Step-3 Computation of average number of packets lost due to monitoring error and of k$_{an}$

> Let $k_{NCM\_avg\_i} \leftarrow 0$;
> **for** $k \leftarrow 1{:}No$
> $\qquad k_{NCM\_avg\_i} \leftarrow \left(k_{NCM\_avg\_i} + k.P_{NCM\_i}\left(\frac{k}{No}\right)\right)$;
> **end**
> $\qquad k_{an\_NCM\_i} = 1 + k_{NCM\_avg\_i}$;

Step-4 Computation of the total number of packets to be transmitted after the emergence of PU

$$k_{TAEPU} = (1 - \rho) \times No;$$

Step-5 Complete Data-loss in the HTCRN

> **If** $\{(\rho \leftarrow 0) \,\&\&\, (P_{me\_i} \leftarrow \text{any value})\}$ **then**
> $\qquad k_{comp\_i} \leftarrow 0$ packet;
> **elseif**$\{(0 < \rho \le 1)\,\&\&\, (P_{me\_i} \leftarrow 0)\}$ **then**
> $\qquad k_{comp\_i} \leftarrow 1$ packet;
> **elseif**$\{(0 < \rho \le 1)\,\&\&\, (0 < P_{me\_i} \le 1)\}$ **then**
> $\qquad$ **if** $(k_{TAEPU} \ge k_{an\_NCM\_i})$ **then**
> $\qquad\qquad k_{comp\_i} \leftarrow k_{an\_NCM\_i}$;
>
> $\qquad$ **else** **then**
> $\qquad\qquad k_{comp\_i} \leftarrow k_{TAEPU}$;
>
> $\qquad$ **end**
> **end**
> **end**
> **end**
> **}END**

Due to imperfection in the SM, the CU is unable to detect the emergence of PU quickly, and continue the data transmission, even after the emergence of PU which results in the interference at PU receiver. The CU transmission starts interfering with the PU transmission when the PU emerges and continues it up to the detection of emergence. As the detection of emergence of the PU during the data transmission period is the function of P_{me} and Q_{me} for noncooperative and cooperative SM case, respectively, therefore the interference at the PU is also the function of P_{me} and Q_{me}.

For the noncooperative SM case, the number of packets lost after the emergence of PU relies on the P_{me_i} for the ith CU. The starting time (I_s) and ending time (I_E) of the interference at PU depends on the traffic intensity (ρ) of PU and on the P_{me_i} and Q_{me} for the cooperative and noncooperative SM cases, respectively, which are computed as:

$$I_s = \lfloor\{(1-\rho)\times(T-(\tau_s))\} + \{(\tau_s)\}\rfloor, \tag{6.13}$$

$$I_{E\_NCM\_i} = I_s + \left(k_{comp\_NCM\_i}\times PT\right), \tag{6.14}$$

$$I_{E\_CM} = I_s + \left(k_{comp\_CM}\times PT\right), \tag{6.15}$$

where PT is the packet duration and defined as: $PT = (T - \tau_s)/No$. Now, the various performance metrics have to be computed by using the preceding analysis of the data loss in Algorithm 6.1. The performance metrics exploited further are the achieved throughput and data loss, interference efficiency, and energy efficiency which are computed as follows.

6.4.1 Computation of Achieved Throughput and Data Loss

The throughput obtained due to collision-free data transmission is defined as the achieved throughput (RA), whereas the throughput obtained during collision is considered as data loss of the network. In the proposed HCRN, the achieved throughput will be the same for all the CUs since the time of collision-free data transmission $(I_s - \tau_s)$ remains same. Therefore, the achieved throughput of the proposed HCRN with N number of CUs is computed as:

$$RA = N \times ((I_s - \tau_s)/T) \times log_2\left(1 + \frac{P_1 h_{ss}}{N_{CU}}\right). \tag{6.16}$$

In the noncooperative SM, $k_{comp\_NCM\_i}$ number of packets are lost which relies on the probability of SM error due to which the total data loss time is: $k_{comp}\times PT$ for the ith CU. Therefore, the average data loss in the complete CRN is the sum of data loss of N number of CUs and computed as:

$$DL_{NCM} = \frac{1}{N}\sum_{i=1}^{N}\left(\frac{\left(k_{comp_{NCM_i}}\times PT\right)}{T}\right) \times log_2\left(1 + \frac{P_1 h_{ss}}{N_{CU}}\right). \tag{6.17}$$

Similarly, in case of cooperative SM, the data loss (DL_{CM}):

$$DL_{CM} = \frac{N}{N}\times\left(\frac{\left(k_{comp_{CM}}\times PT\right)}{T}\right) \times log_2\left(1 + \frac{P_1 h_{ss}}{N_{CU}}\right). \tag{6.18}$$

6.4.2 Computation of Interference Efficiency

The interference efficiency (*IE*) [13, 31] is a prominent performance metric if the CU introduces interference at the PU which is defined as the number of bits transmitted per unit of energy imposed on the PU. In the proposed CRN, it is ratio of the achieved throughput to the power received at the PU receiver when the original state of PU is active and due to which the unit is bits/joule/Hz. Now, the average power received at the PU receiver due to CU transmission is considered as the interference to the PU communication (*IF*) and for the noncooperative and cooperative SM denoted as IF_{NCM} and IF_{CM}, respectively, which are computed as:

$$IF_{NCM} = \frac{1}{N}\sum_{i=1}^{N}\left(\left(\frac{k_{comp_{NCM_i}} \times PT}{T}\right) \times P_1 \times h_{sp}\right), \tag{6.19}$$

$$IF_{CM} = \frac{N}{N}\left(\left(\frac{k_{comp_{CM}} \times PT}{T}\right) \times P_1 \times h_{sp}\right). \tag{6.20}$$

Further, the interference efficiency for the noncooperative and cooperative SM case is computed as follows:

$$IE_{NCM} = \frac{RA}{IF_{NCM}}, \tag{6.21}$$

$$IE_{CM} = \frac{RA}{IF_{CM}}. \tag{6.22}$$

6.4.3 Computation of Energy Efficiency

The energy-efficient nature of CRN supports the green communication concept and liberates the customers from futile power consumption. Therefore, the effect of noncooperative and cooperative SM over the energy efficiency of the proposed CRN is analyzed. The energy efficiency (*EE*) [32–36] is defined as the ratio of achieved throughput to the power consumed by the system and its unit is bits/joule/Hz. The time of data transmission with power P_1 is $(I_{E_{NCM\_i}} - \tau_s)$ for the ith CU and $(I_{E_{CM}} - \tau_s)$ for all the CUs in the noncooperative and cooperative SM, respectively. Therefore, the average power consumption in the noncooperative and cooperative SM is defined as:

$$PC_{NCM} = \frac{1}{N}\sum_{i=1}^{N}\left(\left(\frac{\left(I_{E_{NCM_i}} - \tau_s\right)}{T} \times P_1\right) + P_{si} + P_{Pri}\right), \tag{6.23}$$

$$PC_{CM} = \frac{N}{N}\left(\left(\frac{(I_{E_{CM}} - \tau_s)}{T} \times P_1\right) + P_s + P_{Pr}\right), \tag{6.24}$$

where P_{Pri} and P_{si} are the powers required for the spectrum prediction and sensing techniques in the ith CU; however, $\forall i : P_{Pri} = P_{Pr}$ and $\forall i : P_{si} = P_s$. Therefore, the EE for the noncooperative and cooperative SM case is computed as follows:"

$$EE_{NCM} = \frac{RA}{PC_{NCM}}, \tag{6.25}$$

$$EE_{CM} = \frac{RA}{PC_{CM}}. \tag{6.26}$$

6.5 Results and Discussion

This section presents the numerically simulated results of the data loss, interference efficiency, and energy efficiency for the proposed homogeneous as well as heterogeneous cooperative SM CRN system model and have been compared with the noncooperative SM CRN. The IEEE 802.22 is the first wireless standard that relies on cognitive radio [37] used to form the wireless regional area network (WRAN); therefore, the simulation parameters selected in the proposed CRN are inspired by WRAN standard and are presented in the Table 6.1. Moreover, in order to validate the proposed CRN, we have exploited the randomness in emergence of PU using traffic intensity of PU and the Monte-Carlo simulation for the 10 000 runs [38]."

6.5.1 Homogeneous Cognitive Radio Network

The simulation results for the homogeneous CRN are illustrated in this section. The variations of data loss (in the form of packets) for cooperative and noncooperative SM are presented in Figure 6.2. The cooperative SM outperforms the

Table 6.1 The simulation parameters for the proposed HCRN.

Parameter	Value	Parameter	Value	Parameter	Value
T	100 ms	No	100	h_{ss}	0.8
N	10	P_1	6 W	h_{sp}	0.2
τ_s	2.5 ms	N_{PPU}	0.4 W	N_{PCU}	0.4 W
P_{Pr}	0.2 W	P_S	0.2 W		

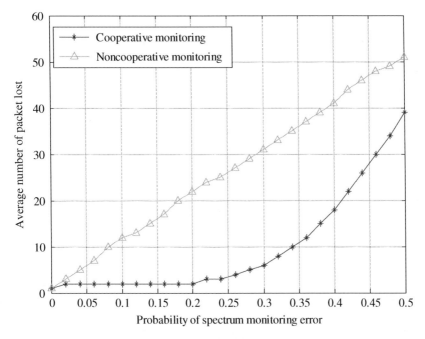

Figure 6.2 The variations of data loss with probability of SM error.

noncooperative SM in terms of data loss. Moreover, it is palpable that the data loss for noncooperative SM increases almost linearly; however, in case of cooperative SM, it remains constant for certain value of P_{me} till the Q_{me} achieved the value which causes packet loss greater than one. The relation between the interference efficiency and probability of SM error for cooperative and noncooperative SM with different traffic intensities is illustrated in Figure 6.3. The interference efficiency is an improved metric in case of the CM as compared to that of the noncooperative SM for all considered values of the traffic intensity. In addition to this, the noncooperative SM depicts the exponential decay of interference efficiency with increase of the probability of SM error; however, in case of cooperative SM, the interference efficiency follows the fixed value till the Q_{me} causes data loss greater than one packet and then starts decaying.

Moreover, the increase in ρ results in the decrease in achieved throughput for certain value of P_{me} as well as Q_{me} and interference efficiency is the ratio of achieved throughput to the interference introduced at PU. Therefore, the interference efficiency is large for small values of ρ and reduces with its increment in both the cases of cooperative and noncooperative SM.

The variations of interference efficiency with channel gain from CU transmitter to the PU receiver (h_{sp}) for various values of the traffic intensity is depicted in

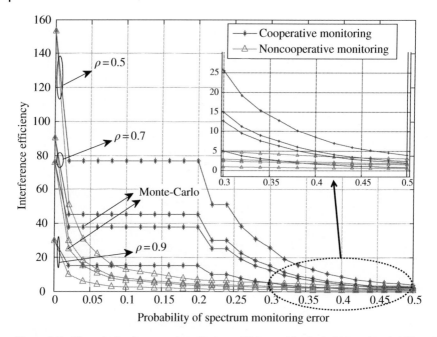

Figure 6.3 The relation between interference efficiency (bits/joule/Hz) and probability of SM error.

Figure 6.4. The rate of change of interference efficiency is significantly large for small values of h_{sp}; however, it decays with increase in the h_{sp}. Moreover, the cooperative SM outperforms the noncooperative SM in terms of interference efficiency. As the channel gain from CU transmitter to PU receiver increases for fixed value of the channel gain from CU transmitter to CU receiver, the interference efficiency decays almost exponentially. The relation between the energy efficiency and probability of SM error for various values of the traffic intensity is presented in Figure 6.5.

The energy efficiency is significantly more in cooperative SM as compared to that of the noncooperative SM; however, it decreases with increase in the probability of SM error for both the cases, i.e. noncooperative and cooperative SM. In addition to this, the energy efficiency shows inversely proportional relation with the traffic intensity which means energy efficiency decreases with increase of the traffic intensity."

6.5.2 Heterogeneous Cognitive Radio Networks

"This section comprises the simulation results for the heterogeneous CRNs. The variations of data loss (in the form of packets) for cooperative and noncooperative

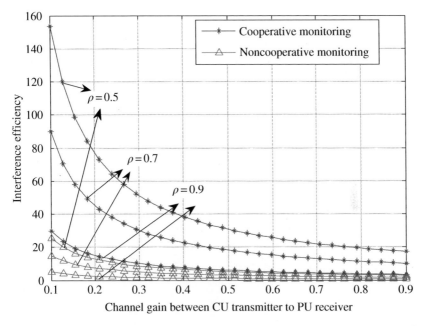

Figure 6.4 The variations of interference efficiency (bits/joule/Hz) and channel gain from CU transmitter to PU receiver [h_{ss} = 0.8].

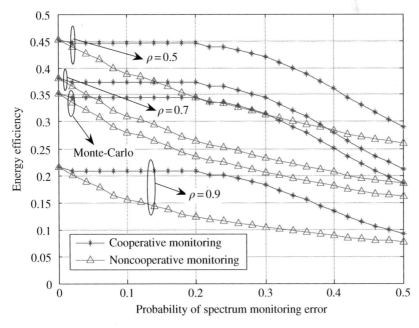

Figure 6.5 The relation between energy efficiency (bits/Joule/Hz) and probability of SM error.

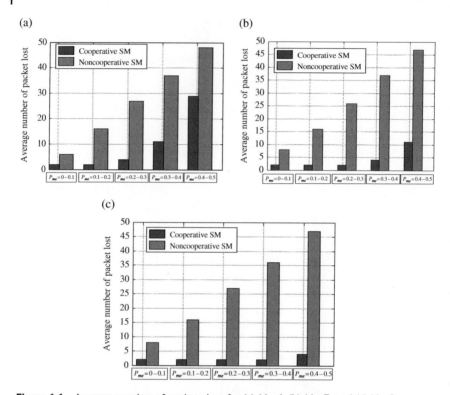

Figure 6.6 Average number of packets lost for (a) $M = 6$, (b) $M = 7$, and (c) $M = 8$.

SM in the proposed HCRN for different values of M in the MOON rule are presented in Figure 6.6 when the traffic intensity of CU is 0.5. The cooperative SM outperforms the noncooperative SM in terms of data loss. Moreover, it is palpable that the data loss for noncooperative SM increases with P_{me}; however, in case of cooperative SM, it remains constant for certain value of P_{me} till the Q_{me} achieved the value which causes packet loss greater than one. Moreover, there is improvement in the data loss in the cooperative case with increase in the value of M as shown in Figure 6.6. The variations of average interference efficiency in cooperative and noncooperative HCRN for different values of M are depicted in Figures 6.7–6.9. The interference efficiency is significantly more in the cooperative SM as compared to that of the noncooperative SM in all the cases ($M = 6$, $M = 7$, and $M = 8$). The interference efficiency decreases with increase in the value of the range of the P_{me} as well as ρ as depicted in Figures 6.7–6.9. In addition to this, there is significant improvement in the interference efficiency for the cooperative SM for large values of probability of SM error when we increase the value of M as shown from Figures 6.7–6.9. In Figure 6.7, for $M = 6$, $\rho = 0.5$, and $P_{me} = 0.5$, the value of

(a)

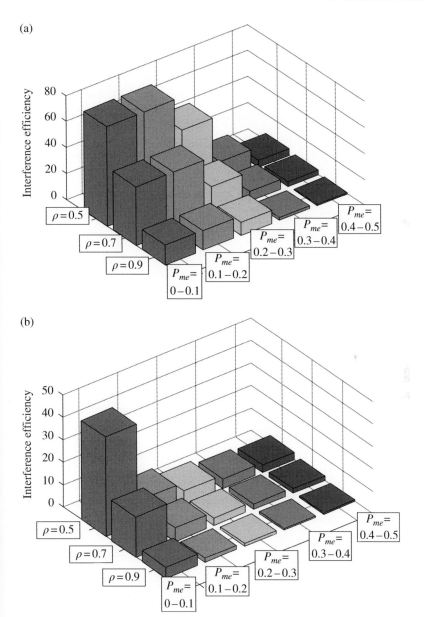

(b)

Figure 6.7 The interference efficiency (bits/joule/Hz) for $M = 6$ in (a) cooperative SM and (b) noncooperative SM.

(a)

(b)

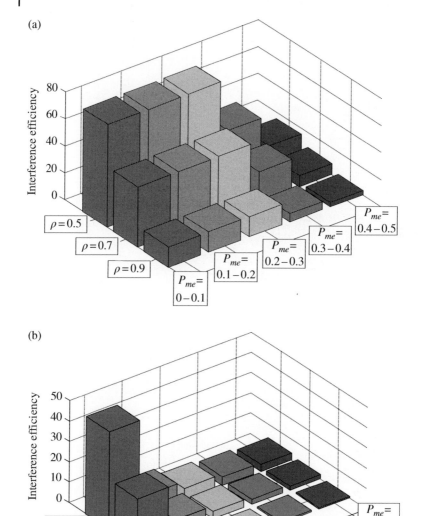

Figure 6.8 The interference efficiency (bits/joule/Hz) for *M* = 7 in (a) cooperative SM and (b) noncooperative SM.

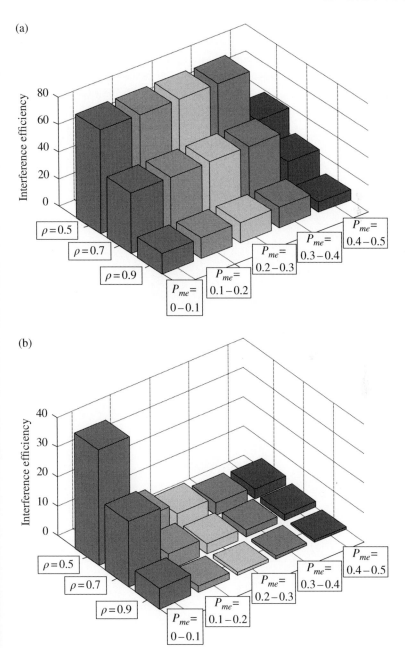

Figure 6.9 The interference efficiency (bits/joule/Hz) for *M* = 8 in (a) cooperative SM and (b) noncooperative SM.

interference efficiency is 5.11 which increases to approximately 15.33 for $M = 7$ (Figure 6.8) and attains the value near 38.34 for $M = 8$ (Figure 6.9). This means, for large value of P_{me}, the cooperative SM with optimal value of M plays a significant role. The average energy efficiency in cooperative and noncooperative HCRN for different values of M, P_{me}, and ρ is depicted in Figures 6.10–6.12. The cooperative SM outperforms the noncooperative SM as compared to that of the

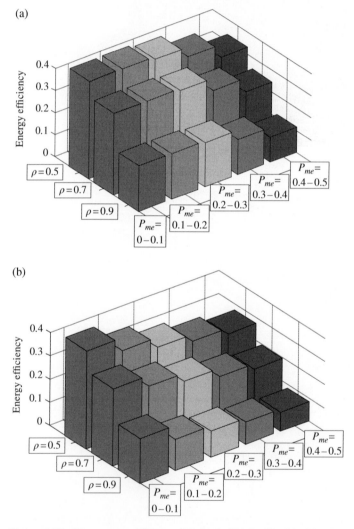

Figure 6.10 The energy efficiency (bits/joule/Hz) for $M = 6$ in (a) cooperative SM and (b) noncooperative SM.

(a)

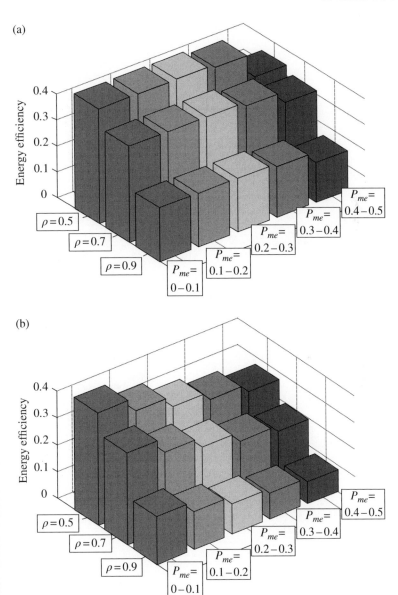

(b)

Figure 6.11 The energy efficiency (bits/joule/Hz) for $M = 7$ in (a) cooperative SM and (b) noncooperative SM.

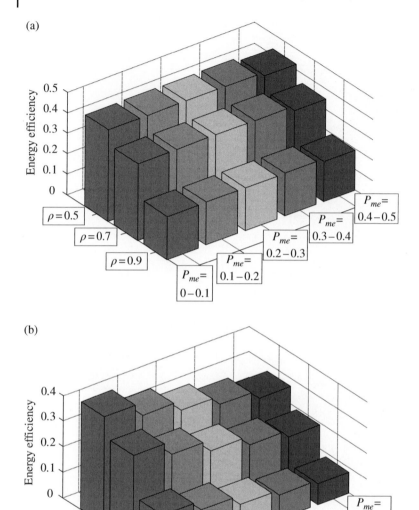

Figure 6.12 The energy efficiency *(bits/joule/Hz)* for *M* = 8 in (a) cooperative SM and (b) noncooperative SM.

noncooperative SM which is palpable from Figures 6.10–6.12. Moreover, there is significant improvement in the energy efficiency for the cooperative SM for large values of P_{me} when M has sufficient value which can be analyzed from Figures 6.10–6.12. In Figure 6.10, for $M = 6$, $P_{me} = 0.5$, and $\rho = 0.5$, the value of energy efficiency is approximately 0.33 which increases to 0.34 for $M = 7$ (Figure 6.11) and 0.42 for $M = 8$ (Figure 6.12). This signifies that for large value of P_{me}, the cooperative SM is an effective approach; however, selection of M plays a key role.

6.6 Summary

In this chapter, we have exploited the cooperative SM in the homogeneous as well as heterogeneous CRN and have analyzed the data loss, achieved throughput, interference efficiency, and energy efficiency. It is concluded that the cooperative SM outperforms the noncooperative SM in terms of the aforementioned performance metrics for the homogeneous as well as heterogeneous CRNs. Further, the Monte-Carlo simulations have been exploited to validate the numerically simulated results in the homogeneous CRN. This chapter explores the centralized architecture where all the CUs are controlled by the centralized CU (CCU). The distributed architecture-based CRN will be explored in Chapter 8.

References

1 Thakur, P., Kumar, A., Pandit, S. et al. (2017). Spectrum mobility in cognitive radio network using spectrum prediction and monitoring techniques. *Phys. Commun.* **24**: 1–8.

2 Boyd, S.W., Frye, J.M., Pursley, M.B., and IV, T.C.R. (2012). Spectrum monitoring during reception in dynamic spectrum access cognitive radio networks. *IEEE Trans. Commun.* **60** (2): 547–558.

3 Thakur, P., Kumar, A., Pandit, S. et al. (2018). Analysis of high-traffic cognitive radio network with imperfect spectrum monitoring technique. *Comput. Commun.* **147**: 27–37.

4 Li, Q., Hu, R.Q., Qian, Y., and Wu, G. (2012). Cooperative communications for wireless networks: techniques and applications in LTE-advanced systems. *IEEE Wireless Commun.* **19** (2): 22–29.

5 Akylidiz, A.F., Lo, B.F., and Balakrishna, R. (2011). Cooperative spectrum sensing in cognitive radio networks: a survey. *Phys. Commun.* **4** (1): 40–62.

6 Rathika, M. and Sivakumar, P. (2018). A survey on co-operative communication in 4G-LTE wireless networks. *International Conference on Recent Trends in Advance Computing (ICRTAC)*, Chennai, India (10–11 September 2018), 193–199.

7 Liu, T., Qu, X., Tan, W., and Cheng, Y. (2019). An energy efficient cooperative communication scheme in ambient RF powered sensor networks. *IEEE Access* **7**: 86545–86554.

8 Afghah, F., Shamsoshoara, A., Njilla, L.L., and Kamhoua, C.A. (2020). Cooperative spectrum sharing and trust management in IoT networks. In: *Modeling and Design of Secure Internet of Things* (eds. C.A. Kamhoua, L.L. Njilla, A. Kott and S. Shetty), 79–109. IEEE.

9 Luo, X. (2020). Secure cooperative spectrum sensing strategy based on reputation mechanism for cognitive wireless sensor networks. *IEEE Access* **8**: 131361–131369.

10 Chu, T.M.C., Phan, H., and Zepernick, H.-J. (2014). Hybrid interweave-underlay spectrum access for cognitive cooperative radio networks. *IEEE Trans. Commun.* **62** (7): 2183–2197.

11 Thakur, P., Kumar, A., Pandit, S. et al. (2019). Performance analysis of cooperative spectrum monitoring in cognitive radio network. *Wireless Netw.* **25**: 989–997.

12 Awoyemi, B.S., Maharaj, B.T., and Alfa, A.S. (2017). Resource allocation in heterogeneous buffered cognitive radio networks. *Wireless Commun. Mobile Comput.* **2017**: 1–12.

13 Thakur, P., Kumar, A., Pandit, S. et al. (2018). Spectrum monitoring in heterogeneous cognitive radio network: how to cooperate? *IET Commun.* **12** (17): 2110–2118.

14 Nath, A. and Sarma, N. (2017). A distributed solution for cooperative spectrum sensing scheduling in multi-band cognitive radio networks. *J. Netw. Comput. Appl.* **94**: 69–77.

15 Jiao, Y., Yin, P., and Joe, I. (2016). Clustering scheme for cooperative spectrum sensing in cognitive radio networks. *IET Commun.* **10** (13): 1590–1595.

16 Liu, X., Jia, M., and Tan, X. (2013). Threshold optimization of cooperative spectrum sensing in cognitive radio networks. *Radio Sci.* **48** (1): 23–32.

17 Banavathu, N.R. and Khan, M.Z.A. (2017). Optimal number of cognitive users in K -out-of- M rule. *IEEE Wireless Commun. Lett.* **6** (5): 606–609.

18 Kay, S.M. (1998). *Fundamentals of Statistical Signal Processing: Detection Theory*. Prentice-Hall PTR.

19 Wang, Y.H. (1993). On the number of successes in independent trials. *Stat. Sin.* **3** (2): 295–312.

20 Chen, X.-H., Dempster, A.P., and Liu, J.S. (1994). Weighted finite population sampling to maximize entropy. *Biometrika* **81** (3): 457–469.

21 Fernandez, M. and Williams, S. (2010). Closed-form expression for the Poisson-Binomial probability density function. *IEEE Trans. Aerosp. Electron. Syst.* **46** (2): 803–817.

22 MathWorks MathWorks introduceert Release 2015b van de MATLAB en Simulink productseries. https://nl.mathworks.com/company/newsroom/mathworks-announces-release-2015b-of-the-matlab-and-simulink-product-families.html (accessed 19 February 2018).

23 Yarkan, S. and Arslan, H. (2007). Binary time series approach to spectrum prediction for cognitive radio. *66th IEEE Vehicular Technology Conference*, Baltimore, MD (30 September to 3 October 2007), 1563–1567.

24 Man, B.P., Kaneko, M., and Taparugssanagorn, A. (2020). A deep convolutional neural network based transfer learning method for non-cooperative spectrum sensing. *IEEE Access* **8**: 164529–164545.

25 Kumar, A., Thakur, P., Pandit, S., and Singh, G. (2020). Threshold selection and cooperation in fading environment of cognitive radio network: Consequences on spectrum sensing and throughput. *AEU-Int. J. Electron. Commun.* **117**: 1–11.

26 Kumar, A., Pandit, S., and Singh, G. (2020). Optimisation of censoring-based cooperative spectrum sensing approach with multiple antennas and imperfect reporting channel scenarios for cognitive radio network. *IET Commun.* **14** (16): 2666–2676.

27 Xing, X., Jing, T., Cheng, W. et al. (2013). Spectrum prediction in cognitive radio networks. *IEEE Wireless Commun.* **20** (2): 90–96.

28 Yang, J. and Zhao, H. (2015). Enhanced throughput of cognitive radio networks by imperfect spectrum prediction. *IEEE Commun. Lett.* **19** (10): 1738–1741.

29 Barnes, S.D., Maharaj, B.T., and Alfa, A.S. (2016). Cooperative prediction for cognitive radio networks. *Wireless Pers. Commun.* **89** (4): 1177–1202.

30 Pandit, S. and Singh, G. (2015). Backoff algorithm in cognitive radio mac protocol for throughput enhancement. *IEEE Trans. Veh. Technol.* **64** (5): 1991–2000.

31 Mili, M.R. and Musavian, L. (2017). Interference efficiency: a new metric to analyze the performance of cognitive radio networks. *IEEE Trans. Wireless Commun.* **16** (4): 2123–2138.

32 Yang, Z., Han, R., Chen, Y., and Wang, X. (2018). Green-RPL: an energy-efficient protocol for cognitive radio enabled AMI network in smart grid. *IEEE Access* **6**: 18335–18344.

33 Mishra, M.K., Trivedi, A., and Pattanaik, K.K. (2018). Outage and energy efficiency analysis for cognitive based heterogeneous cellular networks. *Wireless Netw.* **24** (3): 847–865.

34 Ogbebor, J.O., Imoize, A.L., and Atayero, A.A.-A. (2020). Energy efficient design techniques in next-generation wireless communication networks: emerging trends and future directions. *Wireless Commun. Mobile Comput.* **2020**: 7235362.

35 Thakur, P. and Singh, G. (2019). Energy and spectral efficient SMC-MAC protocol in distributed cognitive radio networks. *IET Commun.* **13** (17): 2705–2713.

36 Alarifi, A., Dubey, K., Amoon, M. et al. (2020). Energy-efficient hybrid framework for green cloud computing. *IEEE Access* **8**: 115356–115369.

37 IEEE-SA – Contact Us. http://standards.ieee.org/contact/form.html (accessed 18 February 2018).

38 Raychaudhuri, S. (2008). Introduction to Monte Carlo simulation. *2008 Winter Simulation Conference*, Miami, FL (7–10 December 2008).

7

Spectrum Mobility in Cognitive Radio Networks Using Spectrum Prediction and Monitoring Techniques

7.1 Introduction

The cognitive user (CU) senses its radio-frequency environment to perceive the unutilized bands of the primary user (PU), then accesses these bands using spectrum accessing techniques, and finally establishes communication on these bands. However, the protection of PU communication is the key function of spectrum accessing techniques [1, 2]. The spectrum sensing acts as the backbone for the complete framework of the CRN, as it detects the idle/unused bands of the spectrum, therefore its performance must be reliable [3–5]. The performance metrics of the spectrum sensing technique are the probability of false alarm (P_f) and probability of detection (P_d) in which the value of P_f and P_d must be low and high, respectively, for the better sensing performance [6–8]. The CU needs to vacate the spectrum on the emergence of PU during the data transmission; however, to continue the CU communication, the CU needs to switch on another idle channel and this process is known as spectrum-handoff or mobility which are well explored in literature [9–20]. Arshid et al. [11] have proposed a PU traffic pattern-based opportunistic spectrum handoff (PUTPOSH) approach that allows the CU to sense the arrival of PU and use an opportunistic handoff scheme. The opportunistic handoff scheme firstly detects the arrival of the PU by energy detection sensing and secondly, it allows a CU to decide whether to do handoff or not contingent upon the overall service time to reduce the unused handoffs. The handoffs can either be reactive or proactive based on the arrival rate of the PU. The superiority of the presented approach is perceived using simulation results in terms of the reduced number of handoffs and overall service time, as well as maintains the channel utilization and throughput. Aggarwal et al. [12] have exploited a centralized device for spectrum handoff to decrease the handoff latency of the CU where the adopted approach is dual processor based that helps in

Spectrum Sharing in Cognitive Radio Networks: Towards Highly Connected Environments,
First Edition. Prabhat Thakur and Ghanshyam Singh.
© 2021 John Wiley & Sons, Inc. Published 2021 by John Wiley & Sons, Inc.

increasing the accuracy of the target channel selection process. The potential performance parameters that are improved using the proposed technique are accuracy, handoff latency, processor speed, and memory. Shekhar et al. [13] have analyzed the spectrum handoff delay by exploiting the finite queuing theory. The authors have investigated the effect of queue length on the performance metrics that are cumulative handoff delay (CHD) and total service time (TST). Cao and Qian [14] have proposed a spectrum handoff approach where a spectrum handoff success rate is introduced into an optimal spectrum resource allocation model in order to ensure the reliability of spectrum handoff, and the closed-form expression for the spectrum handoff success rate is obtained based on the Poisson distribution. Hoque et al. [15] have modeled a key handoff performance measuring metrics that are link maintenance probability, link failure probability, and service completion probability of a nonstationary CU and analyze the impact of mobility parameters to investigate the characteristics and behavior of CRNs. Further, the authors have analyzed the effect of general service time distributions on the service completion probability. Chengyu et al. [16] have proposed a spectrum handoff scheme based on recommended channel sensing sequence (RCSS) that intends to optimize the spectrum handoff delay subject to the sensing reliability and link maintenance constraints. The authors have investigated two cases that are CU performing a spectrum handoff in the current frame and CU performing the spectrum handoff using several frames. The numerical results confirm that the proposed RCSS handoff scheme can achieve better handoff delay performance than others when the received signal-to-noise ratios of the PUs' signals on different channels are nonidentical. Kumar et al. [17] have presented a detailed classification of existing spectrum handoff schemes for CRNs. Usman et al. [18] have illustrated an energy-efficient channel handoff scheme for sensor network-assisted CRNs while in [19], the authors have proposed a three-phase target channel allocation for multiuser CRNs.

The spectrum mobility techniques are generally of two types, namely, the reactive and proactive. In the reactive technique, the CU switches its communication after the emergence of PU; however, the CU switches its communication before the emergence of PU in case of the proactive technique. The detailed discussion of both the techniques is presented as follows. In the conventional cognitive radio networks as reported in [6, 10, 21], the CU senses the channel for time τ_s and start data transmission for time $(T - \tau_s)$, if the channel is sensed idle, otherwise it senses another channel. Once the data transmission has started on the idle sensed channel, the CU periodically examines the state of channel (either active or idle) and if the emergence of PU on that channel is detected in the next sensing interval, the CU switches its communication on another available channel as shown in Figure 7.1a. As in this approach, the data transmission ceases during the spectrum sensing which reduces the achieved throughput of CU; however, the detection of

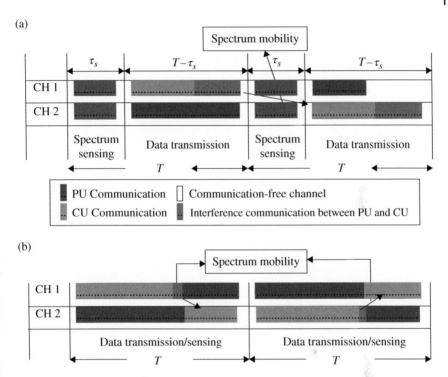

Figure 7.1 The spectrum mobility in cognitive radio networks using (a) periodic sensing and (b) spectrum monitoring with two transceivers at CU.

emergence of PU is possible after certain time, which results in the "interference-at-PU" and "data loss of the CU" [22].

Therefore, in order to exploit the entire time period (T) for data transmission by escaping the spectrum sensing period and to detect the emergence of PU during that period, the researchers have presented an approach in which the spectrum sensing and data transmission are parallel phenomena as depicted in Figure 7.1b. For this functionality, the CU comprises two transceiver units, one for the spectrum sensing and other for the data transmission. Here, this spectrum sensing process concurrent to the data transmission is named as spectrum monitoring as discussed in Chapter 5, Section 5.2.2. In the spectrum monitoring, the emergence of PU is perceived after a certain time (detection delay) and certain switching time (switching delay) is also required to switch the channel. Thus, the total time in which the data loss occurs is the sum of detection and switching delay. Since the detection and switching delay are undetachable entities, therefore in order to avoid this issue, Yang et al. [22] have proposed a novel approach for the spectrum mobility, in which the spectrum prediction technique is exploited to forecast the future states of the channel. On the basis of predicted information, the CU switches its communication before the true emergence of PU, due to which the problem of data loss and interference at PU receiver gets resolved [23].

The conventional spectrum monitoring technique requires two transceivers at every CU and causes extra cost and power consumption. Recently, a new technique for the spectrum monitoring is proposed by Boyd et al. [24], in which the emergence of PU is detected during data transmission period by using only single transceiver. Therefore, this spectrum monitoring technique is preferable over pre-discussed monitoring technique and in the entire discussion, we have considered this. However, for the effective spectrum mobility, the emergence of PU needs to be detected with a reliable system. The available techniques for this process are the spectrum prediction and spectrum monitoring. In the spectrum prediction [25], the CU predicts the emergence of PU on the basis of pre-available information about the channel states; however, the spectrum monitoring [22] is a recently developed technique in which the CU detects the emergence of PU during data transmission period on the basis of received packets' statistics such as receiver error count (REC). The prediction technique requires the information about the prestates of channel, whereas in the spectrum monitoring, no such prerequisite/prestates because the decision is taken only on the current received packets' statistics. However, both the techniques may suffer from the system and/or procedure error known as prediction error [26] and monitoring error [27]; therefore, the detection of the emergence of PU cannot be detected with probability 1. Till now, various researchers have exploited spectrum prediction and monitoring techniques both independently for the detection of emergence of PU during the data transmission for spectrum mobility [28, 29]; however, as per the best of authors' knowledge, the joint effect of these techniques simultaneously is not reported. Therefore, we have exploited these techniques simultaneously to improve the performance of the proposed system during the spectrum mobility by the truthful detection of the spectrum availabilities/holes. The potential contribution in this chapter is summarized as follows:

- The spectrum prediction and monitoring techniques are exploited, simultaneously for the spectrum mobility, in order to improve the performance of CRN by detecting the emergence of the PU quickly and correctly.
- For the simultaneous use of the spectrum prediction and monitoring techniques as well as to combine the results of both, AND and OR fusion rules are used.
- The closed-form expressions for the resource wastage, achieved throughput, interference power at PU, and data loss are derived for the proposed (AND and OR) as well as prediction and monitoring approaches.

- The simulation results for the proposed approaches are presented and compared with those of the reported prediction and monitoring technique.
- The proposed approach is validated by exploiting the special case.

In the next section, the system model of the proposed approach is illustrated. The performance analysis of the proposed approach is presented in Section 7.3. The simulation results and discussion is presented in Section 7.4. Section 7.5 summarizes this chapter.

7.2 System Model

The proposed CRN comprises a transceiver-pair and N number of PUs, in which each PU has a licensed channel for communication. The complete framework of the proposed model is presented in Figure 7.2. Initially, the CU predicts the states (active or idle) of the N number of channels and senses only the idle predicted channels to improve the performance of spectrum sensing (the Nth is sensing free (not-sensed) as it is already predicted active). The activities on each channel are assumed to be independent of each other, as the channels may belong to different networks. Further, the CU establishes communication on the idle sensed channel (as CH 2 is sensed as idle) using an interweave spectrum access technique [1, 30]. Moreover, during the data transmission, the CU starts monitoring the status of PU by exploiting the received packets' statistics. As on the emergence of PU in the channel, the CU needs to switch the communication on another available channel. However, the detection of emergence of PU is the function of spectrum prediction and spectrum monitoring techniques. As in CH 2, the emergence of the PU is already predicted at that time and also confirmed by the spectrum monitoring technique. Moreover, the idle nature of the CH 1 is already predicted for that time. Therefore, the switching occurs, from CH 2 to CH 1 as shown in Figure 7.2. For the spectrum sensing, and to know the true (original) state of the PU, the binary hypothesis considered for the received signal $r(t)$ are H_0 and H_1, which signify the presence and absence of the PU, respectively. There are mainly following two hypothesis testing methods for spectrum sensing [31]: (i) Neyman-Pearson (NP) test [32–36] and (ii) Bays test [37–40]. In the NP test, the potential purpose is to maximize the P_d for given value of P_f whereas in the Bays test, the aim is to minimize the average cost which is also known as Bays risk and defined as

$R = \sum\limits_{i=0}^{1} \sum\limits_{j=0}^{1} C_{ij} P\left(\frac{H_i}{H_j}\right) P(H_j)$, where C_{ij} is the cost, $P\left(\frac{H_i}{H_j}\right)$ are the probabilities of

declaring H_i when H_j is true, $P(H_j)$ is the prior probability of the hypothesis H_j, and $i, j \in (0, 1)$. Here, $P_d = P\left(\frac{H_1}{H_1}\right)$, $P_f = P\left(\frac{H_1}{H_0}\right)$, the probability of miss-detection

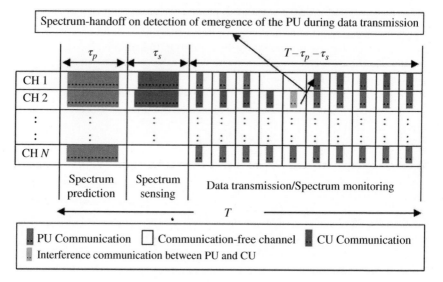

Figure 7.2 The proposed cognitive radio network system model.

($P_{md} = 1 - P_d$). The Bays risk is defined as the sum of all possible costs weighted by the two untrue detection probabilities (false-alarm and miss-detection) and two true-detection probabilities. Since the Bays test minimizes the cost function, however, in the NP test, only P_d maximizes on the cost of P_f. Therefore, in the proposed system model, we have considered the Bays test.

$$X(r) = \frac{f(r/H_1)}{f(r/H_0)} \overset{H_1}{\underset{H_0}{\gtrless}} \frac{P(H_0)(C_{10} - C_{00})}{P(H_1)(C_{01} - C_{11})} = \lambda.$$

The hypothesis H_1 is selected if $X(r) > \lambda$, otherwise, hypothesis H_0 is opted.

In order to examine the effects of spectrum monitoring error and spectrum prediction error irrespective of all other metrics, we have considered the perfect spectrum sensing, i.e. the results of spectrum sensing are 100% accurate or the probability of false-alarm is zero and probability of detection is one [31]. The traffic intensity (ρ) is modeled equivalent to the probability of the channel to be active, which means $\rho = P(H_1) = \mu/\lambda$ as discussed in Section 3.3.1. Moreover, the signal-to-noise ratio (SNR) at the CU receiver due to CU and PU transmissions are denoted as SNR_s and SNR_P, respectively. The CUs' transmitted power is denoted by P_s, whereas h_{sp} denotes the channel power gain from CU transmitter to the PU receiver. This proposed framework requires spectrum prediction and monitoring abilities in a system which increases the cost and complexity. However, due to sufficient exploration of these techniques, it is feasible to use both in a system where the cost and complexity will be within tolerable limits.

7.3 Performance Analysis

In the proposed CRN, initially, the CU predicts the states of channel which is a binary hypothesis and the probability of wrong prediction P_{pe} is used to consider the imperfect spectrum prediction. The predicted and true channel states are assumed to be independent and the probability distribution of these states is presented in Table 7.1. The probability of a channel to be predicted idle is denoted by P_p^0 and defined as:

$$P_p^0 = (1 - P_{pe})P(H_0) + (P_{pe})P(H_1) \tag{7.1}$$

whereas the probability of channel to be predicted active is denoted by P_p^1 and defined as:

$$P_p^1 = (P_{pe})P(H_0) + (1 - P_{pe})P(H_1). \tag{7.2}$$

Further, the CU senses the idle predicted channels and start data transmission on the idle sensed channel. The spectrum monitoring is the parallel phenomenon to the data transmission and P_{me} denotes the spectrum monitoring error, which is used to consider the imperfect spectrum monitoring. Similar to the prediction case, the true channel and monitoring states are also assumed as independent and their probability distribution is depicted in Table 7.2.

Moreover, using aforementioned assumptions, the spectrum prediction, spectrum monitoring, and true channel states are independent phenomenon to each other and the possible combination of the probability distribution of these states is presented in Table 7.3. The performance metrics of CRN for spectrum handoff are considered as resource wastage (RW), achieved throughput (RA), data loss

Table 7.1 The probability distribution of true and predicted channel states.

True channel state (TCS)	Prediction state (PS)	Probability
0	0	$P_{1p} = (1 - P_{pe})\,P(H_0)$
0	1	$P_{2p} = P_{pe}\,P(H_0)$
1	0	$P_{3p} = P_{pe}\,P(H_1)$
1	1	$P_{4p} = (1 - P_{pe})P(H_1)$

Table 7.2 The probability distribution of true and monitoring channel states.

True channel state (TCS)	Monitoring state (MS)	Probability
0	0	$P_{1m} = (1 - P_{me})P(H_0)$
0	1	$P_{2m} = (P_{me})P(H_0)$
1	0	$P_{3m} = (P_{me})P(H_1)$
1	1	$P_{4m} = (1 - P_{me})P(H_1)$

Table 7.3 The probability distribution of the combination of the true channel, prediction and monitoring states.

True channel state	Prediction state	Monitoring state	Probability
0	0	0	$P_{1c} = (1 - P_{me})P(H_0)(1 - P_{pe})$
0	0	1	$P_{2c} = (P_{me})P(H_0)(1 - P_{pe})$
0	1	0	$P_{3c} = (1 - P_{me})P(H_0)(P_{pe})$
0	1	1	$P_{4c} = (P_{me})P(H_0)(P_{pe})$
1	0	0	$P_{5c} = (1 - P_{me})P(H_1)(P_{pe})$
1	0	1	$P_{6c} = (P_{me})P(H_1)(P_{pe})$
1	1	0	$P_{7c} = (1 - P_{me})P(H_1)(1 - P_{pe})$
1	1	1	$P_{8c} = (P_{me})P(H_1)(1 - P_{pe})$

(*DL*), and interference power at the PU (*IP*). The resource wastage is considered in the form of throughput that the CU has wasted due to false-detection of the emergence of PU and switches the channel even it is still idle. The achieved throughput is the total throughput of CU in the presence and absence of the PU. The interference power at PU is considered in the form of total interference produced at PU receiver by the CU transmission. The data loss occurs in the CRN when the CU is unable to detect the emergence of PU and continue data transmission even after its emergence. In the considered CRN, the data loss is considered as the total throughput that can be achieved by the CU if the PU remains absent in that period.

The metrics in prediction technique are evaluated using Table 7.1. As the resource wastage occurs in the network when truly idle channels are predicted as active, the probability of this event is P_{2p}. Therefore, the expression for the resource wastage in prediction technique is RW_P, which is defined as:

$$RW_P = log_2(1 + SNR_s)P_{2p}. \qquad (7.3)$$

The CU continues data transmission and achieves the throughput if the channel is predicted as idle; however, the probabilities of the channel to be predicted idle are P_{1p} and P_{3p}. P_{1p} is the probability when the true state of the PU is idle which means no interference from the PU communication; however, in case of P_{3p}, the PU interferes the CU communication since its true state is active. Therefore, the achieved throughput in the prediction technique is R_P, which is defined as:

$$RA_P = log_2(1 + SNR_s)P_{1p} + log_2\left(1 + \frac{SNR_s}{1 + SNR_P}\right)P_{3p}. \tag{7.4}$$

Moreover, the CU interferes the PU communication when the true state of channel is active; however, it is predicted as idle, and probability of this event is P_{3p}. Therefore, the interference power at PU receiver in prediction technique is denoted as IP_P, which is defined as:

$$IP_P = \left(P_s h_{sp}\right)P_{3p}. \tag{7.5}$$

Furthermore, the data loss in CRN is the function of event, i.e. if the actually active channels are predicted as idle. In this case, the CU transmits data which collides with the PUs' data and thus data lost. The probability of this event is P_{3p} and therefore, the data loss in the prediction technique is DL_P, which is defined as:

$$DL_P = log_2(1 + SNR_s)P_{3p}. \tag{7.6}$$

Similarly, the performance metrics in the spectrum monitoring technique only are computed using Table 7.2.

$$RW_M = log_2(1 + SNR_s)P_{2m}, \tag{7.7}$$

$$RA_M = log_2(1 + SNR_s)P_{1m} + log_2\left(1 + \frac{SNR_s}{1 + SNR_P}\right)P_{3m}, \tag{7.8}$$

$$IP_P = \left(P_s h_{sp}\right)P_{3m}, \tag{7.9}$$

$$DL_P = log_2(1 + SNR_s)P_{3m}. \tag{7.10}$$

The proposed approach fuses the results of prediction and monitoring techniques using AND and OR fusion rule. In the AND fusion rule, the CU only switches its communication if both the techniques (spectrum prediction and spectrum monitoring) confirm the emergence of PU. However, in the OR rule, if one of the technique confirms the emergence of the PU, the CU switches its communication. The performance metrics for the AND-rule (denoted using subscript A) are evaluated using Table 7.3, similar to the way used for prediction technique.

$$RW_A = log_2(1 + SNR_s)P_{4c}, \tag{7.11}$$

$$RA_A = log_2(1 + SNR_s)(P_{1c} + P_{2c} + P_{3c})$$

$$+ log_2\left(1 + \frac{SNR_s}{1 + SNR_P}\right)(P_{5c} + P_{6c} + P_{7c}), \qquad (7.12)$$

$$IP_A = (P_s h_{sp})(P_{5c} + P_{6c} + P_{7c}), \qquad (7.13)$$

$$DL_A = log_2(1 + SNR_s)(P_{5c} + P_{6c} + P_{7c}). \qquad (7.14)$$

Similarly, the performance metrics for the OR-rule are denoted using subscript O, and computed using Table 7.3.

$$RW_O = log_2(1 + SNR_s)(P_{2c} + P_{3c} + P_{4c}), \qquad (7.15)$$

$$RA_O = log_2(1 + SNR_s)(P_{1c}) + log2\left(1 + \frac{SNR_s}{1 + SNR_P}\right)(P_{5c}), \qquad (7.16)^\bullet$$

$$IP_O = (P_s h_{sp})(P_{5c}), \qquad (7.17)$$

$$DL_O = log_2(1 + SNR_s)(P_{5c}). \qquad (7.18)$$

7.4 Results and Discussion

In this chapter, the spectrum prediction and monitoring techniques are exploited simultaneously in order to detect the emergence of PU during data transmission period and the combination of two approaches is denoted by Comb. The AND and OR rules are used to fuse the results of two approaches and various performance metrics are analyzed for Comb-OR and Comb-AND rule, and have been compared with that of the independent prediction and monitoring techniques. The simulation metrics are presented in Table 7.4.

The relation between the resource wastage and prediction error for the proposed approaches (Comb-OR and Comb-AND rule) as well as for the prediction approach is illustrated in Figure 7.3a by considering the different values of the

Table 7.4 The simulation metric values.

Metric	Value	Metric	Value	Metric	Value
h_{sp}	0.1	N_o	10	P_s	6 W
SNR_s	0 dB	$P(H_0)$	0.9		
SNR_p	15 dB	$P(H_1)$	0.1		

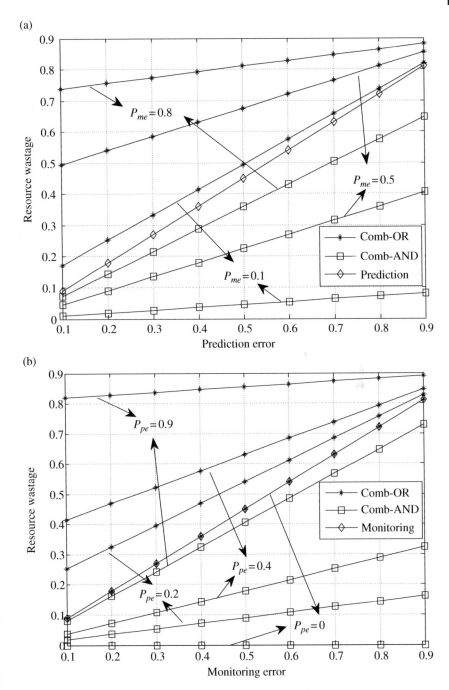

Figure 7.3 The behavior of resource wastage (bits/second/Hz) with reference to (a) prediction error and (b) monitoring error.

spectrum monitoring error. The resource wastage is increasing with increase in the prediction error for all approaches. It is also apparent that the Comb-AND provides a significant improvement in the resource wastage as compared to that of the prediction approach because the CU switches the spectrum on the confirmation by both the spectrum prediction and monitoring approaches. However, the Comb-OR wasted the resources by immediately switching its communication, as the emergence of the PU is confirmed only by one of them (prediction or monitoring); even the true state of the PU is idle. Figure 7.3b depicts the behavior of the resource wastage with reference to the monitoring error, for several values of the prediction error. The variations of the resource wastage with reference to the monitoring error are almost same; however, in Figure 7.3b, a special case of the proposed model is depicted, i.e. the prediction error is zero which means the prediction approach is 100% accurate.

In this special case, for the Comb-AND rule, the resource wastage should be zero as the prediction accuracy is 100%; however, for the Comb-OR rule, only the error occurs due to monitoring error. Figure 7.3b completely supports aforementioned scenarios as for the Comb-AND rule; the resource wastage is zero whereas for the Comb-OR rule, it overlaps with the monitoring approach. Therefore, due to this special case, the proposed approach gets validated. The variations of achieved throughput with the prediction error for the proposed and prediction approaches are presented in Figure 7.4a for several values of the spectrum monitoring error. Figure 7.4a witnesses the superior performance of the Comb-AND approach over the prediction approach and it is due to efficient resource utilization by the CU; however, the vice-versa in case of Comb-OR. The behavior of achieved throughput versus monitoring error for different values of the prediction error is depicted in Figure 7.4b.

Similar to Figure 7.4a, the Comb-AND approach outperforms the monitoring only approach. In the special case, for the Comb-AND approach, the achieved throughput becomes maximum and constant, i.e. independent of the monitoring error. However, for the Comb-OR approach, the maximum achieved throughput overlaps the throughput for the monitoring case. Figure 7.5a depicts the interference power at PU versus prediction error. The interference at PU shows linearly increasing relationship for all approaches with the prediction error. Moreover, the Comb-OR approach improves the interference at PU significantly as compared to that of the prediction approach and it happens because of switching the spectrum even if a single approach results in the emergence of PU. However, Comb-AND uses the confirmation of emergence results by both the approaches, where the single approach errors may result in non-switching case even on the emergence of PU and cause interference at the PU. Figure 7.5b presents the relation between the interference power at PU and spectrum monitoring error for several values of the prediction error. Due to similar cause as in Figure 7.5a, the Comb-OR

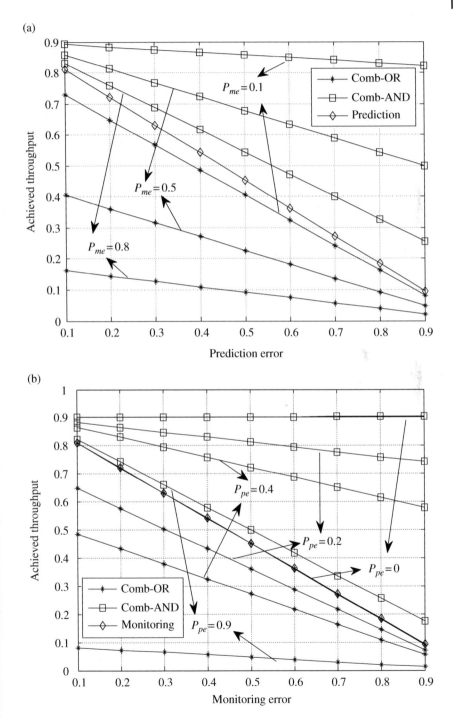

Figure 7.4 The variations of the achieved throughput (bits/second/Hz) with (a) prediction error and (b) monitoring error.

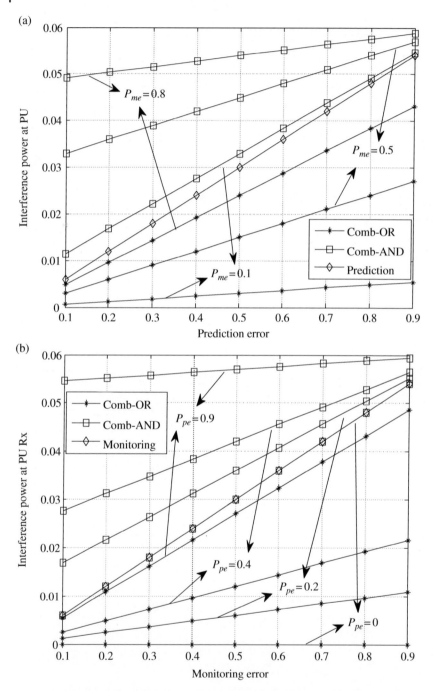

Figure 7.5 The interference power at PU receiver (Watt) with reference to (a) prediction error and (b) monitoring error.

(a)

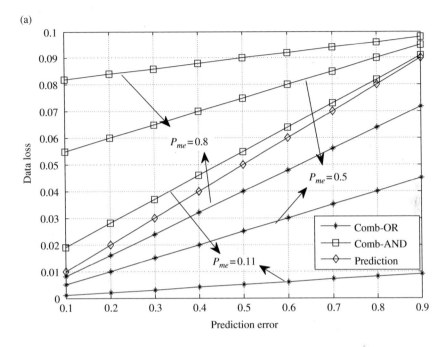

(b)

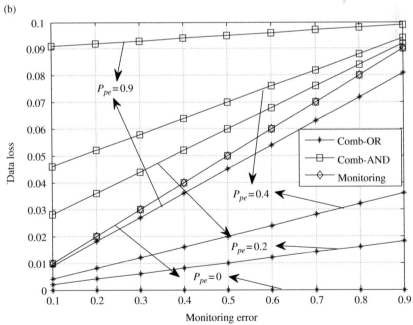

Figure 7.6 The variations of data loss (bits/second/Hz) with (a) prediction error and (b) monitoring error.

outperforms the monitoring approach and vice-versa in case of the Comb-AND. In the special case, the interference becomes zero for the Comb-OR approach; however, it overlies the monitoring approach for the Comb-AND approach.

Figure 7.6a illustrates the relationship between the data loss and prediction error for different values of the monitoring errors using proposed and prediction approaches. The data loss shows the directly proportional relationship with the spectrum prediction error. It is also clear that the Comb-OR approach performs well as compared to that of the prediction approach and it is due to immediate detection of the PU, and switching of its communication by the CU. However, in the Comb-AND approach, the detection of emergence of the PU may be time delayed due to the approval of the two (monitoring and prediction) approaches. Figure 7.6b presents the variations of data loss versus monitoring error for several values of the prediction error, where the data loss shows the same behavior as in Figure 7.6a. In the special case, for the Comb-OR approach, the data loss is constant (zero), i.e. independent to the monitoring error; however, for the Comb-AND approach, it partly covers the data loss in the monitoring approach. From the afore-discussion, we conclude that both the proposed approaches outperform over the prediction and monitoring approaches in different scenarios. The Comb-AND outperforms when resources are limited for the CU, such as high-traffic CRNs. By efficiently utilizing the resources, the achieved throughput of the CU is improved. However, if PU cannot tolerate small interference from the CU, in this case, the CU needs to detect the emergence of PU very keenly, so that the interference and data loss can be improved. Moreover, in the special case, every metric follows the monitoring approach or shows independent nature with reference to the monitoring error. This behavior has verified the validation of the proposed approaches as it works suitably, when limited to the special cases.

7.5 Summary

This chapter has exploited the spectrum monitoring and spectrum prediction techniques concurrently to improve the performance of the spectrum mobility/handoff. The AND and OR fusion rules are used to combine the results of both the spectrum monitoring and spectrum prediction techniques. The numerically simulated results of the proposed approaches (AND and OR) are compared to that of the prediction approach. Moreover, the comparison of the proposed approaches to that of the monitoring approach is also presented. The validation of mathematical techniques used for the proposed approaches is verified by its desired outcomes in the special cases. It is concluded that the Comb-AND outperforms the prediction approach in terms of the achieved throughput by reducing the resource wastage.

However, the Comb-OR performs well with reference to the data loss by immediate switching of communication on the emergence of PU. Therefore, the selection of the particular fusion rule appears as a function of the CU requirements and the PUs' tolerance levels. If the prime goal of the CRN to enhance the network throughput and PU has certain interference tolerability, then we need to opt the Comb-AND approach. Further, if the CRN is unable to tolerate data loss due to the critical nature of data, then the Comb-OR approach is the best option to employ. Moreover, the channel power gain is considered as already known metric to the CU transmitter, i.e. perfect channel state information (CSI) in the entire discussion. However, practically the effect of multipath fading, path loss, and shadowing restrict to yield the perfect CSI, which introduces the potential issues of imperfect CSI. For example, in the spectrum monitoring, the presence of PU is detected on the basis of a number of RECs and threshold value for the detection is selected on the basis of pre-available CSI. The imperfect nature of CSI may increase the values of probabilities of false-alarms and misdetections since the sudden worsening of channel conditions may increase the RECs and the spectrum monitoring system may consider it as the presence of PU which affects the proposed model. This potential issue of imperfect CSI will be reported in future communications.

References

1 Thakur, P., Kumar, A., Pandit, S. et al. (2017). Advanced frame structures for hybrid spectrum access strategy in cognitive radio communication systems. *IEEE Commun. Lett.* **21** (2): 410–413.
2 Yang, C., Lou, W., Fu, Y. et al. (2016). On throughput maximization in multichannel cognitive radio networks via generalized access strategy. *IEEE Trans. Commun.* **64** (4): 1384–1398.
3 Yucek, T. and Arslan, H. (2009). A survey of spectrum sensing algorithms for cognitive radio applications. *IEEE Commun. Surv. Tutorials* **11** (1): 116–130.
4 Bagwari, A., Tuteja, S., Bagwari, J., and Samarah, A. (2020). Spectrum sensing techniques for cognitive radio: a re-examination. *IEEE 9th International Conference on Communication Systems and Network Technologies (CSNT)*, Gwalior, India (10–12 April 2020), 93–96.
5 Babu, S.K. and Venkataramanan, C. (2020). BLSBA-ED-SSO: Bivariate Lévy-stable bat algorithm-based energy detection for spectrum sensing optimization in cognitive radio networks (CRNs). *Int. J. Commun. Syst.* **33** (7): 1–18.
6 Liang, Y.C., Zeng, Y., Peh, E.C.Y., and Hoang, A.T. (2008). Sensing-throughput tradeoff for cognitive radio networks. *IEEE Trans. Wireless Commun.* **7** (4): 1326–1337.

7 Kumar, A., Thakur, P., Pandit, S., and Singh, G. (2020). Threshold selection and cooperation in fading environment of cognitive radio network: consequences on spectrum sensing and throughput. *AEU-Int. J. Electron. Commun.* **17**: 1–11.

8 Kumar, A., Thakur, P., Pandit, S., and Singh, G. (2020). Intelligent threshold selection in fading environment of cognitive radio network: advances in throughput and total error probability. *Int. J. Commun. Syst.* **33** (1): 1–15.

9 Lee, W.Y. and Akyildiz, I.F. (2012). Spectrum-aware mobility management in cognitive radio cellular networks. *IEEE Trans. Mobile Comput.* **11** (4): 529–542.

10 Thakur, P., Kumar, A., Pandit, S. et al. (2017). Spectrum mobility in cognitive radio network using spectrum prediction and monitoring techniques. *Phys. Commun.* **24**: 1–8.

11 Arshid, K., Hussain, I., Bashir, M.K. et al. (2020). Primary user traffic pattern based opportunistic spectrum handoff in cognitive radio networks. *Appl. Sci.* **10** (5): 1–19.

12 Aggarwal, M., Velmurugan, T., and Nandakumar, S. (2020). Dual processor based centralized device for spectrum handoff in cognitive radio networks. *J. Electr. Eng. Technol.* **15** (2): 833–842.

13 Shekhar, S., Hoque, S., and Arif, W. (2020). Analysis of spectrum handoff delay using finite queuing model in cognitive radio networks. *Int. J. Commun. Netw. Distrib. Syst.* **25** (3): 249–264.

14 Cao, K. and Qian, P. (2020). Spectrum handoff based on DQN predictive decision for hybrid cognitive radio networks. *Sensors* **20** (4): 1–11.

15 Hoque, S., Arif, W., and Sen, D. (2020). Assessment of spectrum handoff performance in cognitive radio cellular networks. *IEEE Wireless Commun. Lett.* **9** (9): 1403–1407.

16 Chengyu, W., Chen, H., and Lingge, J. (2013). Spectrum handoff scheme based on recommended channel sensing sequence. *China Commun.* **10** (8): 18–26.

17 Kumar, K., Prakash, A., and Tripathi, R. (2016). Spectrum handoff in cognitive radio networks: a classification and comprehensive survey. *J. Netw. Comput. Appl.* **61**: 161–188.

18 Usman, M., Khan, S., V-Van, H., and Insoo, K. (2015). Energy-efficient channel handoff for sensor network-assisted cognitive radio network. *Sensors* **15** (8): 18012–18039.

19 Chakraborty, T. and Misra, I.S. (2020). A novel three-phase target channel allocation scheme for multi-user cognitive radio networks. *Comput. Commun.* **154**: 18–39.

20 Huang, S., Liu, X., and Ding, Z. (2008). Opportunistic spectrum access in cognitive radio networks. *IEEE INFOCOM 2008 – The 27th Conference on Computer Communications*, Phoenix, AZ (13–18 April 2008).

21 Sahai, A., Hoven, N., and Tandra, R. (2004). Some fundamental limits on cognitive radio. *42nd Allerton Conference on Communication, Control, and Computing*, Monticello, IL (29 September to 1 October 2004), 1–11.

22 Yang, L., Cao, L., and Zheng, H. (2008). Proactive channel access in dynamic spectrum networks. *Phys. Commun.* **1** (2): 103–111.

23 Zhao, Y., Hong, Z., Luo, Y. et al. (2018). Prediction-based spectrum management in cognitive radio networks. *IEEE Syst. J.* **12** (4): 3303–3314.

24 Ali, A. and Hamouda, W. (2015). Spectrum monitoring using energy ratio algorithm for OFDM-based cognitive radio networks. *IEEE Trans. Wireless Commun.* **14** (4): 2257–2268.

25 Yarkan, S. and Arslan, H. (2007). Binary time series approach to spectrum prediction for cognitive radio. *66th IEEE Vehicular Technology Conference*, Baltimore, MD (30 September to 3 October 2007), 1563–1567.

26 Yang, J. and Zhao, H. (2015). Enhanced throughput of cognitive radio networks by imperfect spectrum prediction. *IEEE Commun. Lett.* **19** (10): 1738–1741.

27 Thakur, P., Kumar, A., Pandit, S. et al. (2018). Analysis of high-traffic cognitive radio network with imperfect spectrum monitoring technique. *Comput. Netw.* **147**: 27–37.

28 Xing, X., Jing, T., Cheng, W. et al. (2013). Spectrum prediction in cognitive radio networks. *IEEE Wireless Commun.* **20** (2): 90–96.

29 Thakur, P., Kumar, A., Pandit, S. et al. (2019). Performance analysis of cooperative spectrum monitoring in cognitive radio network. *Wireless Netw.* **25**: 989–997.

30 Akyildiz, I.F., Lo, B.F., and Balakrishnan, R. (2011). Cooperative spectrum sensing in cognitive radio networks: a survey. *Phys. Commun.* **4** (1): 40–62.

31 Kay, S.M. (1998). *Fundamentals of Statistical Signal Processing: Detection Theory*. Prentice-Hall PTR.

32 Watanabe, S. (2017). Neyman-Pearson test for zero-rate multiterminal hypothesis testing. *IEEE International Symposium on Information Theory (ISIT)*, Aachen, Germany (25–30 June 2017), 1–6.

33 Watanabe, S. (2018). Neyman–Pearson test for zero-rate multiterminal hypothesis testing. *IEEE Trans. Inf. Theory* **64** (7): 4923–4939.

34 Scott, C. and Noval, R. (2005). A Neyman-Pearson approach to statistical learning. *IEEE Trans. Inf. Theory* **51** (11): 4923–4939, 3806–3819.

35 Levitan, E. and Mervah, N. (2002). A competitive Neyman-Pearson approach to universal hypothesis testing with applications. *IEEE Trans. Inf. Theory* **48** (8): 2215–2229.

36 Li, Z. and Oechtering, T.J. (2017). Privacy-constrained parallel distributed Neyman-Pearson test. *IEEE Trans. Signal Inf. Process. Netw.* **3** (1): 77–90.

37 Golz, M., Muma, M., Halme, T. et al. (2019). Spatial inference in sensor networks using multiple hypothesis testing and bayesian clustering. *27th European Signal Processing Conference (EUSIPCO)*, A Coruna, Spain (2–6 September 2019), 1–6.

38 Zhang, J., Fillatre, L., and Nikiforov, I. (2017). Bayesian test with quadratic criterion for multiple hypothesis testing problem. *International Conference on Industrial*

Informatics – Computing Technology, Intelligent Technology, Industrial Information Integration (ICIICII), Wuhan, China (2–3 December 2017), 1–6.

39 Halme, T., Golz, M., and Koivunen, V. (2019). Bayesian multiple hypothesis testing for distributed detection in sensor networks. *IEEE Data Science Workshop (DSW)*, Minneapolis, MN (2–5 June 2019), 105–109.

40 Golz, M., Mumma, M., Halme, T. et al. (2019). Spatial inference in sensor networks using multiple hypothesis testing and bayesian clustering. *27th European Signal Processing Conference (EUSIPCO)*, A Coruna, Spain (2–6 September 2019), 1–5.

8

Hybrid Self-Scheduled Multichannel Medium Access Control Protocol in Cognitive Radio Networks

8.1 Introduction

The complete cognitive engine comprises the following functional units, namely (i) spectrum sensing [1–6], (ii) spectrum analysis and decision [4, 7–9], (iii) spectrum accessing/sharing [10–14], and (iv) spectrum mobility [15–20] which are well illustrated in Chapter 1. In the spectrum sensing, the cognitive user (CU) scans the electromagnetic environment to perceive the idle channels and then select the most suitable channel among all the idle channels for communication as well as shares this information with all other CUs. Further, the CU accesses the selected idle channel using appropriate spectrum accessing technique. In addition to this, if the PU resumes its communication, then the CU needs to stop the communication and switches on another idle channel that is known as spectrum mobility [15]. After perceiving the idle channels using spectrum sensing, the CU accesses these channels using suitable medium access control (MAC) protocol where the decision on optimal spectrum sensing, transmission time, and proper coordination with other CUs are the key characteristics of the MAC protocol [21]. Several studies have been reported on the cognitive radio medium access control (CR-MAC) protocols for the distributed cognitive radio networks (DCRNs) because the distributed network shows the superior nature over centralized networks due to architecture-free frameworks and less information exchange requirement [22–24], where the CUs are permitted to sense their environment in order to perceive the idle channels, select the most suitable channels, and access these channels without using any centralized framework [25].

In the conventional MAC protocol, the CUs exploit the common control channel (CCC) for the spectrum sensing information exchange in order to negotiate the channel access. After the spectrum sensing, each CU contends for the access of idle channel using various spectrum access protocols. A potential protocol among these is reported in [26] where the authors have proposed a self-scheduling

Spectrum Sharing in Cognitive Radio Networks: Towards Highly Connected Environments,
First Edition. Prabhat Thakur and Ghanshyam Singh.
© 2021 John Wiley & Sons, Inc. Published 2021 by John Wiley & Sons, Inc.

multichannel-MAC (SMC-MAC) protocol, which allows multiple CUs to transmit data through the sensed idle channels using two cooperative channel sensing algorithms, i.e. fixed channel sensing (FCS) and adaptive channel sensing (ACS). Moreover, the slotted contention mechanism is used to exchange the channel request information for self-scheduling. It is reported that the proposed protocol outperforms in allowing multiple CUs to transmit data frames effectively on the multichannel and adaptively in response to the PUs' traffic dynamics. In this protocol, the contention interval (CI) is fixed and due to this feature, the number of collisions increases if the CI is set small for the large number of CUs. Similarly, the data transmission interval decreases if the CI is set large for the small number of CUs. In order to avoid these issues, Pandit and Singh [21, 27] have proposed a Backoff algorithm for the SMC-MAC protocol in which the CI is flexible and selects the value according to the number of contending users. The time frame (T) of SMC-MAC protocol reported in [21] comprises four subintervals, namely, the idle interval (T_i), the sensing and sharing interval (T_{ss}), contention interval (T_{ct}), and data transmission interval (T_{tr}). In the idle interval, all the CUs get ready to sense the channel. Further, in the spectrum sensing and sharing interval, the CUs sense the channel and share the spectrum sensing information with all other CUs using CCC. In the CI, the CUs contend for the idle periods using Backoff algorithm and the winning CUs start data transmission in the data transmission interval. Since merely the transmission time of the frame is devoted for data transmission, the sensing-sharing and contention intervals also consume significant time of the frame, which results in the inefficient utilization of the spectrum. On the other hand, the CUs who lose the contention need to wait for the next frame which causes major issue regarding the seamless communication. The seamless communication is most preferable in certain applications such as voice-telephony. Further, in [21], the authors have illustrated the protocol for perfect spectrum sensing scenarios; however, in practice, the spectrum sensing is an imperfect phenomenon [28]. Therefore, the proposed protocols must be analyzed for the real scenario, i.e. the imperfect spectrum sensing. By keeping in view the aforementioned facts, for the efficient utilization of spectrum and to ensure seamless communication, we have proposed an improved frame structure for the DCRN so that hybrid spectrum accessing techniques can be implemented for the SMC-MAC protocol, therefore the proposed protocol is named as hybrid SMC-MAC (HSMC-MAC). The authors' potential contribution in this chapter is summarized as follows:

- An improved frame structure for the DCRN is proposed in order to enhance the throughput by exploiting the spectrum efficiently.
- The HSMC-MAC protocol is proposed with the improved frame structure where the idle and active sensed channels are accessed via the interweave and underlay spectrum accessing strategy, respectively.

- The performance of proposed protocol is analyzed for the perfect and imperfect spectrum sensing as well as for more feasible scenarios.
- The closed-form expressions of throughput and spectral utilization for the proposed framework in case of perfect spectrum sensing are derived numerically.
- Further, in case of imperfect spectrum sensing, the closed-form expressions of spectral utilization, data loss, interference efficiency, and resource wastage are derived.
- A novel performance metric is introduced to consider the switching events such as data transmission mode and channel switching, which is named as throughput efficiency and derived numerically. The throughput efficiency is defined as the data rate per unit of time, i.e. bits/second square/Hz. For certain data rate as well as time critical applications (such as live telesurgery where the information exchange rate needs to be maintained) where the data rate needs to be greater than the threshold value, the throughput efficiency is an important metric.
- Moreover, the Monte-Carlo simulations are presented in order to consider the effect of random events of PUs' appearance to validate the numerically simulated results.

The next section comprises the work related to already existing MAC protocols and spectrum accessing techniques.

8.2 Related Work

8.2.1 CR-MAC Protocols

The design of MAC protocols for CR technology is a challenging task as compared to that of the conventional wireless networks due to the simultaneous existence of the PUs and CUs as well as pre-requisition of the spectrum sensing phenomenon. Therefore, the prime objectives of the CR-MAC are as follows: (i) optimization of spectrum sensing and accessing decisions, (ii) control the multiuser access in the multichannel network, (iii) scheduling of traffic transmission, and (iv) supporting of spectrum trading functions. Various researchers have designed different CR-MAC protocols in order to fulfill the aforementioned objectives as reported in [21–27, 29–32]

In [29], the authors have presented a hardware-constrained-MAC (HC-MAC) protocol without network synchronization in order to achieve the efficient spectrum sensing and sharing decisions. Since this protocol suffers from multichannel hidden terminal problem, therefore, a distributed spectrum agile MAC protocol is proposed which is multichannel carrier sense multiple access (CSMA)-based protocol [30]. This protocol supports single channel/multichannels as well as single user/multiusers with a prerequisite of two transceiver units at CUs so that the CU tunes at control and idle channel simultaneously. Thus, the network synchronization is required due to the need of two transceiver units. Cormio and Chowdhury [22] have presented a survey on various MAC protocols in the CRNs and typically categorize them as: (i) random access protocols such as CSMA/collision avoidance (CA) [33, 34], (ii) time-slotted protocols, and (iii) hybrid protocols. An opportunistic-multichannel MAC (OMC-MAC) protocol for DCRNs has been proposed which provides significant quality-of-service (QoS) for the delay-sensitive applications by assigning higher priorities to such applications [23]. Kwon et al. [24] have proposed a pre-emptive opportunistic-MAC (PO-MAC) protocol in which the sensing and pre-emption mechanism are utilized to transmit data and to report sensing results without collisions. It is also reported that the PO-MAC outperforms the existing CR-MAC protocols in terms of the end-to-end delay and aggregate throughput of the CUs. Debroy et al. [25] have proposed a contention-based multichannel MAC protocol which allows the collision-free access of the idle PUs' channels. Moreover, in [26], the authors have proposed a SMC-MAC protocol, which allows multiple CUs to transmit data through the sensed idle channels using two cooperative channel sensing algorithms. However, the major limitation of this protocol is static CI and in order to overcome this, the authors [27] have presented a modified SMC-MAC protocol where CI is flexible and have exploited a Backoff algorithm.

Recent work on advanced MAC protocols is presented in [35–37]. Khanian et al. [35] have proposed a distributed opportunistic MAC scheme for maximizing the expected aggregate throughput of the multichannel wireless network in which each user attempts to send only its best channel. The CU transmits data if the best-channel gain is greater than the given threshold, which is dynamically updated depending on previous idle and collision situations. In [36], the authors have presented a new analytical model for performance analysis of IEEE 802.11ad employing a three-dimensional Markov chain by exploiting all the features of IEEE 802.11ad medium access mechanisms including the presence of non-contention access. Further, the dependencies on contention period and the number of sectors on the MAC delay and throughput are also analyzed. Yao et al. [37] have proposed an analytical model to compute the optimal bandwidth resource allocation and presented a flexible multichannel coordination MAC (FMC-MAC) protocol for vehicular ad-hoc networks (VANETs). The FMC-MAC protocol allows safety messages broadcasted on service channel (SCH) and non-safety data transmitted on the CCC in a flexible way.

In [38], the authors have presented a dynamic spectrum sharing MAC protocol in order to improve the network throughput improvement by avoiding the data collision. A back-off algorithm (BOA) is proposed where two accessing channels techniques for the proposed algorithm are presented that are fixed channel allocation (FCA) and dynamic channel allocation (DCA) schemes. Karaca [39] has analyzed the performance maximization of an m-channel distribution of CUs with harvesting capability in hybrid CRNs under the SINR constraint on PU receivers. Further, an algorithm is proposed which is used to allocate multiple channels in a hybrid CRN where the CU transmitter can harvest from RF signals. In [40], the authors have investigated the performance of reservation-based and contention-based MAC protocols in delay-tolerant sensor networks with reference to the throughput and energy consumption, respectively. The effectiveness of the proposed scheme is perceived using simulation result.

Further, it is palpable from the presented literature that hybrid spectrum accessing techniques are unexplored in the CR-MAC protocols for DCRNs.

8.2.2 Interference at PU

The CU exploits the spectrum of PU for communication by using opportunistic and/or spectrum sharing modes. In the opportunistic mode, the CU senses the state of channels and establishes communication on the idle sensed channel whereas in the spectrum sharing mode [41], the CU accesses the spectrum of PU in the sharing mode in such a way that the CU communication does not affect the PU communication. As both the modes protect the PU communication from interference ideally, however, practically the imperfection of the systems introduced interference at PU. In the opportunistic mode, the interference at PU appears when there are imperfections in the spectrum sensing especially when the active channels are sensed as idle (misdetection) and CU transmits data in the presence of PU. Further, in the spectrum sharing mode, the CU selects the transmission powers so that the received power at PU receiver is within the constraining limits and for this perfect channel, state information must be available at the CU transmitter. However, the formation of perfect channel's state information is a potential challenge due to random nature of the channel; therefore, in practice, imperfect channel state information is available which introduces the interference at PU [42, 43]. Thus, it is palpable that interference at PU receiver is also a potential metric to analyze the performance of cognitive radio frameworks. To this point of view, various researchers have analyzed the interference performance of cognitive radio frameworks [44, 45]. In [44], the authors have exploited the probability of interference at PU receiver to analyze the performance of Additive Links On-line Hawaii Area (ALOHA) as well as distributed coordination function (DCF) protocol in CRNs. Mili and Musavian [45] have introduced a novel performance metrics to evaluate the interference effect at PU receiver which is named as interference efficiency. In addition to the introduction of interference at PU, the imperfections in the system result in the data loss as well as resource wastage phenomenon which are unexplored till now for CR-MAC protocols. Therefore, we have reported these performance metrics in addition to the interference efficiency and spectral utilization.

Further, this chapter is structured as follows. In the next section, the system model and proposed HSMC-MAC protocol are illustrated. Section 8.4 consists of the performance analysis of proposed protocol for the perfect and imperfect spectrum sensing scenarios. In Section 8.5, we have presented the simulation results for the perfect and imperfect spectrum sensing scenarios. Finally, Section 8.6 summarizes the chapter and highlights the future perspectives.

8.3 System Model and Proposed Hybrid Self-Scheduled Multichannel MAC Protocol

8.3.1 System Model

We have considered a primary network comprising N_{pu} number of PUs, in which every PU has a licensed channel for communication and a cognitive radio network having N_{cu} number of CUs, where each CU has two transceiver units (half-duplex mode) for the spectrum sharing and data transmission [46–48]. The use of two transceivers improves the spectral utilization; however, increases the cost of CU units. The proposed model is suitable for the applications where the spectral utilization and throughput improvement is the prime preference over the cost such as military and medical application. The traffic intensity of primary network is assumed as Poisson distribution on the basis of cellular network [27]. The CU accesses the idle channels via interweave spectrum access strategy and active channels using underlay spectrum access. We have considered the two cases of sensing as follows: (i) spectrum sensing reliability of DCRN is 100% (the probability of false alarm is zero and probability of detection is one) which means perfect spectrum sensing and (ii) spectrum sensing reliability is less than 100% (the probability of false alarm is greater than zero and probability of detection is less than one) which means imperfect spectrum sensing [49]. In addition to this, there is one CCC to exchange the information between CUs, which is assumed to be always available to the DCRN. For the perfect spectrum sensing, since the sensing of more number of channels increases the complexity as well as power consumption, therefore it is desirable and sufficient that each user senses only one channel. Based on this, we have considered that each CU senses only one channel and share the spectrum sensing information with other CUs on control channel so that more information about PU channels' states at each CU is available. The signal-to-noise ratio (SNR) at CU receiver due to the interweave and underlay transmission are denoted as SNR_{s1} and SNR_{s2}, respectively. The received SNR at the CU receiver due to PU transmission is denoted as SNR_P. In the proposed model, the energy detection technique is used for the spectrum sensing phenomenon due to its simple and cost-effective nature [46]. To know the true (original) state of the PU, the binary hypothesis exploited for the received signal $r(t)$ are H_0 and H_1, which connote the presence and absence of the PU, respectively.

$$r(t) = \begin{cases} h.s(t) + w(t) & H_1 \qquad (\textit{Presence of the PU}) \\ w(t) & H_0 \qquad (\textit{Absence of the PU}) \end{cases},$$

where h, $s(t)$, and $w(t)$ signify the channel gain coefficient, PUs' transmitted signal, and additive white Gaussian noise (AWGN), respectively. The traffic intensity of PU is assumed to be $\rho = P(H_1) = \mu/\lambda$ as discussed in Section 3.2.1.

8.3.2 Proposed HSMC-MAC Protocol

In the proposed hybrid SMC-MAC protocol, the entire frame time T comprises five sub-time intervals, namely, the idle interval/period (T_i), spectrum sensing interval (T_s), sharing interval (T_{sh}), contention interval (T_{ct}), and data transmission interval (T_{tr}) as shown in Figure 8.1. It is assumed that in the idle interval, all the CUs get synchronized and are ready to perform spectrum sensing on a particular channel using channel number. As the number of CUs is equal to the number of PUs and a number from 1 to N_{pu} is assigned to each CU and PUs' channel, where ith CU will

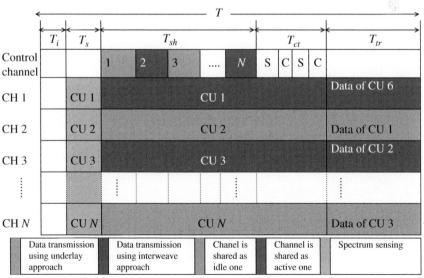

S-Successful contention slot, C -Collided contention slot

Figure 8.1 The proposed improved frame structure for HSMC-MAC protocol.

sense only the ith channel. Further, all the CUs perform spectrum sensing in the sensing interval, simultaneously.

The sharing interval consists of $N_{pu} = N_{cu}$ number of slots and the spectrum sensing results of the ith CU are shared with the other CUs on the ith slot of control channel; however, the CUs compete for the idle PUs' channels in the CI. In addition to this, the CUs start data transmission on the sharing and contention intervals according to the spectrum sensing results, i.e. the idle and active sensed channels are accessed via the interweave and underlay access strategy, respectively. In the sharing interval, the data transmission and sensing results' sharing with other CUs are the parallel phenomenon as shown in Figure 8.2. In the conventional structure, the CUs who won the contention get new idle channels for data transmission period; however, the CUs who lost the contention have to wait for the next frame to transmit the data. However, in the proposed HSMC-MAC protocol, since the number of CUs is equal to the number of channels, therefore, we sort the active sensed channels and the CUs who lost the contention, and the jth active sensed channel is assigned to the jth CU. The CUs who sensed their channel as idle as well as won the contention and get idle sensed channel, continue their data transmission using the interweave spectrum access strategy in the data transmission period, whereas the CUs who sensed their channels as active, however, lose the contention, continue their data transmission via the underlay spectrum access strategy. Moreover, the CUs who sensed their channels as idle, however, lost the contention, switch their communication from the interweave to underlay spectrum access and vice-versa, in the data transmission period.

8.4 Performance Analysis

In the conventional approach [21], the authors have presented a technique of throughput maximization by increasing the number of CUs who gets the idle channel after the contention. Moreover, the CU transmits data merely on the idle sensed channels during data transmission period only; however, the spectrum sharing and contention intervals also consume significant time, in which the CUs are unable to transmit the data. Therefore, in the proposed HSMC-MAC protocol, the CU starts data transmission on the idle and active sensed channels using the interweave and underlay spectrum access strategy, respectively, in the spectrum sharing and contention intervals. Further, the following four possible cases have appeared when the transmission switches from the contention period to the data transmission period.

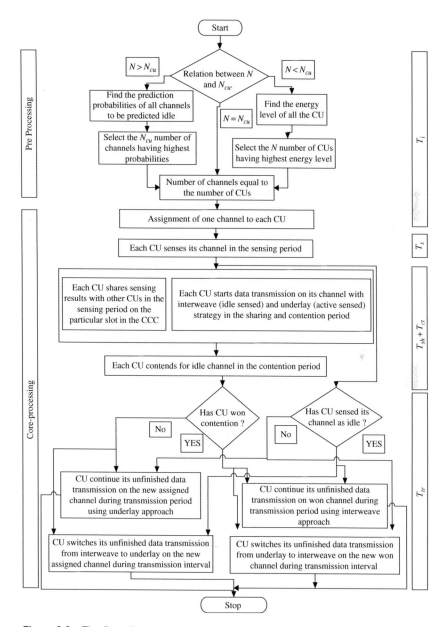

Figure 8.2 The flow diagram of proposed HSMC-MAC protocol.

- The CU who sensed its channel as idle as well as won the contention also, continues its data transmission using the interweave approach on the new won channel (after contention) during the data transmission period.
- The CU who sensed its channel as idle and lost the contention, switches its data transmission from the interweave to underlay approach on the assigned active sensed channel (after contention) during the data transmission period.
- The CU who sensed its channel as active and won the contention, switches its data transmission from the underlay to interweave approach on the new won channel (after contention) during the data transmission period.
- The CU who sensed its channel as active and lost the contention, continues its data transmission using underlay approach on the new assigned active sensed channel (after contention) during the data transmission period.

As in the proposed HSMC-MAC model, we have assumed that the CU who sensed its channel as an idle, starts data transmission in the sharing and contention intervals using the interweave approach and via the underlay approach on the active sensed channels.

8.4.1   With Perfect Spectrum Sensing

In this chapter, we develop and analyze a novel performance metric, called interference efficiency, which shows the number of transmitted bits per unit of interference energy imposed on the PUs in an underlay CRN. The N_{cua} and N_{cui} denotes the number of CUs sensed their channels as active and idle, respectively. The traffic intensity of PU network (ρ) is a very prominent phenomenon to compute the number of active sensed channel, as sensing is a perfect phenomenon in the proposed network. Thus, the relationship between the N_{cua} and ρ is presented as:

$$N_{cua} = \lceil \rho \times N_{cu} \rceil. \tag{8.1}$$

The N_{cui} is related to N_{cua} and defined as:

$$N_{cui} = (N_{cu} - N_{cua}). \tag{8.2}$$

The total number of idle channels (N_{pui}) is equal to the number of CUs who sensed their channel as idle since the number of CUs is same as the number of PUs channels is: $N_{pui} = N_{cui}$. The total number of active channels (N_a) is obtained by subtracting the number of idle sensed channels from the total number of PUs channels which means $N_a = (N_{pu} - N_{pui})$. Further, the throughput of CU using the interweave approach in the sharing and contention period (R_{isc}) for channel with bandwidth (B) is:

$$R_{isc} = \frac{T_{sh} + T_{ct}}{T} (B \log_2(1 + SNR_{s1})) \tag{8.3}$$

and the throughput of CU using the underlay approach in the spectrum sharing period and a contention period (R_{usc}) is:

$$R_{usc} = \frac{T_{sh} + T_{ct}}{T} \left(Blog_2 \left(1 + \frac{SNR_{s2}}{(1 + SNR_p)} \right) \right). \tag{8.4}$$

Moreover, the throughput of CU using the interweave approach in the data transmission period (R_{it}) is:

$$R_{it} = \frac{T_{tr}}{T} (Blog_2(1 + SNR_{s1})) \tag{8.5}$$

and the throughput of CU using the underlay approach in the data transmission period (R_{ut}) is:

$$R_{ut} = \frac{T_{tr}}{T} \left(Blog_2 \left(1 + \frac{SNR_{s2}}{(1 + SNR_p)} \right) \right). \tag{8.6}$$

As the number of CUs is equal to the number of channels and N_{pui} number of channels is idle, the same number of CUs will get idle channel after contention and $(N_{cu} - N_{pui})$ number of CUs unable to find the channel; however, in the proposed model, these $(N_{cu} - N_{pui})$ CUs start data transmission on the active channels using the underlay approach. Thus, it is clear that in the sharing, contention, and data transmission interval, N_{cui} number of CU will transmit data using the interweave approach; however, $(N_{pu} - N_{pui})$ CUs will transmit data using the underlay approach. Therefore, the total throughput of DCRN for the proposed HSMC-MAC protocol (R_{prop}) is:

$$R_{prop} = \left(N_{cui} \left(\frac{T_{sh} + T_{ct} + T_{tr}}{T} (Blog_2(1 + SNR_{s1})) \right) \right) +$$

$$\left((N_{cu} - N_{cui}) \left(\frac{T_{sh} + T_{ct} + T_{tr}}{T} \left(Blog_2 \left(1 + \frac{SNR_{s2}}{(1 + SNR_p)} \right) \right) \right) \right). \tag{8.7}$$

However, in the conventional approach, the data are transmitted on the N_{pui} number of idle sensed channels by the N_{cui} number of CUs and the throughput of DCRN for the conventional approach (R_{conv}) is:

$$R_{conv} = \left(N_{cui} \left(\frac{T_{tr}}{T} (B log_2(1 + SNR_{s1})) \right) \right). \tag{8.8}$$

The spectral utilization of a network is the throughput achieved per unit of bandwidth; therefore, the spectral utilization (SU) for the proposed HSMC-MAC and conventional SMC-MAC protocol is defined as follows:

$$SU_{prop} = R_{prop}/B, \tag{8.9}$$

$$SU_{conv} = R_{conv}/B. \tag{8.10}$$

8.4.2 With Imperfect Spectrum Sensing

In the previous section, we have analyzed the proposed HSMC-MAC protocol by assuming sensing as a perfect phenomenon, i.e. probability of false alarm (P_f) is zero and probability of detection is one. However, in the practical frameworks, the spectrum sensing is an imperfect phenomenon which means the P_f and P_d have values greater than zero and less than one, respectively. Therefore, in order to consider the practical scenarios, we have analyzed the proposed HSMC-MAC protocol for imperfect spectrum sensing. The CUs transmit data on idle sensed channels with full-power; however, some of them are actually active which are sensed as idle due to false alarm and results in the data loss as well as introduces interference at PU. Similarly, the CUs transmit data with constrained power on the active sensed channels; however, some of them are actually idle which are sensed as active due to misdetection and results in the resource wastage. Therefore, in this chapter, the explored performance metrics are spectral utilization (SU), data loss (DL), interference efficiency (IE), and resource wastage (RW) in case of imperfect spectrum sensing. The computation of the performance metrics relies on the sensing and original/true states of the channel and probability of these states.

The selection of data transmission modes (interweave and underlay) relies on spectrum sensing states and interference as PU transmission depends on the original state of channel and there are four possible cases as shown in Table 8.1. As shown in the previous section, the total number of idle sensed channels transmits data with full power on the sharing, contention, and data transmission intervals whereas the active sensed channels transmit constrained power on the same intervals in the complete frame time (T).

Table 8.1 The data rates of CU for various conditions.

True state (TS)	Sensing state (SS)	Throughput
0	0	$c_0 = \frac{T_{sh} + T_{ct} + T_{tr}}{T} \cdot log_2(1 + SNR_{s1})$
0	1	$c_1 = \frac{T_{sh} + T_{ct} + T_{tr}}{T} \cdot log_2(1 + SNR_{s2})$
1	0	$c_2 = \frac{T_{sh} + T_{ct} + T_{tr}}{T} \cdot log_2\left(1 + \frac{SNR_{s1}}{1 + SNR_p}\right)$
1	1	$c_3 = \frac{T_{sh} + T_{ct} + T_{tr}}{T} \cdot log_2\left(1 + \frac{SNR_{s2}}{1 + SNR_p}\right)$

- As in row 1 of Table 8.1, the c_0 is the achieved data rate when the true and sensing states are idle where the CU transmits data with full power when there is zero interference at CU receiver from the PU communication.
- As in row 2 of Table 8.1, the c_1 is achieved data rate when the true and sensing states are idle and active, respectively, where the CU transmits data with constrained power when there is zero interference at CU receiver from the PU communication.
- As in row 3 of Table 8.1, the c_2 is achieved data rate when the true and sensing states are active and idle, respectively, where the CU transmits data with full power when the CU receiver experiences interference from the PU communication.
- As in row 4 of Table 8.1, the c_3 is achieved data rate when the true and sensing states are active where the CU transmits data with constrained power when the CU receiver experiences interference from the PU communication.

For the aforementioned cases, the number of channels (*NC*) relies on the traffic intensity, probability of detection, and probability of false alarm as presented in Table 8.2. Now, the spectral utilization of network is defined as the achieved data rates by all the CUs per unit of bandwidth, which is numerically defined as:

$$SU = \{(NC1 \times c_0) + (NC2 \times c_1) + (NC3 \times c_2) + (NC4 \times c_3)\}. \tag{8.11}$$

The data loss in the proposed HSMC-MAC is defined as the data transmitted on the idle sensed channels when they are actually active and numerically represented as:

$$DL = NC1 \times B \times c_0. \tag{8.12}$$

The performance metric of the interference caused at PU receiver due to CU transmission is the probability of interference (P_{in}) for the CR networks in which the spectrum is accessed by interweave spectrum access technique. Since the CU communication introduces interference at PU receiver if the misdetection occurs and CU transmits data even in the presence of PU, the probability of interference is

Table 8.2 The number of channels in the CRN for various conditions.

TS	SS	Probability	Number of channels
0	0	$(1 - P_f)$	$NC1 = N \times (1 - \rho) \times (1 - P_f)$
0	1	(P_f)	$NC2 = N \times (1 - \rho) \times (P_f)$
1	0	$(1 - P_d)$	$NC3 = N \times (\rho) \times (1 - P_d)$
1	1	(P_d)	$NC4 = N \times (\rho) \times (P_d)$

considered equivalent to the probability of misdetection, i.e. $P_{in} = (1 - P_d)$. However, in the proposed HSMC-MAC protocol, we have exploited the hybrid spectrum access technique due to which two modes (interweave and underlay) cause interference with different power levels; therefore, the probability of interference due to interweave and underlay modes will be equivalent to probability of misdetection $(1 - P_d)$ and probability of detection (P_d), respectively. As the probability of detection is also contributing the probability of interference at PU receiver, this is unsuitable performance metric for the interference analysis in the proposed HSMC-MAC. Therefore, we have exploited one recently explored interference performance metric known as interference efficiency. The interference efficiency [50, 51] is defined as the number of bits transmitted per unit of energy imposed on the PU and in the proposed HSMC-MAC protocol, it is the ratio of achieved throughput to the power received at PU receiver when the original state of PU is active, which is numerically presented as:

$$IE = \frac{(NC3 \times c_2) + (NC4 \times c_3)}{(NC3 \times P_1 \times h_{sp}) + (NC3 \times P_2 \times h_{sp})}. \tag{8.13}$$

Further, the resource wastage is defined as the data rate that can be achieved at the idle channels when they are sensed as active, however, having idle states actually. Since the data are transmitted with constrained power on the active sensed channels, the resource wastage, in this case, is the difference between data rate achieved on the idle channels with full power and constrained power, which is numerically represented as:

$$RW = NC1 \times B \times (c_0 - c_1). \tag{8.14}$$

8.4.3 More Feasible Scenarios

The spectral utilization of the proposed HSMC-MAC protocol is defined for the idle events in which the sensing, sharing, contention, and data transmission intervals are idle and no switching event is considered. However, in the practical scenarios, the CUs need switching time as they move from sensing to sharing interval for data transmission and this time is denoted as ST_{ssh}. In addition, the CUs switch their communication mode, i.e. from interweave to underlay or vice versa when moving from CI to data transmission interval and simultaneously CUs switch the channel also after contention period. The transmission mode switching and channels' switching time is denoted as ST_{tm} and ST_{ch}, respectively, and since both the events occur simultaneously, the switching time which affects the system performance is the larger value between ST_{tm} and ST_{ch} which is known as effective switching time (EST) and numerically presented as: $EST = max(ST_{tm}, ST_{ch})$. Considering these events, Table 8.1 is modified as depicted in Table 8.3.

Table 8.3 The data rates of CU for various conditions.

True state (TS)	Sensing state (SS)	Throughput
0	0	$c_{00} = \dfrac{T_{sh} + T_{ct} + T_{tr} - EST}{T} \cdot log_2(1 + SNR_{s1})$
0	1	$c_{01} = \dfrac{T_{sh} + T_{ct} + T_{tr} - EST}{T} \cdot log_2(1 + SNR_{s2})$
1	0	$c_{10} = \dfrac{T_{sh} + T_{ct} + T_{tr} - EST}{T} \cdot log_2\left(1 + \dfrac{SNR_{s1}}{1 + SNR_p}\right)$
1	1	$c_{11} = \dfrac{T_{sh} + T_{ct} + T_{tr} - EST}{T} \cdot log_2\left(1 + \dfrac{SNR_{s2}}{1 + SNR_p}\right)$

Now the spectral utilization of the proposed HSMC and conventional SMC protocols is defined as:

$$SU_{mp} = \{(NC1 \times c_{00}) + (NC2 \times c_{01}) + (NC3 \times c_{10}) + (NC4 \times c_{11})\}. \tag{8.15}$$

However, the spectral utilization remains same for the conventional framework because there is no channel and power switching for the data transmission.

Further, for the data rate critical applications in which a predefined data rate must be achieved either by compromising or without compromising the other metrics. One such application is live telesurgery where the surgeon remotely performs surgery and data rate reduction even for a small time results in the loss of human life. For such applications, the throughput efficiency (*TI*) will play a vital role which is defined as the throughput achieved per unit of time consumed and for the proposed HSMC-MAC protocol, it is numerically represented as:

$$TI_{Prop} = \frac{SU_{prop}}{T_i + T_s + T_{sh} + T_{ct} + T_{tr} + max\left(ST_{tm}, ST_{ch}\right)}. \tag{8.16}$$

However, for the conventional framework, the *TI* is numerically presented as:

$$TI_{conv} = \frac{SU_{conv}}{T_i + T_s + T_{sh} + T_{ct} + T_{tr}}. \tag{8.17}$$

The throughput efficiency shows the rate of change of throughput and measured in bps^2/Hz.

8.5 Simulations and Results Analysis

In this section, the simulation results are presented for the proposed HSMC-MAC protocol and have been compared with that of the conventional SMC-MAC protocol [27]. The simulation environment of the proposed protocol is simulated using the MATLAB 2010a [52]. The performance of the proposed HSMC-MAC protocol is analyzed for the following two scenarios of spectrum sensing, namely, the perfect and imperfect spectrum sensing. Therefore, the simulation results for both the scenarios for various performance metrics are illustrated as follows.

8.5.1 With Perfect Spectrum Sensing

The total throughput of DCRN for the proposed HSMC-MAC and conventional SMC-MAC is presented by Eqs. (8.5) and (8.6), respectively. In the considered DCRN, the maximum throughput is achieved if all the PU channels are sensed as idle consequently, all the CUs transmit data with the interweave approach in the sharing, contention, and data transmission intervals. Thus, the maximum achieved throughput of the DCRN (R_{max}) is:

$$R_{max} = \left(N_{cu} \left(\frac{T_{sh} + T_{ct} + T_{tr}}{T} \left(log_2(1 + SNR_{s1}) \right) \right) \right). \quad (8.18)$$

The normalized throughput for the proposed HSMC-MAC (R_{propn}) and conventional SMC-MAC (R_{convn}) protocols are defined as follows:

$$R_{propn} = \frac{R_{prop}}{R_{max}}, \quad (8.19)$$

$$R_{convn} = \frac{R_{conv}}{R_{max}}. \quad (8.20)$$

The IEEE 802.22 is the first cognitive radio-based standard which is used for wireless regional area network (WRAN) in order to exploit the television (TV) white-spaces [53, 54]. Therefore, the numerical values of simulation parameters are inspired by IEEE 802.22 and are tabulated in Table 8.4. The variations of normalized throughput with the traffic intensity of primary network are illustrated in Figure 8.3. The normalized throughput of DCRN decreases linearly with increase in the traffic intensity for both the proposed HSMC-MAC and conventional SMC-MAC protocols. The throughput of proposed approach is significantly high as compared to that of the conventional approach.

Moreover, the normalized throughput of DCRN for the proposed HSMC-MAC approach is 1 when the traffic intensity is 0, however, has certain value even when the traffic intensity is 1 (means all the PUs channels are active) which causes due to the underlay transmission on all the active channels. On the other hand, in the

Table 8.4 The numerical values of simulation metrics.

Metric	Value	Metric	Value	Metric	Value
T	100 ms	N_{pu}	20	SNR_{s1}	20 dB
T_i	2 ms	T_{sh}	20 ms	SNR_{s2}	0 dB
T_s	5 ms	T_{ct}	10 ms	B	1 unit
N_{cu}	20	SNR_P	0 dB	ST_{tr}	0.1 ms

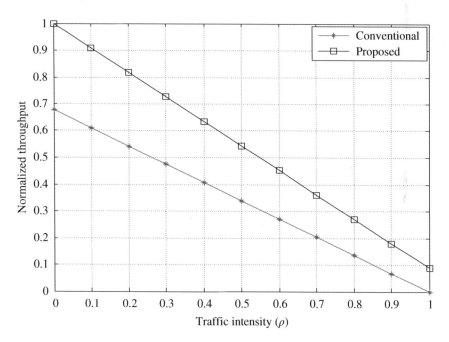

Figure 8.3 The normalized throughput variations with traffic intensity of PU.

conventional approach, the normalized throughput is even below 0.7 when the traffic intensity is 0 and reduces to zero as traffic intensity reaches to one.

The normalized throughput reduces to zero because the traffic intensity 1 means all the PUs channels are active and CU transmits data only on the idle channels in the conventional approach. The variations of spectral utilization with the traffic intensity are presented in Figure 8.4 where the proposed HSMC-MAC protocol

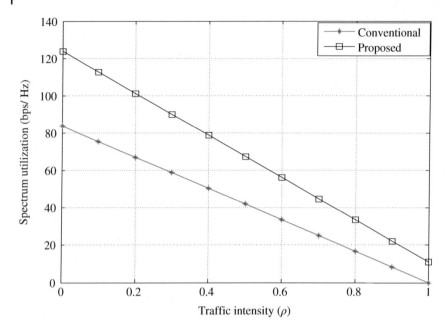

Figure 8.4 The spectral utilization versus traffic intensity of PU.

outperforms the conventional one, due to the same reason as discussed for Figure 8.3. Due to the random nature of traffic intensity (PUs appearance), we have randomly selected the value of ρ between 0 and 1, and have presented the Monte-Carlo simulation results for 10 000 runs as depicted in Figures 8.5 and Figure 8.6. The relationship between the normalized throughput and SNR at CU receiver due to the underlay transmission for various values of the traffic intensity is presented in Figure 8.5. It is apparent that the normalized throughput of proposed HSMC-MAC protocols is significantly high as compared to that of the conventional approach and increases with the increase of SNR. However, the normalized throughput of conventional approach remains constant with reference to the "SNR at CU receiver due to the underlay transmission" and it happens since there is only the interweave transmission mode in the conventional approach, whereas the proposed HSMC-MAC protocol has exploited both the interweave and underlay modes.

Thus, it is clear that the normalized throughput in conventional approach is independent of the "SNR at CU due to underlay transmission." The variations of normalized throughput with reference to the "SNR at CU due to interweave transmission" for different values of the traffic intensity are illustrated in

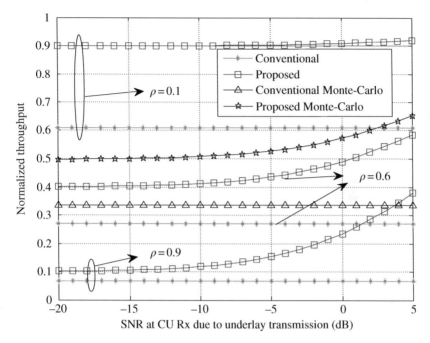

Figure 8.5 Normalized throughput versus SNR at CU due to underlay transmission.

Figure 8.6. The proposed HSMC-MAC protocols outperform the conventional SMC-MAC protocols in terms of the normalized throughput; however, the proposed protocol and conventional protocol show the steady nature with respect to the SNR. This steady nature is because of the normalized throughput which is the ratio of achieved throughput to the maximum throughput; however, the increase in SNR results in the same increase in achieved and maximum throughput which provides the steady nature of the normalized throughput. Moreover, the Monte-Carlo simulation results for the proposed and conventional protocols also follow same nature in Figures 8.5 and 8.6 which further validates the numerically simulated results.

8.5.2 With Imperfect Spectrum Sensing

The simulation parameters for the imperfect spectrum sensing are same as shown in Table 8.4, except the probability of false alarm and probability of detection. The variations of the spectral utilization with traffic intensity for the conventional HSMC-MAC and SMC-MAC protocols with perfect as well as imperfect spectrum sensing are presented in Figure 8.7. It is already shown that the proposed protocol outperforms the conventional one; however, the imperfect spectrum sensing

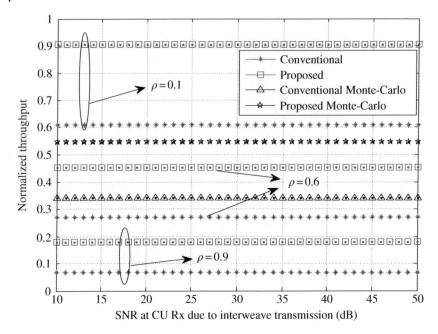

Figure 8.6 Normalized throughput versus SNR at CU due to interweave transmission.

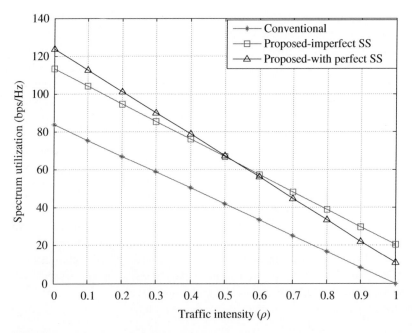

Figure 8.7 The variations of spectral utilization with traffic intensity. (P_d = 0.9, P_f = 0.1) for imperfect spectrum sensing.

reduces the spectral utilization of the proposed protocol. Further, the proposed framework outperforms for imperfect spectrum sensing as compared to that of the perfect spectrum sensing when traffic intensity is greater than 0.5. This happens because as depicted in Eq. (8.11), in the computation of spectrum utilization for perfect spectrum sensing, only *NC1* and *NC4* contribute; however, for imperfect spectrum sensing, *NC1, NC2, NC3,* and *NC4* contribute. *NC1* and *NC3* show the inversely and directly proportional relationship, respectively, with the traffic intensity. Therefore, when the traffic intensity goes greater than 0.5, *NC1* reduces and *NC3* increases, due to which the *SU* of the imperfect spectrum sensing increases as compared to that of the perfect spectrum sensing. The relationship between the resource wastage and traffic intensity for different values of the P_d and P_f is depicted in Figure 8.8. It is apparent that the more imperfections in the spectrum sensing phenomenon, higher will be the resource wastage due to underutilization of the idle channels.

Moreover, the resource wastage reduces linearly with increase in the traffic intensity and it happens because the increase in traffic intensity implies a reduced number of idle channels and resource wastage is defined for the actually idle channel which is sensed as active due to imperfect spectrum sensing. The variations of the data loss with the traffic intensity for several values of the P_d and P_f are presented in Figure 8.9. It is clear that the data loss linearly increases with increase in

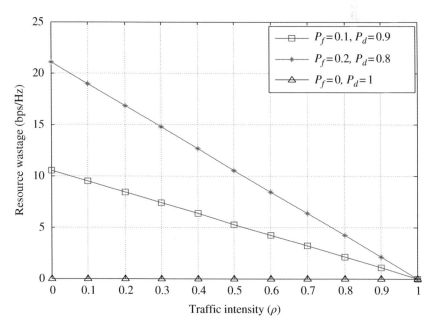

Figure 8.8 The resource wastage versus traffic intensity of PU.

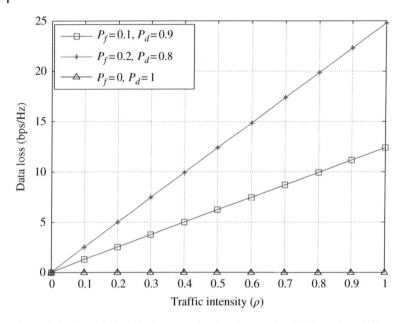

Figure 8.9 The relationship between the data loss and traffic intensity of PU.

the traffic intensity and it happens since large traffic intensity implies higher number of active channels and data loss is computed for the actually active channels which are sensed as idle.

In addition to this, the large-value imperfection metrics result in the higher data loss. The relationship between the interference efficiency and channel gain between CU transmitter and the PU receiver (h_{sp}) is depicted in Figure 8.10. It is perceptible that the interference efficiency decreases with the increase of h_{sp} since higher the channel gain, more the interference at the PU. In addition to this, the imperfection in spectrum sensing also affects the interference efficiency of the proposed system. The more imperfection implies the large value of probability of misdetection results in data transmission even in the presence of PU which introduces the interference at the PU and reduces the interference efficiency.

The variations of throughput efficiency with channel switching time for conventional SMC-MAC as well as proposed HSMC-MAC protocols with perfect and imperfect spectrum sensing are illustrated in Figure 8.11. It is apparent that the throughput efficiency decreases with increase in the channel switching time for the proposed HSMC-MAC protocols in both the cases, however, follows constant nature for the conventional approach. This constant nature appears since the role and effect of channel switching time is null. It is clear that the switching speed of the CU for data transmission mode and channel also affects the performance of the framework.

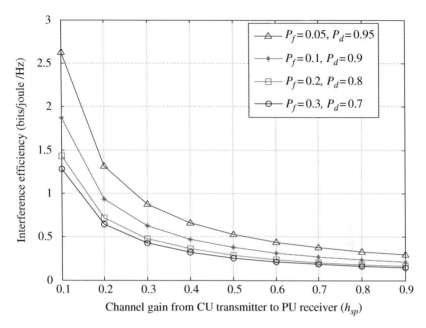

Figure 8.10 The variations of interference efficiency to the channels gains from CU transmitter to PU receiver.

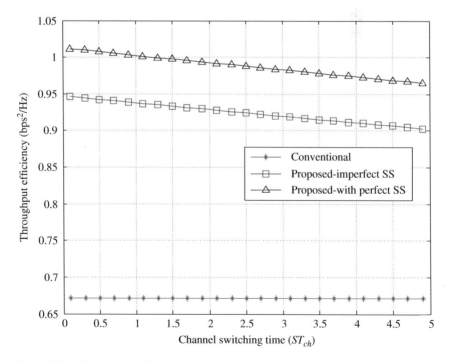

Figure 8.11 Throughput efficiency versus channel switching time ($\rho = 0.2$).

8.6 Summary

In this chapter, we have proposed an improved frame structure for the HSMC-MAC protocol in the DCRN to achieve the enhanced throughput and spectral utilization. The HSMC-MAC protocol is illustrated using improved frame structure for feasible DCRN by exploiting the perfect as well as imperfect spectrum sensing scenarios. The simulation results are presented for various scenarios of the traffic intensity of PU and SNR level due to underlay as well as interweave transmission, which are witnesses of significant improvement in the normalized throughput and spectral utilization of the DCRN. Further, the imperfections in spectrum sensing degrade the performance of proposed protocol by increasing the data loss and resource wastage as well as reduce the spectral utilization and interference efficiency. Further, the less switching speed of the CU degrades the performance of proposed HSMC-MAC. The optimization of various performance metrics is also a challenging and important issue which needs to be explored in future research works. Moreover, the energy efficiency is also a potential phenomenon which was well explored by us in [55].

References

1 Yucek, T. and Arslan, H. (2009). A survey of spectrum sensing algorithms for cognitive radio applications. *IEEE Commun. Surv. Tutorials* 11 (1): 116–130.

2 Ali, A. and Hamouda, W. (2015). Spectrum monitoring using energy ratio algorithm for OFDM-based cognitive radio networks. *IEEE Trans. Wireless Commun.* 14 (4): 2257–2268.

3 Akyildiz, I.F., Lo, B.F., and Balakrishnan, R. (2011). Cooperative spectrum sensing in cognitive radio networks: a survey. *Phys. Commun.* 4 (1): 40–62.

4 Khan, M.S., Gul, N., Kim, J. et al. (2020). A genetic algorithm-based soft decision fusion scheme in cognitive iot networks with malicious users. *Wireless Commun. Mobile Comput.* 2020: 1–10.

5 Pan, G., Li, J., and Lin, F. (2020). A cognitive radio spectrum sensing method for an ofdm signal based on deep learning and cycle spectrum. *Int. J. Digital Multimedia Broadcast.* 2020: 1–10.

6 Zaimbashi, A. (2020). Spectrum sensing in a calibrated multi-antenna cognitive radio: exact LRT approaches. *AEU – Int. J. Electron. Commun.* 113: 1–16.

7 Masonta, M.T., Mzyece, M., and Ntlatlapa, N. (2013). Spectrum decision in cognitive radio networks: a survey. *IEEE Commun. Surv. Tutorials* 15 (3): 1088–1107.

8 Wang, C.W. and Wang, L.C. (2009). Modeling and analysis for proactive-decision spectrum handoff in cognitive radio networks. *IEEE International Conference on Communications*, Dresden, Germany (14–18 June 2009), 1–6.

9 Cao, K. and Qian, P. (2020). Spectrum handoff based on DQN predictive decision for hybrid cognitive radio networks. *Sensors* 20 (4): 1–11.

10 Bandhari, S., Bandhari, M., and Joshi, N. (2019). Spectrum sharing in cognitive radio networks for 5G vision. *2nd International Conference on Advanced Computational and Communication Paradigms (ICACCP)*, Gangtok, India (25–28 February 2019), 1–6.

11 Thakur, P. and Singh, G. (2020). Power management for spectrum sharing in cognitive radio communication system: a comprehensive survey. *J. Electromag. Waves Appl.* 34 (4): 407–461.

12 Ahmad, W.S.H.M.W., Radzi, N.A.M., Samidi, F.S. et al. (2020). 5G technology: towards dynamic spectrum sharing using cognitive radio networks. *IEEE Access* 8: 14460–14488.

13 Thakur, P., Kumar, A., Pandit, S. et al. (2017). Advanced frame structures for hybrid spectrum access strategy in cognitive radio communication systems. *IEEE Commun. Lett.* 21 (2): 410–413.

14 Yang, C., Lou, W., Fu, Y. et al. (2016). On throughput maximization in multichannel cognitive radio networks via generalized access strategy. *IEEE Trans. Commun.* 64 (4): 1384–1398.

15 Christian, I., Moh, S., Chung, I., and Lee, J. (2012). Spectrum mobility in cognitive radio networks. *IEEE Commun. Mag.* 50 (6): 114–121.

16 Lee, W.Y. and Akyildiz, I.F. (2012). Spectrum-aware mobility management in cognitive radio cellular networks. *IEEE Trans. Mobile Comput.* 11 (4): 529–542.

17 Thakur, P., Kumar, A., Pandit, S. et al. (2017). Spectrum mobility in cognitive radio network using spectrum prediction and monitoring techniques. *Phys. Commun.* 24: 1–8.

18 Thomas, J. and Menon, P.P. (2017). A survey on spectrum handoff in cognitive radio networks. *2017 International Conference on Innovations in Information, Embedded and Communication Systems (ICIIECS)*, Coimbatore, India (17–18 March 2017), 1–4.

19 Salgado, C., Hernandez, C., Molina, V., and Beltran-Molina, F.A. (2016). Intelligent algorithm for spectrum mobility in cognitive wireless networks. *Procedia Comput. Sci.* 83: 278–283.

20 Yewada, P.S. and Dong, M.T. (2019). Intelligent process of spectrum handoff/ mobility in cognitive radio networks. *J. Electri. Comput. Eng.* 2019: 1–12.

21 Pandit, S. and Singh, G. (2015). Backoff algorithm in cognitive radio mac protocol for throughput enhancement. *IEEE Trans. Veh. Technol.* 64 (5): 1991–2000.

22 Cormio, C. and Chowdhury, K.R. (2009). A survey on MAC protocols for cognitive radio networks. *Ad Hoc Netw.* 7 (7): 1315–1329.

23 Jha, S.C., Phuyal, U., Rashid, M.M., and Bhargava, V.K. (2011). Design of OMC-MAC: an opportunistic multi-channel mac with qos provisioning for distributed cognitive radio networks. *IEEE Trans. Wireless Commun.* 10 (10): 3414–3425.

24 Kwon, S., Kim, B., and Roh, B.H. (2014). Preemptive opportunistic MAC protocol in distributed cognitive radio networks. *IEEE Commun. Lett.* 18 (7): 1155–1158.

25 Debroy, S., De, S., and Chatterjee, M. (2014). Contention based multichannel mac protocol for distributed cognitive radio networks. *IEEE Trans. Mobile Comput.* 13 (12): 2749–2762.

26 Lim, S. and Lee, T.-J. (2011). A self-scheduling multi-channel cognitive radio mac protocol based on cooperative communications. *IEICE Trans. Commun.* E94.B (6): 1657–1668.

27 Pandit, S. and Singh, G. (2017). *Spectrum Sharing in Cognitive Radio Networks: Medium Access Control Protocol Based Approach*. Springer International Publishing.

28 Wang, J., Wang, F., and Li, W.W. (2017). Strategic behavior and admission control of cognitive radio systems with imperfect sensing. *Comput. Commun.* 113: 53–61.

29 Jia, J., Zhang, Q., and Shen, X.S. (2008). HC-MAC: a hardware-constrained cognitive mac for efficient spectrum management. *IEEE J. Sel. Areas Commun.* 26 (1): 106–117.

30 Motamedi, A. and Bahai, A. (2007). MAC protocol design for spectrum-agile wireless networks: stochastic control approach. *2007 2nd IEEE International Symposium on New Frontiers in Dynamic Spectrum Access Networks*, Dublin, Ireland (17–20 April 2007), 448–451.

31 Hossain, M.A. and Sarkar, N.I. (2018). A distributed multichannel MAC protocol for rendezvous establishment in cognitive radio ad hoc networks. *Ad Hoc Netw.* 70: 44–60.

32 Hu, P. and Ibnkahla, M. (2014). A MAC protocol with mobility support in cognitive radio ad hoc networks: protocol design and analysis. *Ad Hoc Netw.* 17: 114–128.

33 Shah, G.A. and Akan, O.B. (2014). Performance analysis of CSMA-based opportunistic medium access protocol in cognitive radio sensor networks. *Ad Hoc Netw.* 15: 4–13.

34 Jung, B.C. and Lee, W. (2018). Performance analysis of opportunistic CSMA schemes in cognitive radio networks. *Wireless Netw.* 24 (3): 833–845.

35 Khanian, Z.B., Rasti, M., Salek, F., and Hossain, E. (2016). A distributed opportunistic MAC protocol for multichannel wireless networks. *IEEE Trans. Wireless Commun.* 15 (6): 4263–4276.

36 Chandra, K., Prasad, R.V., and Niemegeers, I. (2017). Performance analysis of IEEE 802.11ad MAC protocol. *IEEE Commun. Lett.* 21 (7): 1513–1516.

37 Yao, Y., Zhang, K., and Zhou, X. (2017). A flexible multi-channel coordination MAC protocol for vehicular ad hoc networks. *IEEE Commun. Lett.* 21 (6): 1305–1308.

38 Abeer, A., Hussin, S., and Fouad, M. (2020). Performance analysis of different channel allocation schemes of random access (RA) MAC protocol with back-off algorithm (BOA). *Wireless Pers. Commun.* 112: 1981–1993.

39 Karaca, H.M. (2020). Throughput maximization of multichannel allocation mechanism under interference constraint for hybrid overlay/underlay cognitive radio networks with energy harvesting. *Wireless Netw.* 26: 3905–3928.

40 Boukerche, A. and Zhou, X. (2020). A novel hybrid MAC protocol for sustainable delay-tolerant wireless sensor network. *IEEE Trans. Sustainable Comput.* https://doi.org/10.1109/TSUSC.2020.2973701.

41 Hoang, A.T., Liang, Y.C., Wong, D.T.C. et al. (2009). Opportunistic spectrum access for energy-constrained cognitive radios. *IEEE Trans. Wireless Commun.* 8 (3): 1206–1211.

42 Thakur, P., Singh, G., and Satasia, S.N. (2016). Spectrum sharing in cognitive radio communication system using power constraints: a technical review. *Perspect. Sci.* 8: 651–653.

43 Bharti, B. and Singh, G. (2017). Analysis of capacity limits over fading environment with imperfect channel state information for cognitive radio network. *Ann. Telecommun.* 72 (7–8): 469–482.

44 Yang, W. and Zhao, X. (2017). Robust resource allocation for orthogonal frequency division multiplexing-based cooperative cognitive radio networks with imperfect channel state information. *IET Commun.* 11 (2): 273–281.

45 Zhang, S., Zhao, H., Wang, S., and Hafid, A.S. (2017). Impact of access contention on cooperative sensing optimisation in cognitive radio networks. *IET Commun.* 11 (1): 94–103.

46 Sharma, S.K., Bogale, T.E., Le, L.B. et al. (2018). Dynamic spectrum sharing in 5G wireless networks with full-duplex technology: recent advances and research challenges. *IEEE Commun. Surv. Tutorials* 20 (1): 674–707.

47 Ozfatura, M.E., ElAzzouni, S., Ercetin, O., and ElBatt, T. (2018). Optimal throughput performance in full-duplex relay assisted cognitive networks. *Wireless Netw.* 25: 1–17.

48 Thakur, P., Kumar, A., Pandit, S. et al. (2019). Performance analysis of cooperative spectrum monitoring in cognitive radio network. *Wireless Netw.* 25: 989–997.

49 Kay, S.M. (1998). *Fundamentals of Statistical Signal Processing: Detection Theory.* Prentice-Hall PTR.

50 Mili, M.R. and Musavian, L. (2017). Interference efficiency: a new metric to analyze the performance of cognitive radio networks. *IEEE Trans. Wireless Commun.* 16 (4): 2123–2138.

51 Thakur, P., Kumar, A., Pandit, S. et al. (2018). Spectrum monitoring in heterogeneous cognitive radio network: how to cooperate? *IET Commun.* 12 (17): 2110–2118.

52 MathWorks introduceert release 2015b van de MATLAB en Simulink productseries [Online]. https://nl.mathworks.com/company/newsroom/mathworks-announces-release-2015b-of-the-matlab-and-simulink-product-families.html (accessed 19 February 2018).

53 IEEE-SA – Contact Us. http://standards.ieee.org/contact/form.html (accessed 18 February 2018).

54 Mody, A., Saha, A., Reede, I. et al. IEEE 802.22/802.22.3 cognitive radio standards: theory to implementation. https://link.springer.com/content/pdf/10.1007%2F978-981-10-1394-2_54.pdf (accessed 08 August 2020).

55 Thakur, P. and Singh, G. (2020). Performance analysis of MIMO-based CR–NOMA communication systems. *IET Commun.* 14 (16): 2676–2686.

9

Frameworks of Non-Orthogonal Multiple Access Techniques in Cognitive Radio Networks

9.1 Introduction

The next-generation communication systems demand huge spectral and energy efficiency, low latency, high scalability, improved connectivity and reliability, as well as advanced security due to massive growth in the wireless connected devices such as internet-of-things (IoT), wireless sensor networks (WSNs), wireless body area networks (WBANs), etc. The prominent and effective approaches to fulfill these demands are the cognitive radio (CR), non-orthogonal multiple access (NOMA), multi-antennas or multiple-input-multiple-output (MIMO), cooperative communication (CC), network function virtualization (NFV), millimeter-wave communication (MMC), ultra-classification, etc. [1]. As we have discussed in Chapter 1, that the CR is a well-explored technique in literature to manage the issue of spectrum scarcity by exploiting the unutilized spectrum opportunities whose basics as well as advances are discussed in detail in [2–14]. However, the NOMA is a recently explored technique for spectral efficient communication by creating the new spectrum opportunities and the briefs about NOMA technologies are given as follows. The NOMA is an emerging technique for the 5G communication systems in order to fulfill the demand for ultrahigh data rate and low latency. In literature, numerous techniques are reported for the feasibility of NOMA mechanism which are conventionally categorized as power domain multiplexing (PDM) and code domain multiplexing (CDM) [15–17] which further includes power division multiple access (PDMA), sparse code multiple access (SCMA), low density spreading – code division multiple access (LDS-CDMA), and pattern division multiple access (PTDMA). The PDMA is popular technique which is explored more for NOMA as compared to all other techniques [16]; therefore, we have chosen this for further discussion. The PDMA is a potential multiple access technique which relies on the two prerequisites at transmitter and receiver, namely, the superposition coding (SC) and successive interference cancellation (SIC), respectively. The SC is a technique which allows simultaneous transmission

Spectrum Sharing in Cognitive Radio Networks: Towards Highly Connected Environments,
First Edition. Prabhat Thakur and Ghanshyam Singh.

of multiple users' information on the same channel from the transmitter end and the reception at receivers is performed using SIC technique.

9.1.1 Related Work

The multiple access (MA) is a very important as well as integral part for any generation of communication systems among which the popular ones are time division multiple access (TDMA), frequency division multiple access (FDMA), CDMA, orthogonal frequency division multiple access (OFDMA), etc., till the fourth generation (4G). The interesting point which needs to mention here is that till 4G, the MA techniques used are orthogonal multiple access (OMA). However, for the next-generation communication systems, researchers are seeking for the NOMA which is recently explored. Therefore, the MA techniques are generally classified as OMA and NOMA which are further classified as shown in Figure 9.1. The OMA technique has mainly three variants that are single carrier orthogonal multiple access (SCOMA), TDMA, and OFDMA. On the other hand, as discussed in the previous section, the NOMA has different techniques such as PDMA, multi-user shared access (MUSA), SCMA, PTDMA, and resource shared multiple access (RSMA). Various researchers have presented the potential review and tutorial articles to explain the concepts of NOMA over OMA and its various possible varieties [16–21]. Islam et al. [16] have presented a review which primarily focused on power-domain NOMA that utilizes SC at the transmitter and SIC at the receiver.

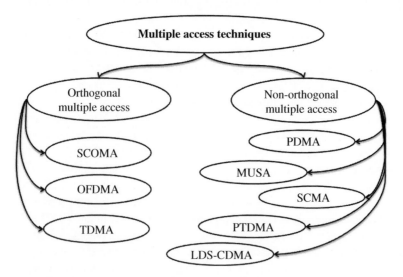

Figure 9.1   Classification of multiple access techniques.

The authors have presented the recent progress of NOMA in 5G systems where the key intent is on the state-of-the-art capacity analysis, power allocation strategies, user fairness, and user-pairing schemes in NOMA. Further, the interplay of NOMA with other existing 5G techniques such as CC, MIMO, beam-forming, space-time coding, and network coding among users are illustrated. In addition, several important issues on NOMA implementation and some avenues for future research are highlighted. Ding et al. [17] have presented a review on NOMA techniques where these techniques are classified as a single carrier (power domain-NOMA (PD-NOMA) and CR-NOMA) and multi-carrier (LDS, SCMA, and PDMA), MIMO-NOMA, cooperative-NOMA, and millimeter-wave NOMA. Further, the practical implementation challenges for NOMA such as imperfect channel state information (CSI), cross-layer resource allocation, and coding and modulation for NOMA are discussed. In addition, the future research challenges such as simultaneous wireless information and power transfer (SWIPT)-NOMA, CR-NOMA, and security in NOMA are also explored. Liu et al. [18] have provided a comprehensive survey of the state-of-the-art PDM-aided NOMA, with a focus on the theoretical NOMA principles, multiple-antenna-aided NOMA design, on the interplay between NOMA and cooperative transmission, on the resource control of NOMA, and on the coexistence of NOMA with other emerging potential 5G techniques. The key intent is on the superiority of the power-domain multiplexing NOMA compared to other NOMA techniques. Further, the research challenges of existing NOMA techniques with their potential solutions are discussed. Yunzheng et al. [19] have presented the briefs about the various existing NOMA technique including SCMA, MUSA, PTDMA, and some key waveforms including filter-bank-based multicarrier (FBMC), universal filtered multi-carrier (UFMC), and generalized frequency division multiplexing (GFDM). The potential challenges and research directions in the field of NOMA are presented. Wang et al. [20] have presented the briefs about various NOMA techniques and have highlighted the challenges for the standardization of NOMA techniques by the International Telecommunication Union (ITU).

Wani et al. [21] have proposed a downlink multiuser NOMA with full and partial CSI feedback where a beam design and user clustering from the throughput-fairness trade-off perspective. Further, to enhance this trade-off, two proportional fairness (PF)-based scheduling algorithms are proposed; each has two stages. The first algorithm is based on integrating the maximum product of effective channel gains and the maximum signal to interference ratio with the PF principle (PF-MPECG-SIR), to select the strong users in the first stage and the weak users in the second stage. In [22], Sena et al. have presented a multiuser multi-cluster massive MIMO system with NOMA where in the downlink scenario, the impact of imperfect SIC is carried out. Further, the authors have derived the closed-form expressions for the outage probability and ergodic rates. Qiu et al. [23] have presented a NOMA system with partial CSI for downlink and uplink transmission in mobile scenarios where the users are deployed randomly with their random movement around the base station (BS). The authors have proposed dynamic power allocation (DPA) for downlink NOMA and dynamic power control (DPC) for uplink NOMA scenarios where the outage performance is optimized by using the distance information. In [24], Saggese et al. have exploited a resource allocation problem for multiuser MIMO-NOMA (MU-MIMO-NOMA) downlink transmissions where the users are organized in clusters of strong/weak pair. The objective of the proposed system is to find an optimal clustering, beamforming, and power allocation scheme in order to minimize the power transmitted. Nguyen et al. [25] have presented a comprehensive survey where the key emphasis is on the innovative approaches for NOMA in terms of the spectral efficiency and energy efficiency. Further, the emerging technologies involved with NOMAs with the potential research challenges and future research directions are described. Abrardo et al. [26] have investigated the problem of power and channel allocation for multicarrier NOMA full duplex systems. It is revealed that for such systems, the allocation task is a non-convex and extremely challenging problem due to multiple interfering users transmitting over the same channel. To resolve this issue, the authors have exploited the lock-coordinated descent approach and have proposed two algorithms based on the decomposition of the original allocation problem in lower-complexity subproblems, whose solutions can be find using the Lagrangian dual domain method. In [27], Maraqa et al. have presented a comprehensive survey for the integration of power-domain-NOMA with the enabling communications technologies in order to fulfill the demands of next-generation communication systems. In this survey, the authors have emphasized over the different rate optimization scenarios when power-domain-NOMA is combined with other communication technologies such as MISO, MIMO, massive-MIMO, advanced antenna architectures, mmWave and THz, Coordinated Multi Point, CCs, CR, VLC, UAV, and others.

The CR is well-explored technique in the literature [2–7]; however, the practical implementation and reliable communication is still a challenging issue. The recent literatures which took one step ahead toward the practical implementation and reliable communication of PU and CU are illustrated as follows. Wang et al. [28] have presented the most feasible framework for spectrum management in wireless networks using the software-defined networking. The authors have illustrated the design principles and key challenges in realizing the software-defined wireless networking (SDWN)-enabled spectrum management architecture. By considering these principle and design challenges, the authors have developed a general architecture with a new baseband virtualization design. Further, a prototype is designed that seamlessly integrates with the IEEE 802.11 protocol stack and commodity RF front-end. It is reported that the proposed architecture increases the spectrum efficiency significantly.

9.1.2 Motivation

Since the CR and NOMA support the concept of spectral efficiency, which is major concern for the next-generation communication systems, therefore, simultaneous exploitation of the CR and NOMA in order to develop the more spectral efficient systems suitable for next-generation communication systems is the potential demand. The comparison of review articles presented in the literature is tabulated in Table 9.1. From the discussion in Section 9.1.1, it is perceptible that Ding and Liu [17, 18] have discussed the concept of CR-NOMA in brief where the CR is used to select the power allocation strategy in the NOMA. However, the spectral efficient nature of both the techniques, i.e. CR and NOMA, encourages exploring the simultaneous exploitation of the CR and NOMA to achieve higher spectral efficiency. Therefore, in this chapter, we have proposed the potential CR-NOMA frameworks with the implementation challenges and future research directions.

9.1.3 Organization

The rest of the chapter is structured as follows. In Section 9.2, the spectrum accessing strategies in CR are presented. Section 9.3 comprises the fundamentals of NOMA. In Section 9.4, we have presented the prominent frameworks of NOMA in CR for uplink and downlink scenarios. Section 9.5 comprises the challenging issues in CR-NOMA frameworks and their recommended solutions. Finally, Section 9.6 illustrates the summary of the chapter.

9.2 CR Spectrum Accessing Strategies

The spectrum accessing techniques are exploited in the CR communication to avoid the interference at PU receiver due to CU communication. The spectrum accessing techniques are categorized as follows: (i) Interweave spectrum access, (ii) Underlay spectrum access, (iii) Hybrid spectrum access, and (iv) Overlay spectrum access [29]. In the interweave spectrum access, the CU performs spectrum sensing to detect the idle channels/bands of the spectrum known as spectrum holes/white spaces and starts data transmission with full power on the suitable idle channel [30]. However, if all the channels are detected as active, then CU needs to stop its data transmission which is a major milestone for seamless communication. Therefore, in order to achieve the seamless communication, the underlay

Table 9.1 Comparison of various multiple accessing strategies.

References	NOMA technique	Prerequisites	Performance analysis using	Integration of NOMA with other techniques	NOMA+ CR	Partial or imperfect CSI	Remarks
Islam et al. [16]	PD-NOMA	SC and SIC techniques	Capacity analysis, power allocation strategies, user fairness, and user-pairing schemes	COPC, MIMO, beamforming, space-time coding, and network coding among users	No	Yes	Perfect SC at the transmitter and error-free SIC at the receiver, optimum power allocation, QoS-oriented user fairness, appropriate user pairing, and good link adaptation are also required to obtain the maximum benefits offered by NOMA.
Ding et al. [17]	PDMA, LDS, SCMA	SC and SIC techniques	—	CC, SWIPT, CR, millimeter wave	Yes (Brief)	Yes	NOMA is an enabling technology to achieve high throughput, low latency, and massive connectivity.
Liu et al. [18]	PDM	SC and SIC techniques	—	Multiple antennae, CC, SWIPT, CR, MMC	Yes (Brief)	Yes´	Highlights the main advantages of power-domain multiplexing NOMA compared to other existing NOMA techniques.

Reference	Techniques	Method	Parameters				Description
Yunzheng et al. [19]	SCMA, MUSA, PTDMA and FBMC, UFMC, GFDM.	SC and SIC techniques	—	No	No	No	Have discussed various non-orthogonal multiple access and non-orthogonal modulation techniques.
Wang et al. [28]	PD-NOMA, SCMA, PTDMA, MUSA	SC and SIC techniques	Sum throughput, Target SNR, complexity	No	No	No	Presented the relationship among OMA, PD-NOMA, SCMA, PDMA, and MUSA. Illustrates the progress and challenges on standardization.
Proposed	PD-NOMA, CR-PD-NOMA	SC and SIC techniques	Sum throughput of user-1 and user-2	CR	Yes (in detail with proposed frameworks)	Yes	The simultaneous use of CR and NOMA results in the improved spectral efficiency and allows more users to access the same spectrum.

spectrum access technique is proposed by the researchers in which the CU transmits the data parallel to the PU on the same channel, time, and space [31]. The interference from CU communication to PU is avoided by constraining the transmitted power so that it does not interfere with the PU receiver. However, the major limitation of the underlay spectrum access technique is the limited channel capacity due to constrained power transmission. As both the techniques have specific inadequacies, therefore, in order to avoid these inadequacies and to enjoy the adequacies of both the techniques, the hybrid spectrum accessing technique is explored by several researchers [32–35]. In this technique, the CU accesses the idle sensed channels using an interweave spectrum access technique, i.e. data transmission with full power whereas the active sensed channels are accessed via the underlay spectrum access technique, i.e. data transmission with the constrained-power transmission. In [32], the authors have presented the capacity analysis of interweave, underlay, as well as a combination of both (hybrid) spectrum accessing techniques and reported that the hybrid technique outperforms both the techniques. In addition to this, a simple power allocation scheme for the hybrid spectrum access strategy is illustrated and claims that its achieved capacity is very close to the maximum achievable capacity of the CU. Jiang et al. [33] have proposed a hybrid technique which combines the interweave and underlay spectrum access schemes by exploiting a double-threshold energy detection method. Moreover, a Markov chain model is introduced to derive the performance metrics for the proposed technique. In [34], the authors have proposed a hybrid transmission system that exploits both the interweave and underlay techniques using multicarrier code-division multiple accesses due to its interference avoidance capability. To the best of authors' knowledge, the MAC protocols are not exploited with hybrid spectrum accessing techniques in order to analyze the performance of CR networks. From the above discussion, it is palpable that the spectrum accessing strategy is an integral part of the CR communication systems where the power allocation to the CU has a very key role whether the full power or constrained power. In addition to the interweave, underlay, and hybrid spectrum accessing strategies, the recently developed strategy is the overlay in which the CU and PU access the channel simultaneously with full power; however, the interference cancellation with each other is achieved by using the advanced encoding schemes such as dirty-paper coding and Gelfand-Pinsker binning [35]. Due to the requirement of such advance encoding techniques, the overlay technique shows complex nature due to which it is less explored. The comparison of different spectrum accessing strategy is illustrated in Table 9.2. Since the interweave and underlay spectrum accessing strategies seem like the primary spectrum accessing strategies, it is worth analyzing the effect of NOMA implementation in the CRNs for these two strategies. Therefore, in Section 9.4, we have presented the proposed CR-NOMA frameworks.

Table 9.2 Cognitive radio spectrum accessing strategies.

Spectrum accessing strategy	Simultaneous transmission of CU and PU	Prerequisite	Constraints	Interference management	NOMA-CR in literature ref.
Interweave	Not allowed	Spectrum sensing to perceive the idle channels.	—	Interference controlling	Proposed
Underlay	Allowed	Interference power tolerable limit of PU	Power at PU receiver due to CU transmission needs to be below the interference limit.	Interference avoiding	[30, 31], Proposed
Hybrid	Allowed	Spectrum sensing, Interference power tolerable limit of PU	Power at PU receiver due to CU transmission needs to be below the interference limit tolerable by the CU.	Interference controlling and Interference avoiding	—
Overlay	Allowed	Advanced interference cancellation techniques are required at PU and CU.	—	Interference mitigating	—

9.3 Functions of NOMA System for Uplink and Downlink Scenarios

The PDMA is a potential NOMA technique which is explored in the literature [15, 36, 37] and relies on the SC at the transmitter end and SIC at the receiving end. In this section, both the techniques are presented and their implementation in the downlink, as well as uplink scenarios, is illustrated as follows. The SC is a technique which allows simultaneous transmission of information from the transmitter to multiple receivers at the same frequency/code with different power levels. The transmitter transmits a signal S_i with power P_i for ith user ($i = 1, 2, 3 \ldots$ N), where $E[|S_i|^2] = 1 E[|S_i|^2] = 1$, $\sum_{i=1}^{N} P_i = P$ and signals are superposition coded to formulate the transmitted signal as:

$$x = \sqrt{P_1}S_1 + \sqrt{P_2}S_2 + \sqrt{P_3}S_3 \ldots \sqrt{P_i}S_i \ldots \sqrt{P_N}S_N. \tag{9.1}$$

The received signal at the ith user is represented as:

$$y_i = h_i x + w_i, \tag{9.2}$$

where h_i is the complex channel gain between the ith user and transmitter, and w_i denotes the additive white Gaussian noise (AWGN) including intercell interference at the ith user receiver.

Further, the SIC is a technique which is employed at the receiver in order to decode the superimposed signals at the particular receiver. The key approach exploited in the SIC is that the users are successively decoded which means initially one users' signal is decoded and after that, it is subtracted from the combined signal before to decode the next user. In order to execute the process of SIC, initially, the users are sorted in descending order of their signal strengths so that the receiver can decode the high-power signal first, subtract it from the combined signal, and isolate the weaker signal from the residue [14]. The complete detail of SIC and representation using constellation diagrams are presented in [15, 36]. Further, the downlink and uplink scenarios for cellular-NOMA are discussed as follows.

9.3.1 Downlink Scenario for Cellular-NOMA

In the downlink scenario for cellular-NOMA, the BS serves the number of users (subject to the number of channels) in their cellular boundaries. However, at a particular channel, the BS broadcasts the information in order to serve the number of users (N) with different information. The channel gains from the BS to the ith users is h_{Di} where $i = 1, 2 \ldots N$ and $|h_{D1}|^2 > |h_{D2}|^2 > |h_{D3}|^2 > \ldots > |h_{Di}|^2 \ldots > |h_{DN}|^2$. The power allocated to the ith user is P_i such that $P_1 < P_2 \ldots < P_i \ldots < P_N$. The process of the signal detection signal at the ith user in downlink scenario proceeds as follows.

- Decode the signal for Nth user (S_{Nd}) from the received signal y_i by assuming signals from all another user as interference.
- Subtract S_{Nd} from the y_i and yield the resultant signal without S_{Nd} (S_{wN}).
- Decode the signal for $(N-1)$th user ($S_{(N-1)d}$) from the signal S_{we} by considering signals from all another user, i.e. S_i where $i < N-1$ as interference.
- Subtract $S_{(N-1)d}$ from the S_{wN} and yield the resultant signal without S_{Nd} and $S_{(N-1)d}$ ($S_{w(N-1)}$).
- This process continues till the decoding of ith user.

The phenomenon of downlink scenario is depicted as shown in Figure 9.2 for $N = 2$ in Eq. (9.1). In the downlink scenario, the BS serves as a transmitter and the information of all users are superposition coded which results in the transmission of that superimposed signal at same time/frequency/code. The assumed bound in the channel power gains due to considered set up is: $|h_{D1}|^2 > |h_{D2}|^2$, where subscript D is used for downlink metrics. Moreover, the process of decoding the signal at the receivers is shown in Figure 9.2 and proceeds as follows. The users with poor channel conditions (user-2) decode their information by assuming the signals from other users as interference whereas the users with strong channel conditions (user-1), initially, decode the information signal of the partner user and then yield their own signal by subtracting the partners' signal from the combined received signal.

Since the user-1 performs user-2 detection first, then decode its own signal by subtracting user-2 signal from the combined received signal y_{D1}, and due to this, the signal from the user-2 does not interfere with user-1, the throughput of the user-1 is:

$$R_{DN1} = B_D.\log_2\left(1 + {}^{P_1|h_{D1}|^2}\!/\!_{N_{01}}\right). \tag{9.3}$$

On the other hand, at the user-2, the SIC is implemented by treating the signal from user-1 as interference; therefore, the throughput of the user-2 is:

$$R_{DN2} = B_D.\log_2\left(1 + {}^{P_2|h_{D2}|^2}\!/\!_{P_1|h_{D1}|^2 + N_{02}}\right), \tag{9.4}$$

where N_{01} and N_{02} are the noise powers at user-1 and user-2, respectively. B_D denotes the bandwidth of downlink channel. Moreover, in the downlink scenario with OMA implementation using FDMA, the bandwidth must be divided into two parts, i.e. B_{D1} for user-1 and B_{D2} for user-2 such that $B_{D1} + B_{D2} = B_D$. Since the individual channels are allocated to every user in order to avoid interference to each other, the throughput of user-1 and user-2 is derived as:

$$R_{DO1} = B_{D1}.\log_2\left(1 + {}^{P_1|h_{D1}|^2}\!/\!_{N_{01}}\right) \tag{9.5}$$

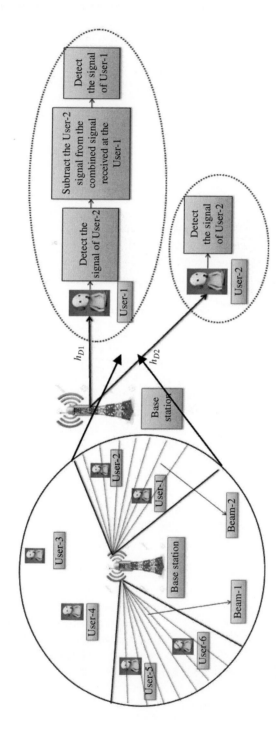

Figure 9.2 The schematic of cellular-NOMA for downlink scenario.

and

$$R_{DO2} = B_{D2}.\log_2\left(1 + P_2|h_{D2}|^2\big/N_{02}\right).\tag{9.6}$$

9.3.2 Uplink Scenario for Cellular-NOMA

In the uplink scenario for cellular-NOMA, all the users transmit their information signal S_{Ui} on the same channel/time/code toward the BS through a channel having gain coefficient h_i; however, the power is allocated to the users in two ways: either full power allocation or controlled power. In the controlled power strategy, the total power P_{Ui} is allocated to ith user on such a way that $\sum_{i=1}^{N} P_{Ui} = P_U$, where P_U is the maximum power that can be transmitted from the BS in order to avoid the interference with the adjacent cells and $P_{U1} = P_{U2}... = P_{UN}$. In the uplink scenario, the BS receives the superimposed signal of all different users.

Therefore, the complete received signal y_U at the BS is the sum of all received signals due to all the transmitted signals which are represented as:

$$y_U = \sqrt{P_{U1}}h_{U1}S_{U1} + \sqrt{P_{U2}}h_{U2}S_{U2}....\sqrt{P_{UN}}h_{UN}S_{UN} + W,\tag{9.7}$$

where W is the AWGN with noise power N_0 at the BS receiver. Further, at the receiving end, the BS must have SIC ability to extract information of each user. The process of users' signal detection is performed as follows:

- Arrange the users (U) according to their received signal strengths at the BS, i.e. $U1 > U2 > U3... UN$.
- Decode the signal for first user (S_{1d}) from the received signal y_U as given in Eq. (9.7) by assuming signals from all other users as interference.
- Now, subtract the decoded signal S_{1d} from the received signal y_U, i.e. $y_{s2} = y_U - S_{1d}$.
- Decode the signal S_{2d} for user 2 from the signal y_{s2} by considering signals from the user having signal strength lower than $U2$.
- This process continues till the detection or decoding of the signal for all users at the BS.

Let us assume the similar case as considered for the downlink scenario, i.e. $N = 2$, as shown in Figure 9.3 the achieved data rates for the user-1 and user-2 are computed as follows. The P_1 and P_2 are assumed to be equal, however, $h_{U1} > h_{U2}$. The received signal from user-1 has high signal strength as compared to that of the user-2. Therefore, at the BS, initially, the signal of user-1 is detected by assuming the signal of user-2 as interference. Further, a resultant signal is yielded by subtracting the user-1s' signal from the combined received signal and the resultant signal is exploited to decode the signal of user-2. Thus, it is apparent that the

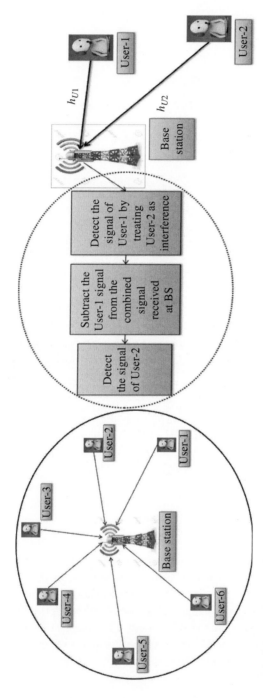

Figure 9.3 The schematic of cellular-NOMA for the uplink scenario.

user-1 gets interference due to user-2; however, the user-2 enjoys the interference-free communication. Therefore, the data rates of user-1 and user-2 are as follows:

$$R_{UN1} = B_U . log_2\left(1 + {P_1|h_{U1}|^2}\big/{\left(P_2|h_{U2}|^2 + N_0\right)}\right) \tag{9.8}$$

and

$$R_{UN1} = B_U . log_2\left(1 + {P_2|h_{U2}|^2}\big/{N_0}\right), \tag{9.9}$$

where B_U is the total bandwidth dedicated to the uplink channel. In the uplink scenario for conventional OMA technique, the bandwidth B_{U1} and B_{U2} are dedicated to the user-1 and user-2 under the constraint $B_{U1} + B_{U2} = B_U$. Therefore, the achieved data rates for the user-1 and user-2 are presented as follows:

$$R_{UO1} = B_{U1} . log_2\left(1 + {P_{U1}|h_{U1}|^2}\big/{N_0}\right) \tag{9.10}$$

and

$$R_{UO2} = B_{U2} . log_2\left(1 + {P_{U2}|h_{U2}|^2}\big/{N_0}\right). \tag{9.11}$$

9.4 Proposed Frameworks of CR with NOMA

It is well known that the CR and NOMA are promising techniques for the next-generation communication systems in order to improve the spectral efficiency. Therefore, recently, researchers have focused on the analysis of simultaneous exploitation of CR and NOMA for the next-generation communication systems and known as CR-NOMA [38, 39]. The existence of CR-NOMA is a challenging issue due to the requirement of interference avoidance technique for NOMA and simultaneous accessing of the same channel for the CU and PUs and in this section, we have proposed two potential frameworks for the CR-NOMA systems.

9.4.1 Framework-1

The framework relies on the underlay spectrum accessing strategy where the CU exploits the spectrum of PU in the presence of PU; however, the power-controlled transmission is achieved via NOMA technique. In the Framework-1, a network-cell comprises the PUs and a BS is considered where the BS uses three antennas to serve the PUs in three different sectors as shown in Figure 9.4a. The CU is allowed to use the spectrum of PUs simultaneously using NOMA techniques as depicted in Figure 9.4b. In every sector, the PU and CU can share the spectrum by employing the NOMA technique and selection of the user for SIC relies on the channel conditions. If we consider only the path loss for channel condition, the far user gets interference due to the information signal of the near user; therefore, the far user decodes its signal by considering the near user as interference. This framework imposes the prerequisite of SIC supporting nature for the PUs

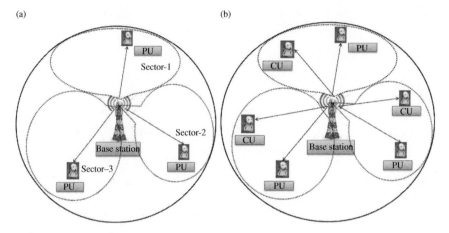

Figure 9.4 Proposed framework-1 for (a) PU network and (b) CU and PU network with NOMA.

and SC ability at the BS. In addition to this, the CR transmitting unit is assumed to be at the BS where both the transmitter (CR and PU) cooperate with each other to formulate the superimposed transmitted signal having information of both the users. In general, this framework violated the fundamental concept of the CR which is as follows. The CU must establish its communication using PUs' spectrum in such a way that its structure and communication remains impervious; however for the next-generation communication system where the spectral efficiency is the prime objective, the proposed framework is a suitable option. The Framework-1 is suitable for the underlay CR networks where the CU is allowed to access the channel of PU subject to avoid the interference at PU and the power allocation plays an important role for this. The implementation of NOMA in underlay CRN strengthens the power allocation strategies and interference management. In addition to this, the interference management allows the CU to transmit the signal with more power which results in the improvement in throughput of CU. The data rate or capacity computation for each sector is similar as discussed in the downlink and uplink scenarios in Section 9.3.

9.4.2 Framework-2

The Framework-2 relies on the interweave spectrum access strategy where CU perceives the idle spectrum through spectrum sensing and have implemented the NOMA technique to support more than one CUs on that spectrum. In the Framework-2, we have assumed a cell for the CU communication where a BS serves the CUs and it is assumed that within that cell there is no PU receiver; however, some CU may experience interference due to the PU transmission as shown in Figure 9.5a. The BS serves CU in the circular cell using three different sectors

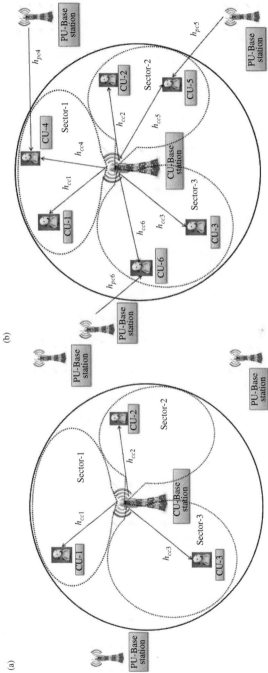

Figure 9.5 Proposed framework-2 CR network with (a) OMA and (b) NOMA.

where the single frequency channel can support only one user at a time in one sector. On the other hand, if the NOMA technique is employed in the CR systems, the number of CU supported in the cell can be increased with their sum rate as depicted in Figure 9.5b. Moreover, some of CUs that lie on the edge of the cell get interference from the PU signals which affects the signal-to-interference-plus-noise ratio (SINR) at the CUs. Therefore, the data rate of CU-1 and CU-4 in Figure 9.5b is:

$$R_{D1N} = B_D . \log_2 \left(1 + {P_1 |h_{cc1}|^2}/{N_{01}}\right) \tag{9.12}$$

and

$$R_{D4N} = B_D . \log_2 \left(1 + {P_2 |h_{D2}|^2}/{P_1 |h_{D1}|^2 + P_{PU} |h_{pc4}|^2 + N_{02}}\right), \tag{9.13}$$

where h_{cci} is the channel gain coefficient between the CUs' in a BS and ith CU in the network, P_1 and P_2 are the powers allocated to CU-1 and CU-4, respectively, $h_{cc1} > h_{cc4}$, due to which $P_2 > P_1$. h_{pci} denotes the channel gain coefficient between the PU BS and ith CU in the network and P_{PU} is the power transmitted by the PU BS.

On the other hand, if two cognitive users need to be served in the same sector without using NOMA, the bandwidth must be divided into two parts, i.e. B_{D1} and B_{D2}, where $B_{D1} + B_{D2} = B_D$. Thus, the data rates of user-1 and user-4 are:

$$R_{D1O} = B_{D1} . \log_2 \left(1 + {P_1 |h_{cc1}|^2}/{N_{01}}\right) \tag{9.14}$$

and

$$R_{D4O} = B_{D2} . \log_2 \left(1 + {P_2 |h_{D2}|^2}/{P_{PU} |h_{pc4}|^2 + N_{02}}\right). \tag{9.15}$$

9.5   Simulation Environment and Results

We have considered the downlink scenario in order to analyze the effect of NOMA, OMA, CR-NOMA (Framework-2), and CR-OMA (Framework-2) techniques on the data rates of the user-1 and user-2. In the simulated environment, the user-1 and user-2 are considered static and mobile, respectively. Therefore, due to static nature of the user-1, the channels' gain $h_{D1} = 0.9$ is assumed to be constant; however, h_{D2} varies due to mobile nature of user-2. The total bandwidth of the channel is assumed to be a unity which is divided into two parts in case of OMA; however, complete bandwidth is assigned to both the users in case of NOMA. The total power is taken as 10 W whereas the power allocated to user-1 and user-2 varies

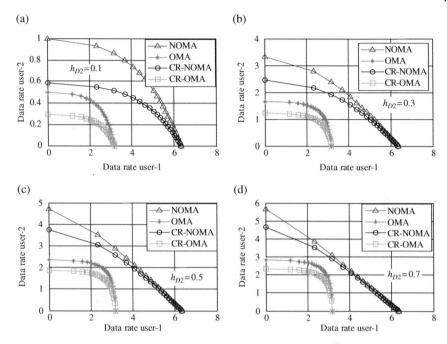

Figure 9.6 The variations of the data rate of user-2 (bits/second) with data rate of user-1 (bits/second) in the downlink scenarios.

between 0 and 10 and vice-versa, respectively. The noise power at both the users is assumed to be 0.1 W. From the presented results in Figure 9.6, it is apparent that there is a significant improvement in the data rate of user-2 for the NOMA technique as compared to that of the OMA technique; however, this improvement decreases with increase in the value of h_{D2} which validates the concept of NOMA.

The fundamental concept of NOMA is that it performs well as compared to that of the OMA if the channel gain difference between two users is high since it improves the performance of the detection using SIC [7]. Moreover, the data rate achieved in case of CR-OMA and CR-NOMA are less as compared to that of the conventional OMA and NOMA data rates and it is due to additional interference at user-2 due to PU communication in Framework-2. However, it is worth mentioning that the effect of cognitive concept on the data rates is not considered here and achieved results are obtained for the Framework-2 in which the spectrum holes are already detected and a user is using that spectrum hole. The proposed CR-framework (Framework-2) supports three users on the same channel in one sector; however, the conventional OMA supports only two users. Conventionally, the CR-NOMA sum-data rates are more as compared to that of the conventional NOMA technique but due to aforementioned reason, its data rates seem less in Figure 9.5. Moreover, the results of CR-NOMA Framework-1 will be similar to that of the NOMA technique because in this case, entire framework is similar to NOMA except the user names since user-1 and user-2 are replaced with CU and PU; however, Framework-1 supports only two users.

9.6 Research Potentials for NOMA and CR-NOMA Implementations

The proposed frameworks are recommended as potential solutions to implement the NOMA techniques in the CR communication systems. However, the implementation of CR-NOMA technique is in the infant stage and still need to be explored to achieve feasible systems by exploiting the following challenging issues that are depicted in Figure 9.7.

9.6.1 Imperfect CSI

The NOMA technique relies on the power allocation to the users on the basis of channel conditions which impose a bound on the transmitter side to obtain the CSI. However, it is very difficult to obtain the perfect CSI at the transmitter side due to random nature (uncertainty) of the channel. By considering this critical issue, Yang et al. [40] have investigated the performance of the NOMA over

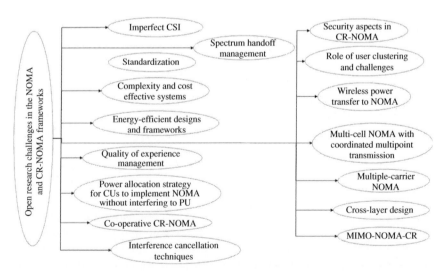

Figure 9.7 The open research challenges in the NOMA and CR-NOMA frameworks.

OMA for two scenarios of the partial channel states as follows: (i) imperfect CSI and (ii) second-order-statistics (SOS). For the first scenario, which is based on imperfect CSI, the authors have presented a simple closed-form approximation for the outage probability and the average sum rate, as well as their high signal-to-noise ratio (SNR) expressions. However, for the second scenario which is based on SOS, the authors have derived a closed-form expression for the outage probability and an approximate expression for the average sum rate. Wei et al. [41] have proposed a power-efficient resource allocation algorithm when the imperfect CSI is available at the transmitter (CSIT). A non-convex optimization problem is formulated in which the optimal SIC decoding policy was determined by the channel-to-noise ratio (CNR) outage threshold. Further, a suboptimal resource allocation scheme was proposed based on the difference of convex (D.C.) programming, which can converge to a close-to-optimal solution rapidly. Through simulation results, it is reported that the proposed resource allocation schemes provide significant transmit power savings and enhanced robustness against channel uncertainty via exploiting the heterogeneity of channel conditions and QoS requirements of users in MC-NOMA systems. Liu et al. [42] have investigated a NOMA downlink multiuser system where the transmitter perceives the CSI through limited feedback channel and have studied the effects of two traditional beam-forming technologies namely, zero-forcing beam-forming and random beam-forming. Based on the CSI available at the transmitter, the authors have proposed a user selection scheme to reduce the interference between the NOMA users. Moreover, a power allocation scheme is proposed to improve the sum-rate of NOMA system. Through simulation results, it is perceived that the NOMA system with limited feedback channel still gains larger system rate than traditional orthogonal multiple access systems. It is also reported that the random beam-forming is more suitable for the NOMA system with limited CSI feedback.

These research efforts to manage the imperfect CSIT in NOMA are appreciable; however, simultaneous exploitation of CR and NOMA in the networks demands further exploration of imperfect CSIT in the NOMA-CR frameworks. Therefore, the analysis of CR-NOMA systems with imperfect/partial CSI needs to be investigated.

9.6.2 Spectrum Hand-off Management

The spectrum hand-off management is a crucial issue in the CR networks; however, the implementation of NOMA in CR networks makes it more challenging since the power allocation relies on the channel gain coefficients. The users need to follow the cellular communication protocols in the cellular region and in addition, have to take care of the power allocation according to the variations of channel gain. Thus, the spectrum hand-off management for CR-NOMA is completely unexplored and potential research area.

9.6.3 Standardization

The standardization of regulatory policies for the CR-NOMA is a challenging task due to its two different frameworks as discussed. In the Framework-1, the BS must have SC and SIC ability for the downlink and uplink scenarios, respectively, which is generally overlooked in the cellular systems. Further, the PU must have the ability of SIC and these prerequisites demand the new standards for the next-generation wireless communication systems so that according to that CR-NOMA standards can be suggested. Moreover, in the Framework-2 as the role of PU is not so significant and therefore the standards for this framework can be suggested.

9.6.4 Less Complex and Cost-Effective Systems

The simultaneous use of CR and NOMA enhances the spectral efficiency significantly but on the cost of complex and costly systems as well as algorithms such as coding and interference cancelation techniques, etc. Thus, it opens a door of research for less complex and cost-effective systems and algorithms for implementation of CR-NOMA systems.

9.6.5 Energy-Efficient Design and Frameworks

The energy-efficient designs are the most preferable in order to support the green communication which is much desired. In addition, most of the communications devices are battery operated which further requires the energy-efficient designs. Therefore, the energy-efficient designs are further open research issues which must be investigated. To the best of the authors' knowledge, recently in [43, 44] a step ahead in this direction is taken where the authors have investigated a CR-NOMA from the perspective of energy efficiency and have reported that it outperforms the CR-OMA approach. Even though, it is interesting as well as open research problem which needs to be investigated. Yang et al. [45] have explored an uplink energy minimization problem for machine-to-machine communications (MMC) with NOMA. The formulated problem is of non-convex nature; therefore, it is proved that transmitting with minimum rate and full time is optimal. Further, the original problem is transformed into an equivalent convex problem, which can be effectively solved by the proposed optimal power control and time scheduling scheme. These efforts to achieve energy efficiency frameworks of NOMA are in infantry stage which needs to explore more for NOMA and CR-NOMA frameworks.

9.6.6 Quality-of-Experience Management

The quality of service (QoS) is the major objective for the next-generation communication systems in order to support the concept of reliability. The next-generation networks appear to be crowded by multiple and heterogeneous devices/users which should serve various applications. At every user, the quality needs to be maintained to serve all the users simultaneously. By considering this, Chen et al. [46] have proposed a new performance metric known as quality of experience (QoE). The QoE is the perceptual QoS from the users' perspective [47]. In CR networks, the prime objective is to enhance the performance of CU network by satisfying the QoS constraints of the PU network. However, in the CR-NOMA, the QoE metrics of both the networks must be improved since both the networks directly affect the performance of each other (Framework-2). To the best of authors' knowledge, this is completely unexplored and interesting as well as broad research area so far.

9.6.7 Power Allocation Strategy for CUs to Implement NOMA Without Interfering PU

In NOMA, researchers have put some efforts on the power allocation schemes [48–50]. The authors in [48] have proposed two power allocation strategies for NOMA which are based on the CSI experienced by NOMA users and on the predefined QoS per NOMA user. Lei et al. [49] have investigated a problem of jointly optimizing the power and channel allocation for NOMA. The key insights on the complexity and optimality have been provided, where an algorithm framework based on Lagrangian dual optimization and dynamic programming is proposed. Shahab et al. [50] have proposed a novel power allocation strategy that can optimize the system ergodic sum capacity constraining the minimum mutual interference between the paired users. It is claimed that the proposed scheme outperforms conventional power allocation in terms of bit error rates at the user ends with negligible effect on the system ergodic sum capacity. In the NOMA technique, the power is allocated to CUs on the basis of channel conditions. However, in the CR-NOMA techniques such as Framework-2, the transmitted power must be allocated to the CUs in such a way that the PU communication remains unaffected, which means the power allocation in CR-NOMA is a more challenging task as compared to that of NOMA technique which is also an open research issue. The potential power allocation techniques such as uniform power allocation, adaptive power allocation, optimal power allocation, etc., are well explored for the visible-light-NOMA communication systems [51].

9.6.8 Cooperative CR-NOMA

The CC is a promising technology for the efficient spectrum utilization and this technology is also eligible for the CR-NOMA (Framework-2). In [38, 39, 52], the authors have presented the concept of cooperative-NOMA in order to improve the performance of NOMA systems. Therefore, the cooperative CR-NOMA is a completely unexplored area which should be investigated for the next-generation communication systems.

9.6.9 Interference Cancellation Techniques

The entire concept of NOMA relies on the SC and interference cancellation techniques which are well explored for wireless communication; however, the latter one has scope to be explored; therefore, researchers are focusing on more efficient interference cancellation techniques [53–55] such as triangular interference cancellation technique. This is also a challenging area of research in order to improve the performance of CR-NOMA and NOMA systems.

9.6.10 Security Aspects in CR-NOMA

From the first view about NOMA, it seems that the security is a major challenge in NOMA due to detection of the other user's information before decoding its own information. However, researchers have worked on this issue and perceived that the security can be provided in NOMA. Ding et al. [56] have explored the security aspects with reference to the uni-casting and multicasting and it is reported that the use of NOMA in uni-casting always improves the uni-casting security when compared with the OMA. Further, the uni-casting secrecy outage probability is investigated in order to further enhance the security of NOMA. In [57], the authors have investigated the physical layer security of NOMA in the large-scale networks with invoking stochastic geometry for the single- and multi-antenna transmission scenarios, where the BS communicates with randomly distributed NOMA users. Moreover, the exact expressions of the security outage probability are derived for both single-antenna and multiple-antenna-aided scenarios. In [58], Qin et al. have studied the aspects of security aspects of NOMA in large-scale networks in which both the NOMA users and eavesdroppers are spatially randomly deployed. To improve the security of the random network, a protected zone around the source node is adopted. The secrecy performance is analyzed through the new deriving asymptotic expression of the security outage probability. Further, it is perceived from simulation results that the secure performance of the NOMA networks can be improved by either enlarging the scope of the protected zone or reducing the scope of the user zone.

These security aspects are presented with reference to the NOMA techniques; however, the CR has its own security challenges such as primary user emulation attack (PUEA), falsifying data, or denial of service (DOS) attack [59]. Therefore, the CR-NOMA will face the security challenges of both the techniques and potential research efforts are required for future communication.

9.6.11 Role of User Clustering and Challenges

Since in the downlink NOMA, multiple data flows are superimposed in the power domain and user decoding is based on SIC, its performance highly relies on the power split among the data flows and the associated power allocation (PA) problem. By keeping this in view, the problem of user fairness arises. In order to study this problem, Timotheou and Krikidis [60] have investigated PA techniques which guarantee the fairness in the downlink for two scenarios of CSI, namely (i) instantaneous CSI at the transmitter and (ii) average CSI. The key feature of NOMA is a trade-off between throughput and user fairness which is studied for the single antenna case in this paper. To consider the multi-antenna concept for this problem, Liu et al. [61] have investigated a dynamic user allocation and power optimization problem by considering the fairness issue in cluster-based MIMO-NOMA systems in which an NP-hard optimization problem is formulated. The authors have proposed a two-step suboptimal method for solving the dynamic user allocation problem for the user fairness. Moreover, to optimize the power allocation coefficients by invoking a bi-section search-based algorithm in order to maximize the SINR of the worst user in each cluster, three efficient user allocation algorithms are designed to seek a trade-off between computational complexity and throughput of the worst user. Further, Ali et al. [62] dynamically group the users' receive antennas into several clusters equal to or more than the number of BS transmit antennas where a beam-forming vector is shared by all the receive antennas in a cluster. A linear beam-forming technique is proposed which significantly cancels the inter-cluster interference by exploiting the all receive antennas. Moreover, for inter- and intra-cluster power allocation, the DPA solutions are provided where the key intent is to maximize the overall cell capacity. Further in [63], the authors have extended this problem for both the uplink and downlink scenarios and have investigated the problem of dynamic user clustering for both uplink and downlink NOMAs, where a sum-throughput maximization problem in a cell is formulated in such a way that the user clustering (i.e. grouping users into a single cluster or multiple clusters) and power allocations in NOMA clusters can be optimized under transmission power constraints, minimum rate requirements of the users, and SIC constraints.

The power allocation techniques with user fairness are explored for NOMA; however, CR-NOMA power allocation strategies are well explored which needs to be synchronized with reference to NOMA users' fairness demand.

9.6.12 Wireless Power Transfer to NOMA

The exponential increase in the tiny battery-operated devices such as IoT, WSNs, body area networks (BANs) demands the energy-efficient communication in addition to the spectral efficient communication. Due to the tiny size of the device, the battery size needs to be small which means the limited battery life and charging the battery is almost an impossible task due to remote and inaccessible locations. The prominent solution for this problem is provided by the SWIPT phenomenon where the information and power to the recipient user are transmitted wirelessly. The concept of SWIPT is well explored in the recent years [64, 65]. The implementation of SWIPT in NOMA is an interesting problem and a step toward this is taken by the researchers [66–68].

The authors have investigated a wireless-powered uplink communication system with NOMA which comprises one BS and multiple energy harvesting users and key intent is on the data rates optimization and fairness increase. The authors have formulated optimization problem and perceived that it can be solved optimally and efficiently by using linear programming methods or convex optimization, which witnesses the ease of practical implementation of the proposed scheme. The proposed scheme outperforms the orthogonal multiple access scheme; however, there is dependence between sum-throughput, minimum data rate, and harvested energy. Moreover, the convergence speed of the proposed greedy algorithm is evaluated, and it is shown that the required number of iterations is linear with respect to the number of users.

Liu et al. [67] have exploited the cooperation for SWIPT in the NOMA networks where users are spatially randomly located. A novel protocol is proposed in which NOMA users close to the source act as energy harvesting relays to help the far NOMA users. The location of the users has a prominent role in this protocol; therefore, three user selection schemes based on the user distances from the BS are proposed. The performance of the proposed protocol is analyzed through the derived closed-form expressions of the outage probability and system throughput. It is concluded that the use of SWIPT does not harm the diversity gain when compared to the conventional NOMA. The SWIPT-NOMA is in its infant stage and needs to be explored more to achieve the mature stage for practical implementation. In addition, the SWIPT-CR-NOMA demands more research efforts for practical implementation.

9.6.13 Multicell NOMA with Coordinated Multipoint Transmission

In the downlink NOMA, the BS allocates transmit power in such a way that the SIC decoding is performed according to an ascending order of the channel gains of the NOMA users [16]. The power allocation strategy results in a low received signal-to-intra-cell-interference ratio for lower channel gain users (e.g. cell-edge users) who are also susceptible to the intercell interference. Therefore, intercell interference management is a crucial phenomenon in multicell downlink NOMA systems. In addition to this, the popularity of phantom cell communication where the cell is divided into small cells in order to accommodate more number of users provokes to manage the effect of intercell interference.

In order to consider this scenario, the authors have proposed a framework to use coordinated multi-point (CoMP) transmission technology in downlink multicell NOMA systems considering distributed power allocation at each cell. The CoMP transmission is used for users experiencing strong receive signals from multiple cells while each cell adopts NOMA for resource allocation to its active users. The numerical results witness the spectral efficiency gain of the proposed CoMP-NOMA models over CoMP-OMA. Finally, this article is concluded by identifying the potential major challenges in implementing CoMP-NOMA in future cellular systems.

9.6.14 Multiple-Carrier NOMA

The multi-carrier access in the communication system is a real and practical scenario which needs to be investigated for the CR-NOMA users also. Recent works presented on multi-carrier NOMA in the literature are as follows. Hsueh and Chen [68] have proposed a CR-NOMA scheme based on multiuser orthogonal frequency division multiplexing scheme in order to reduce the total transmit power. In the proposed framework, each subcarrier is assigned to two users (PU and CU) for data transmission and transmits data with a lower order of a modulation mode. The PU has first preference to use channel and CU accesses some of the channels simultaneous to the PU. The key concern of the proposed scheme is that how the PU and CU perform resource allocation under different subcarriers and channel gains. The proposed scheme outperforms the OFDM scheme in terms of transmitting power or capacity equivalently. The authors in [69] have investigated an optimal resource allocation for multicarrier (MC) multiple-input single-output-NOMA (MISO-NOMA) downlink systems. The resource allocation design for the maximization of the weighted system throughput is formulated as a non-convex optimization problem taking into account the QoS requirements of the downlink receivers. The formulated problem is solved by using monotonic optimization. Chatziantoniou et al. [70] have proposed a novel transmission scheme which combines NOMA and multi-carrier index keying (MCIK). This scheme is proposed as a mechanism to enable multiple access for dense wireless device-to-device (D2D) systems that require high energy efficiency and effective interference management. The practical and effective nature of multi-carrier CR-NOMA further encourages the researchers to explore in this field.

9.6.15 Cross-Layer Design

The CR is a framework which needs to work on almost all layers of the open system interconnection (OSI) model [71]. The spectrum sensing is a process which is physical layer phenomenon. However, the accessing and sharing of spectrum using appropriate spectrum accessing technique is a medium access control (MAC) layer phenomenon. The routing information exchange of path and its reconfiguration are the functions of network layer. The CU pair link establishment and providing the reliable communication without affecting the PU communication become the functions of transport and session layers. In addition to this, the spectrum mobility is a phenomenon which comprises all the operations of cognitive cycle. Therefore, the spectrum mobility is a phenomenon which works on all layers of OSI model. On the other hand, the NOMA appears to work on the physical layer in order to support the SIC at transmitter and SC at the receiver.

On the other hand, the NOMA is assumed to be a Physical and MAC layer mechanism till now in the literature where the single channel/spectrum is utilized by more than one CU non-orthogonally. The data are provided from the end-user to the system which is on the application layer and further decoding of data, clustering, transmission, etc., functions are performed at the lower layers [47].

From here it is perceived that both the CR and NOMA are working on different layers of the OSI model. Therefore, the CR-NOMA needs the cross-layer designing of the protocols so that proper synchronization between two technologies can be achieved. An effort toward providing the cross-layer design for NOMA with the help of software-defined networking is presented in [40] where the functions at different layers are controlled through the software. However, in CR-NOMA, in addition to controlling the functions at different layers, the CU needs to perform all other functions of spectrum accessing and management. Therefore, the cross-layer design for CR-NOMA is an open and potential research issue.

9.6.16 MIMO-NOMA-CR

The researchers have well explored the MIMO-NOMA to improve the spectral and energy efficiency of the network in the recent past. Recently, the efforts to exploit the CR in NOMA are performed [38, 39] to resolve the issue of power allocation to NOMA-users and the improvement in the performance is reported. The 5G demands huge improvement in the spectral efficiency, energy efficiency, etc., and in order to achieve this demand, the simultaneous use of MIMO, CR, and NOMA will be a milestone in the field of communication. Therefore, in MIMO+ CR+NOMA, the open research challenge is how to exploit one technique in another one.

9.7 Summary

The CR and NOMA are the promising candidates to fulfill the demand of high spectral efficiency; however, more spectral efficient frameworks are desired for the next-generation communication systems. The potential way to accomplish this demand is simultaneous use of the CR and NOMA; therefore, the important frameworks of real-time exploitation of both the techniques are proposed in this chapter. The CR-NOMA improves the spectrum efficiency and massive connectivity as compared to that of the CR and NOMA techniques separately. Moreover, the prominent and challenging issues regarding the implementation and feasibility of the proposed frameworks are illustrated.

References

1 Alylidiz, I.F., Nie, S., Lin, S.-C., and Chandrasekaran, M. (2016). 5G roadmap: 10 key enabling technologies. *Comput. Netw.* 106 (4): 17–48.

2 Federal Communications Commission. (2002). Notice of proposed rule-making and order: facilitating opportunities for flexible, efficient, and reliable spectrum use employing cognitive radio technologies. ET Docket No. 03-108.

3 Mitola, J. and Maguire, G.Q. (1999). Cognitive radio: making software radios more personal. *IEEE Pers. Commun. Mag.* 6 (4): 13–18.

4 Alylidiz, I.F., Lee, W.-Y., Vuran, M.C., and Mohanty, S. (2006). NeXt generation/ dynamic spectrum access/cognitive radio wireless networks: a survey. *Comput. Netw.* 50 (13): 2127–2159.

5 Zhao, Q. and Sadler, B.M. (2007). A survey of dynamic spectrum access: signal processing, networking, and regulatory policy. *IEEE Signal Process. Mag.* 24 (3): 79–89.

6 Akyildiz, I.F., Lee, W.-Y., Vuran, M.C., and Mohanty, S. (2006). A survey on spectrum management in cognitive radio networks. *IEEE Commun. Mag.* 50 (13): 2127–2159.

7 Thakur, P., Singh, G., and Satashia, S.N. (2016). Spectrum sharing in cognitive radio communication system using power constraints: a technical review. *Perspect. Sci.* 8: 651–653.

8 Haykin, S. (2005). Cognitive radio: brain-empowered wireless communications. *IEEE J. Sel. Areas Commun.* 23 (2): 201–220.

9 Ali, A. and Homouda, W. (2017). Advances on spectrum sensing for cognitive radio networks: theory and applications. *IEEE Commun. Surv. Tutorials* 19 (2): 1277–1304.

10 Masonta, M.T., Mzyece, M., and Ntlatlapan, M. (2013). Spectrum decision in cognitive radio networks: a survey. *IEEE Commun. Surv. Tutorials* 15 (3): 1088–1107.

11 Thakur, P., Kumar, A., Pandit, S. et al. (2017). Spectrum mobility in cognitive radio network using spectrum prediction and monitoring techniques. *Phys. Commun.* 24: 1–8.

12 Thakur, P., Kumar, A., Pandit, S. et al. (2017). Advanced frame structures for hybrid spectrum access strategy in cognitive radio communication system. *IEEE Commun. Lett.* 21 (2): 410–413.

13 Yu, R., Zhang, C., Zhang, X. et al. (2014). Hybrid spectrum access in cognitive-radio-based smart-grid communications systems. *IEEE Syst. J.* 8 (2): 577–587.

14 Yang, C., Fu, Y., Zhang, Y. et al. (2014). An efficient hybrid spectrum access algorithm in OFDM-based wideband cognitive radio networks. *Neurocomputing* 125 (11): 33–40.

15 Higuchi, K. and Benjebbour, A. (2015). Non-orthogonal multiple access (NOMA) with successive interference cancellation for future radio access. *IEICE Trans. Commun.* E98-B (3): 403–414.

16 Islam, S.M.R., Avazov, N., Dobre, O.A., and Kwak, K.-S. (2017). Power domain non-orthogonal multiple access (NOMA) in 5G systems: potential and challenges. *IEEE Commun. Surv. Tutorials* 19 (2): 721–742.

17 Ding, Z., Lei, X., Karagiannidis, G.K. et al. (2017). A survey on non-orthogonal multiple access for 5G networks: research challenges and future trends. *IEEE J. Sel. Areas Commun.* 35 (10): 2181–2195.

18 Liu, B.Y., Qin, Z., Lan, M.E. et al. (2017). Non-orthogonal multiple access for 5G and beyond. *Proc. IEEE* 105 (12): 2347–2381.

19 Yunzheng, T., Long, L., Shang, L., and Zhi, Z. (2015). A survey: several technologies of non-orthogonal transmission for 5G. *China Commun.* 12 (10): 1–15.

20 Wang, Y., Ren, B., Sun, S. et al. (2016). Analysis of non-orthogonal multiple access for 5G. *China Commun.* 13 (2): 52–66.

21 Wani, M.M.A., Sali, A., Noordin, N.K. et al. Robust beamforming and user clustering for guaranteed fairness in downlink NOMA with partial feedback. *IEEE Access* 7: 121599–121611.

22 Sena, A.S.D., Rafael, F., Lima, M. et al. (2020). Massive MIMO-NOMA networks with imperfect SIC: design and fairness enhancement. *IEEE Trans. Wireless Commun.* 19 (9): 6100–6115.

23 Qiu, H., Gao, S., Tu, G., and Zong, S. (2020). Position information-based NOMA for downlink and uplink transmission in mobile scenarios. *IEEE Access* 8: 150808–150822.

24 Saggese, F., Moretti, M., and Abrardo, A. (2020). A quasi-optimal clustering algorithm for MIMO-NOMA downlink systems. *IEEE Wireless Commun. Lett.* 9 (2): 152–156.

25 Nguyen, H.V., Kim, H.M., Kang, G.-M. et al. (2020). A survey on non-orthogonal multiple access: from the perspective of spectral efficiency and energy efficiency. *Energies* 13 (16): 1–20.

26 Abrardo, A., Moretti, M., and Saggese, F. (2020). Power and subcarrier allocation in multicarrier NOMA-FD systems. *IEEE Trans. Wireless Commun.* https://doi.org/10.1109/TWC.2020.3021036.

27 Maraqa, O., Rajasekaran, S.A., A-Ahmadi, S. et al. (2019). A survey of rate-optimal power domain NOMA with enabling technologies of future wireless networks. *IEEE Commun. Surv. Tutorials* 8 (11): 2192–2235.

28 Wang, W., Chen, Y., Zhao, Q., and Jiang, T. (2016). A software-defined wireless networking enabled spectrum management architecture. *IEEE Commun. Mag.* 54 (1): 33–39.

29 Thakur, P., Kumar, A., Pandit, S. et al. (2017). Frame structures for hybrid spectrum accessing strategy in cognitive radio communication system. *International Conference on Contemporary Computing (IC3)*, Noida, India (11–13 August 2016), 1–6.

30 Kaushik, A., Sharma, S.K., Chatzinotas, S. et al. (2016). Sensing-throughput tradeoff for interweave cognitive radio systems: a deployment-centric viewpoint. *IEEE Trans. Wireless Commun.* 15 (5): 3690–3707.

31 Yang, C., Lou, W., Fu, Y. et al. (2016). On throughput maximization in multichannel cognitive radio networks via generalized access strategy. *IEEE Trans. Commun.* 64 (4): 1384–1398.

32 Khoshkholgh, M.G., Navaie, K., and Yanikomeroglu, H. (2010). Access strategies for spectrum sharing in fading environment: overlay, underlay and mixed. *IEEE Trans. Mobile Comput.* 9 (12): 1780–1793.

33 Jiang, X., Wang, K.K., Zhang, Y., and Edwards, D. (2013). On hybrid overlay-underlay dynamic spectrum access: double-threshold energy detection and Markov model. *IEEE Trans. Veh. Technol.* 62 (8): 4078–4083.

34 Jasbi, F. and Sood, K.C. (2016). Hybrid Overlay/underlay cognitive radio network with MC-CDMA. *IEEE Trans. Veh. Technol.* 65 (4): 2038–2047.

35 Sharma, S.K., Bogale, T.E., Chatzinotas, S. et al. (2015). Cognitive radio techniques under practical imperfections: a survey. *IEEE Commun. Surv. Tutorials* 17 (4): 1858–1888.

36 Han, W., Zhang, Y., Wang, X. et al. (2016). Orthogonal power division multiple access: a green communication perspective. *IEEE J. Sel. Areas Commun.* 34 (12): 3828–3842.

37 Lu, F., Xu, M., Cheng, L. et al. (2016). Non-orthogonal multiple access in large-scale underlay cognitive radio networks. *IEEE Trans. Veh. Technol.* 65 (12): 10152–10157.

38 Lv, L., Chen, J., and Ni, Q. (2016). Cooperative non-orthogonal multiple access in cognitive radio. *IEEE Commun. Lett.* 20 (10): 2059–2062.

39 Yang, Z., Ding, Z., Fan, P., and Karagiannidisg, K. (2016). On the performance of non-orthogonal multiple access systems with partial channel information. *IEEE Trans. Commun.* 64 (2): 656–657.

40 Wei, Z., Ngd, W.K., Yuan, J., and Wang, H.-M. (2017). Optimal resource allocation for power-efficient MC-NOMA with imperfect channel state information. *IEEE Trans. Commun.* 69 (9): 3944–3961.

41 Liu, S. and Zhang, C. (2016). Non-orthogonal multiple access in a downlink multiuser beamforming system with limited CSI feedback. *Eurasip J. Wireless Commun. Netw.* 2016: 1–11.

42 Zhang, Y., Yang, Q., Zheng, T.-X. et al. (2016). Energy efficiency optimization in cognitive radio inspired non-orthogonal multiple access. *27th IEEE International Symposium on Personal, Indoor and Mobile Radio Communication-(PIMRC)*, Valencia, Spain (4–8 September 2016), 1–6.

43 Ding, Z., Liu, Y., Choi, J. et al. (2017). Applications of non-orthogonal multiple access in LTE and 5G networks. *IEEE Commun. Mag.* 55 (2): 185–191.

44 Yang, Z., Xu, W., Xu, H. et al. (2017). Energy efficient non-orthogonal multiple access for machine-to-machine communications. *IEEE Commun. Lett.* 21 (4): 817–820.

45 El-Sayedm, M., Ibrahima, S., and Khairym, M. (2016). Power allocation strategies for non-orthogonal multiple access. *IEEE International Conference on Selected Topics in Mobile and Wireless Networking (MoWNeT)*, Cairo, Egypt (11–13 April 2016), 1–6.

46 Chen, Y., Wu, K., and Zhang, Q. (2015). From QoS to QoE: a tutorial on video quality assessment. *IEEE Commun. Surv. Tutorials* 17 (2): 1126–1165.

47 Wang, W., Liu, Y., Luo, Z. et al. Toward cross-layer design for non-orthogonal multiple access: a quality-of-experience perspective. *IEEE Wireless Commun.* 25 (2): 118–124.

48 Lei, L., Yuan, D., Hoc, K., and Sun, S. (2016). Power and channel allocation for non-orthogonal multiple access in 5G systems: tractability and computation. 15 (12): 8580–8594.

49 Shahabm, B., Kaderm, F., and Shins, Y. (2016). On the power allocation of non-orthogonal multiple access for 5G wireless networks. *International Conference on Open Source Systems and Technologies (ICOSST)*, Lahore, Pakistan (15–17 December 2016), 1–6.

50 Ding, Z., Fan, P., and Poor, H.V. (2016). Impact of user pairing on 5G non-orthogonal multiple access downlink transmission. *IEEE Trans. Veh. Technol.* 65 (8): 6010–6023.

51 Ngene, C.E., Thakur, P., and Singh, G. (2020). Power allocation techniques for visible light – nonorthogonal multiple access communication systems. In:

Enabling Technologies for Next Generation Wireless Communications (eds. M. Usman, M.D. Ansari and M. Wajid), 45–78. CRC Press.

52 Haci, H., Zhu, H., and Wang, J. (2017). Performance of non-orthogonal multiple access with a novel asynchronous interference cancellation technique. *IEEE Trans. Commun.* 65 (3): 1319–1335.

53 Haci, H. (2018). Performance study of non-orthogonal multiple access (NOMA) with triangular successive interference cancellation. *Wireless Netw.* 24: 2145–2163.

54 Jalaianb, A., Yuan, X., Shi, Y. et al. (2018). On the integration of SIC and MIMO DOF for interference cancellation in wireless networks. *Wireless Netw.* 24 (7): 2357–2374.

55 Ding, Z., Zhao, Z., Peng, M., and Poor, H.V. (2017). On the spectral efficiency and security enhancements of NOMA assisted multicast-unicast streaming. *IEEE Trans. Commun.* 65 (7): 3151–3163.

56 Liu, Y., Qin, Z., Elkashlanm, M. et al. (2017). Enhancing the physical layer security of non-orthogonal multiple access in large-scale network. *IEEE Trans. Wireless Commun.* 16 (3): 1656–1672.

57 Qin, Z., Liu, Y., Ding, Z. et al. (2016). Physical layer security for 5G non-orthogonal multiple access in large-scale networks. *IEEE International Conference on Communication*, Kuala Lumpur, Malaysia (22–27 May 2016), 1–6.

58 Jianwul, L., Zebing, F., Zhiyong, F., and Ping, Z. (2015). A survey of security issues in cognitive radio networks. *China Commun.* 12 (3): 132–150.

59 Timotheou, S. and Krikidis, I. (2015). Fairness for non-orthogonal multiple access in 5G systems. *IEEE Signal Process. Lett.* 22 (10): 1647–1651.

60 Liu, Y., Elkashlan, M., Ding, Z., and Karagiannidisg, K. (2016). Fairness of user clustering in MIMO non-orthogonal multiple access systems. *IEEE Commun. Lett.* 20 (7): 1465–1468.

61 Ali, M.S., Hossain, E., and Kim, D.I. (2017). Non-orthogonal multiple access (NOMA) for downlink multiuser MIMO systems: user clustering, beamforming, and power allocation. *IEEE Access* 5: 565–577.

62 Ali, M.S., Tabassum, H., and Hossain, E. (2016). Dynamic user clustering and power allocation for uplink and downlink non-orthogonal multiple access (NOMA) system. *IEEE Access* 4: 6325–6343.

63 Zhang, R. and Hoc, K. (2013). MIMO broadcasting for simultaneous wireless information and power transfer. *IEEE Trans. Wireless Commun.* 12 (5): 1989–2001.

64 Zhou, X., Zhang, R., and Hoc, K. (2013). Wireless information and power transfer: architecture design and rate-energy tradeoff. *IEEE Trans. Commun.* 61 (11): 4754–4767.

65 Diamantoulakisp, D., Pappik, N., Ding, Z., and Karagiannidisg, K. (2016). Optimal design of non-orthogonal multiple access with wireless power transfer. *IEEE International Conference on Communication*, Kuala Lumpur, Malaysia (22–27 May 2016), 1–6.

66 Diamantoulakisp, D., Pappik, N., Ding, Z., and Karagiannidisg, K. (2017). Wireless-powered communications with non-orthogonal multiple access. *IEEE Trans. Wireless Commun.* 15 (12): 1656–1672.

67 Liu, Y., Ding, Z., Ellashlanm, M., and Poor, H.V. (2016). Cooperative non-orthogonal multiple access with simultaneous wireless information and power transfer. *IEEE J. Sel. Areas Commun.* 34 (4): 938–953.

68 Hsueh, W.-H. and Chen, Y.-F. (2017). Resource allocation for NOMA in multiuser multicarrier systems. *International Conference on Applied System Innovation (ICASI)*, Sapporo, Japan (13–17 May 2017), 1–6.

69 Sun, Y., Ng, D.W.K., and Schoberr, R. (2017). Optimal resource allocation for multicarrier MISO-NOMA system. *IEEE International Conference on Communications (ICC)*, Paris France (21–25 May 2017), 1–6.

70 Chatziantoniou, E., Ko, Y., and Choi, J. (2017). Non-orthogonal multiple access with multi-carrier index keying. *23th European Wireless Conference*, Dresden, Germany (17–19 May 2017), 1–6.

71 Towhidlou, V. and Shikh-Bahaei, M. (2017). Cross-layer design in cognitive radio standards, 1–9. https://arxiv.org/ftp/arxiv/papers/1712/1712.05003.pdf

10

Performance Analysis of MIMO-Based CR-NOMA Communication Systems

10.1 Introduction

The rapid proliferation of communication devices due to explosive growth in the applications of Internet-of-Things (IoT) demands high speed, low latency, and more reliability that led to a problem of spectrum scarcity over the radio frequency regime which is a limited resource on the electromagnetic spectrum from 3 kHz to 300 GHz for the next-generation communication systems [1, 2]. Therefore, significantly more spectral efficient designs are desired for the next-generation communication systems and prominent techniques available in the literature are cognitive radio (CR) [3–5], non-orthogonal multiple access (NOMA) [6–13], and multiple-input-multiple-output (MIMO) [14–17]. As we have discussed in Chapter 1 that the CR is a framework that is inspired by the report of Federal Communication Commission (FCC) according to which the problem of spectrum scarcity arises due to underutilized/unutilized licensed spectrum at particular time and space [18–20]. Further, the NOMA is a recently explored spectrum accessing technique that enables more than one user to communicate on the same orthogonal source such as channel/space/time. Various researchers have proposed several techniques for the feasibility of the NOMA mechanism that are usually classified as power domain multiplexing (PDM) and code domain multiplexing (CDM) [21, 22] which further comprise power division multiple access (PDMA), sparse code multiple access (SCMA), low-density spreading-code division multiple access (LDS-CDMA), and pattern division multiple access (PTDMA). The PDMA is a prominent technique that is explored more for NOMA as compared to all other techniques [8]; therefore, we have chosen this for further discussion. The PDMA is a potential multiple access technique which relies on the two prerequisites at transmitter and receiver, namely, the superposition coding (SC) and successive interference cancellation (SIC), respectively. The SC is a technique which allows simultaneous transmission of multiple users' information on

Spectrum Sharing in Cognitive Radio Networks: Towards Highly Connected Environments,
First Edition. Prabhat Thakur and Ghanshyam Singh.

the same channel from the transmitter end and the reception at receivers is performed using SIC technique.

Also, the MIMO is a potential technology that performs well when compared to the single-input-single-output (SISO) in terms of high data rate, low bit error rate (BER) and enhanced signal-interference-plus-noise-ratio (SINR) [23]. The popular frameworks which are exploited in MIMO are as follows: (i) spatial/antenna multiplexing [24] (ii) spatial/antenna diversity [25] and (iii) beamforming technique [26]. The use of spatial multiplexing enables to achieve higher data rates however, the spatial diversity outperforms in terms of BER. Moreover, the beamforming improves the SINR significantly by using smart antenna [27]. The low BER results in the reliable communication which is achieved using spatial diversity and is possible at both transmitting (transmitter diversity [28]) and receiving end (receiver diversity [29]). On the other hand, the antenna multiplexing permits to transmit the different signals from different antennas through the individual channels and results in the improved data rate which is in relation with the number of antennas and individual channels in the MIMO systems.

Various researchers have explored the simultaneous exploitation of two techniques among these three, which are CR-MIMO [30–32], CR-NOMA [33–38], and MIMO-NOMA [39–41], for better spectral efficiency. Moreover, the simultaneous exploitation of all these three techniques is suggested in [7]; however, it is completely unexplored. In [42], the authors have investigated a joint antenna selection (AS) problem for a MIMO-CR-NOMA system and an algorithm is proposed to increase the signal-to-noise ratio (SNR) at each user when satisfying the interference constraints at PU. The closed-form expressions of outage performance for this algorithm are derived. Thus, to the best of author's knowledge, the use of all these techniques for the efficient spectral utilization or analysis of data rate/throughput is unexplored. Therefore, in this chapter, we have investigated the simultaneous exploitation of CR, NOMA, and MIMO for efficient use of spectrum as well as to achieve enhanced throughput. The potential contribution of this chapter is summarized as follows:

- A framework for the simultaneous use of the CR, NOMA, and MIMO is explored for the downlink and uplink scenarios.
- The closed-from expressions for the total throughput at each user due to multiple antennas in different cases such as CR-NOMA, MIMO-NOMA, and MIMO-based CR-NOMA are derived numerically for both the downlink and uplink scenarios.
- Further, the closed-from expression for the computation of total/sum throughput for the CR-OMA, CR-NOMA, CR-MIMO, and CR-MIMO-NOMA frameworks are derived numerically, for both the downlink and uplink scenarios.

- In order to address the interference constraints at PU due to CU-4 transmission in the uplink scenario, we have derived a new interference metric known as interference efficiency in the uplink scenario.
- The proposed frameworks are simulated for different numbers of transmitting and receiving antennas at the base station (BS) and each user, respectively, and the results are compared with the presented frameworks.

The next section comprises related work regarding simultaneous use of CR-NOMA, CR-MIMO, MIMO-NOMA as well as CR-MIMO-NOMA communication systems.

10.2 Related Work for Several Combinations of CR, NOMA, and MIMO Systems

This section presents the recent reported literatures related to the CR-NOMA, CR-MIMO, MIMO-NOMA and CR-MIMO-NOMA/MIMO-CR-NOMA communication systems. In [30], the authors have presented the three cognitive NOMA architectures namely, the underlay, overlay, as well as CR-inspired NOMA networks. It is reported that the interference in the considered approaches limits the reliability of the received signal. Therefore, a cooperative relaying strategy is proposed in order to address inter-network and intra-network interference and have achieved significantly lower value of outage probabilities. Shokair et al. [31] have proposed two equivalent cooperative relay selection schemes for a downlink-controlled CR-based NOMA system and have derived the closed-form expressions for the average achievable sum-rate at high SINR. The authors have claimed the improved performance of proposed system in terms of higher average achievable sum-rate saves the time resources and maximizes the network throughput by using the simulation results. In [32], the authors have presented a NOMA-based CR system where the PU-first-decoding mode (PFDM) and CU-first-decoding mode (SFDM) are proposed at the PU and CU receivers to decode the NOMA signals, respectively. In the PFDM, the improved CU throughput is achieved by controlling the sub-channel power in such a way that the PU throughput is maintained; however, in the SFDM, the CU throughput is decreased due to the PU interference. Therefore, by considering the PFDM and SFDM, the authors have proposed two optimization problems, respectively, which aim to maximize the normalized throughput of CU by jointly optimizing spectrum resource that are the number of sub-channels and sub-channel transmission power. Further, a joint-optimization-algorithm is developed to solve the optimization problems. The simulation results are witness of the improved transmission performance of the NOMA-based CR.

Miridakis et al. [33] have investigated the performance of an underlay MIMO-CR system where power assigned to each antenna of CU is maximized in order to yield the enhanced throughput by constraining the interference power at PU. Further, the closed-form expression of lower bound on the ergodic channel capacity is derived. Hao et al. [34] have investigated a power allocation problem in massive-MIMO CR communication systems in which an orthogonal pilot sharing scheme at pilot transmission phase is proposed, where CUs are permitted to use pilots for channel estimation only when there are temporarily unused orthogonal pilots. Further, an optimization problem for the power allocation to CRN for downlink scenario is formed where the sum-rate of CU is maximized by constraining the total transmit power and SINR at the PU. Satheesan and Sudha [35] have exploited the hybrid spectrum accessing technique in the MIMO-CR communication system and beamforming is used in order to limit the interference in the desired direction. Especially, the power management between the CUs and PUs is controlled by the Nash equilibrium (NE) game model. It is reported that the proposed beamforming technique improves the achieved data rate in the considered system. In [36], the authors have proposed a low-cost reception scheme for MIMO-CR communication systems where the authors have exploited the beamforming and sub-array formation in order to improve the CUs' capacity and to avoid computationally complex processes. Specifically, the antenna sub-array formation (ASF) scheme is used to maximize the SINR by using all antenna elements. Using simulation results, it is illustrated that the ASF schemes outperform when compared with that of the AS scheme. Iwata et al. [37] have put efforts toward the practical implementation of MIMO-CR communication system by using Tx/Rx beamforming and have validated the feasibility by measuring its throughput performance for CUs as well as PUs. It is perceived that the considered system performs well if the PUs' transmit power is large enough when compared to that of the CUs' transmit power. Liu et al. [38] have investigated the cooperative game theory in the MIMO-CR communication systems to maximize the CUs' sum-rate. The problem of CU transmission is modeled as a cooperative game where the strategy of each CU is the transmit covariance matrix; however, the utility is an approximation of the information rate. Each CU controls its transmit covariance matrix to increase the utility under both transmit power and total interference power constraints. The CUs bargain with each other over the assignment of the interference budget until they reach at a unique and Pareto-efficient solution which maximizes the network utility. It is concluded through simulation results that the proposed bargaining solution of the cooperative game outperforms in terms of sum-rate when compared with that of the NE solution of the non-cooperative game.

The authors in [39] have investigated the performance of MIMO-NOMA communication system where an optimal way of power allocation is proposed in order to maximize the sum-rate when a layered transmission scheme is used. Further, the closed-form expressions for the average sum-rate are derived with statistical channel state information (CSI) available at the transmitter and the lower and upper bounds on the average sum-rate are also evaluated. Moreover, via simulation results, it is claimed that the scaling property of MIMO-NOMA with layered transmissions also holds as conventional MIMO does, which means the average sum-rate grows linearly with the number of antennas. In order to specify the role of NOMA in the Massive MIMO communication systems, the authors have investigated massive MIMO-NOMA communication systems and it is perceived that the efficiency of the system is questionable when the BS has many more antennas than the users in a cell [40]. The authors have analyzed and compared the performance achieved by NOMA and multiuser beamforming in both non-line-of-sight as well as line-of-sight scenarios and found that the latter scheme provides highest average sum-rate in massive MIMO systems. Further, the specific cases where NOMA performs well with massive MIMO systems as well as highlights the significant role that NOMA plays in is creating the hybrid of NOMA and multiuser beamforming technique. Lin et al. [41] have investigated a deep learning (DL)-based MIMO-NOMA downlink scenario where a solution of effective signal detection is proposed by using the CSI. In the proposed scheme, a learning method is developed that automatically analyzes the CSI and detects the original transmit sequences. Unlike the existing SIC schemes, which must search for the optimal order of the channel gain and remove the signal with higher power allocation factor while detecting a signal with a lower power allocation factor, the proposed DL method can combine the channel estimation process with recovery of the desired signal suffering from channel distortion and multiuser signal superposition. The simulations for the proposed scheme are executed at extensive scale and it is perceived that the DL method successfully addresses the channel impairments that results in effective detection performance. Unlike the conventional optimization methods, MIMO-NOMA-DL searches for the optimal solution via a neural network (NN) and claimed that the DL is an effective tool.

Yu et al. [42] have explored the MIMO in CR-NOMA communication systems where the joint AS problem is investigated. Further, a computationally efficient joint AS algorithm is proposed in order to maximize the SNR of CU by satisfying the quality of service at PU which is named as subset-based joint AS (SJ-AS). In addition, the authors have derived the closed-form expression of outage performance for SJ-AS as well as proved the minimal outage probability achieved by SJ-AS among all possible joint AS schemes. The provided numerical results demonstrate the superior performance of the proposed scheme. Further, the numerical results are the witness of superior performance of the SJ-AS over both the conventional max-min approach and the random selection scheme. In [43], Givi et al. have proposed a machine-to-machine (M2M) communication underlaying heterogeneous network with mmWave-NOMA transmission, where small cell users and machine-type communication devices share the same resource block in order to improve the network capacity. Further, the authors have presented a low computational complexity suboptimum solution relies on the non-cooperative game and an iterative algorithm with fast convergence speed and a reduced complexity search-based algorithm are, respectively, presented to solve the power control and user selection problems. The improved performance of the proposed technique in terms of spectral efficiency and energy efficiency is illustrated by using simulation results.

From the above discussion, it is perceived that there are very less efforts toward feasible as well as spectral efficient designs using the MIMO-based CR-NOMA communication system. Moreover, the potential frameworks with the sum-rate and throughput analysis are also missing. Therefore, we have proposed the potential frameworks of MIMO-based CR-NOMA communication system and have analyzed its performance for uplink as well as downlink scenarios in terms of throughput at each NOMA user and total/sum throughput in the MIMO-based CR-NOMA framework.

Further, this chapter is structured as follows. The proposed system model for the MIMO-based CR-NOMA framework is illustrated in Section 10.3. In Section 10.4, the analysis of the proposed system model with mathematical derivations is presented. The simulation results and discussion are described in Section 10.5. Finally, Section 10.6 highlights the summary and future scope of the chapter.

10.3 System Model

In this section, the CR, NOMA, and MIMO are simultaneously designed for efficient spectral utilization in the future generation communication systems. We have considered the downlink and uplink scenarios of the CR system where a BS detects the idle channels in the absence of PUs and serves several users depending upon the number of sectors as shown in Figure 10.1. Here, three sectors are considered and due to which, the BS serves three CUs which are CU-1, CU-2, and CU-3. Further, the implementation of NOMA allows the CU-4, CU-5, and CU-6 to share the spectrum of CU-1, CU-2, and CU-3, respectively, by using the PDM with SC and SIC ability at the transmitting and receiving end, respectively. The CU-1, CU-2, and CU-3 are near to the BS and denoted as nearby users; however, CU-4, CU-5, and CU-6 are assumed to be at the edge of cell and known as cell-edge users. Further, we have considered that the BS is equipped with N_{BS} number of antennas; however, each user is equipped with N_{CUi} number of antennas where i denotes the CU number. In addition, the PU comprises N_{PUT} number of antennas. The total channel bandwidth available for the PU transmission is assumed to be B Hz. Moreover, it is assumed that for MIMO multiplexing, each

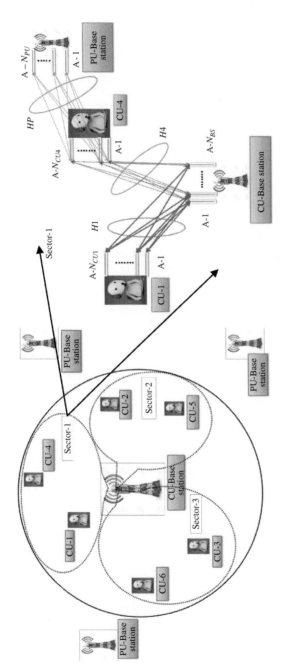

Figure 10.1 The downlink scenario of MIMO-based CR-NOMA communication system.

user as well as BS is able to perform effective detection by already knowing at the receiving end [44]. For the ease of understanding, we have considered only Sector-1, where the CU-1 and CU-4 establish the communication due to NOMA ability. The system specifications are presented for downlink and uplink scenarios as follows.

10.3.1 Downlink Scenarios

The channel matrices from the BS to the CU-1 and CU-4 are denoted as H_D1 and H_D4, respectively, as shown in Figure 10.1 which are further represented as:

$$H_D1 = \begin{bmatrix} h_D1_{11} & h_D1_{12} & & h_D1_{1N_{BS}} \\ h_D1_{21} & h_D1_{22} & & h_D1_{2N_{BS}} \\ : & : & \cdots\cdots & : \\ : & : & & : \\ h_D1_{N_{CU1}1} & h_D1_{N_{CU1}2} & & h_D1_{N_{CU1}N_{BS}} \end{bmatrix}, \qquad (10.1)$$

$$H_D4 = \begin{bmatrix} h_D4_{11} & h_D4_{12} & & h_D4_{1N_{BS}} \\ h_D4_{21} & h_D4_{22} & & h_D4_{2N_{BS}} \\ : & : & \cdots\cdots & : \\ : & : & & : \\ h_D4_{N_{CU4}1} & h_D4_{N_{CU4}2} & & h_D4_{N_{CU4}N_{BS}} \end{bmatrix}. \qquad (10.2)$$

Similar to h_D1_{ij}, h_D4_{ij} signifies the channel gain from the ith antenna of BS to the jth antenna of CU-4.

Moreover, it is worth mentioning that the CU-4 located at the edge of circle due to which it gets interference from the PU transmitter the channel matrix from the PU transmitter having N_{PUT} number of antennas, to the CU-4 is $H_D P$ and mathematically represented as:

$$H_D P = \begin{bmatrix} h_D P_{11} & h_D P_{12} & & h_D P_{1N_{BS}} \\ h_D P_{21} & h_D P_{22} & & h_D P_{2N_{BS}} \\ : & : & \cdots\cdots & : \\ : & : & & : \\ h_D P_{N_{CU4}1} & h_D P_{N_{CU4}2} & & h_D P_{N_{CU4}N_{BS}} \end{bmatrix}. \qquad (10.3)$$

Further, X_D1 is the input signal vectors assigned for the CU-1, which is further represented as:

$$X_D1 = \begin{bmatrix} x_D1_1 & x_D1_2 & x_D1_3 ..... x_D1_{N_{BS}} \end{bmatrix}', \qquad (10.4)$$

where the components of input vectors $x_D1_1, x_D1_2, x_D1_3.....$ and $x_D1_{N_{BS}}$ are assumed to be mutually independent (uncorrelated) and A' denotes the transpose of A. Moreover, assume that the overall power assigned to the transmitted signal is P_D1 that is allocated among all the antennas as per the uniform power allocation rule [44], which is popular and suitable when the CSI is unavailable at the transmitter. Similarly, the input signal vector for the CU-4 and PU transmitter is:

$$X_D4 = \begin{bmatrix} x_D4_1 & x_D4_2 & x_D4_3 ..... x_D4_{N_{BS}} \end{bmatrix}', \qquad (10.5)$$

$$X_DP = \begin{bmatrix} x_DP_1 & x_DP_2 & x_DP_3 ..... x_DP_{N_{BS}} \end{bmatrix}'. \qquad (10.6)$$

Similar to the assumption for eq. (10.4), the components of input vectors $x_D4_1 \, x_D4_2 \, ...x_D4_{N_{BS}}$ and $x_DP_1 \, x_DP_2 \, ...x_DP_{N_{BS}}$ are mutually independent and total power assigned to the transmitted signal for CU-4 and PU are P_D4 and P_DP, respectively.

Further, the Y_D1 and Y_D4 is the output signal vector available at CU-1 and CU-4, respectively, which are further represented as:

$$Y_D1 = \begin{bmatrix} y_D1_1 & y_D1_2 & y_D1_3 ..... y_D1_{N_{CU}} \end{bmatrix}', \qquad (10.7)$$

$$Y_D4 = \begin{bmatrix} y_D4_1 & y_D4_2 & y_D4_3 ..... y_D4_{N_{CU}} \end{bmatrix}'. \qquad (10.8)$$

Moreover, N_D1 and N_D4 are the noise vectors at the CU-1 and CU-4, respectively and are denoted as follows:

$$N_D1 = \begin{bmatrix} n_D1_1 & n_D1_2 & n_D1_3 ..... n_D1_{N_{CU}} \end{bmatrix}', \qquad (10.9)$$

$$N_D4 = \begin{bmatrix} n_D4_1 & n_D4_2 & n_D4_3 ..... n_D4_{N_{CU}} \end{bmatrix}'. \qquad (10.10)$$

Further, the input covariance matrix of input vector X_D1, X_D4, and X_DP is denoted as Q_D1, Q_D2, and Q_DP, respectively. Moreover, the channel gains from BS to the CU-4 are assumed be less as compared to the H_D1; therefore, we can form a relation between the H_D1 and H_D4 using a factor k which is function of path loss exponent, i.e. $H_D4 = H_D1/k$. The σ_D1_i and σ_D4_i represents the parallel channel gain coefficients from the ith antenna of the BS to the ith antenna of CU-1 and CU-4, respectively, which are yielded by using parallel channel decomposition [42]. Similarly, σ_DP_i denotes the parallel channel gain coefficients from the ith antenna of PU to the ith antenna of CU-4. In addition to this, the R_{H_D1}, R_{H_D4}, and R_{H_DP} are the number of nonzero singular values of the channel matrix H_D1, H_D4, and H_DP, respectively.

10.3.2 Uplink Scenario

For the uplink scenario, the channel gains from the CU-1 to the BS are denoted as:

$$
H_U1 = \begin{bmatrix}
h_U1_{11} & h_U1_{12} & & h_U1_{1N_{BS}} \\
h_U1_{21} & h_U1_{22} & & h_U1_{2N_{BS}} \\
\vdots & \vdots & \cdots\cdots & \vdots \\
\vdots & \vdots & & \vdots \\
h_U1_{N_{CU1}1} & h_U1_{N_{CU1}2} & & h_U1_{N_{CU1}N_{BS}}
\end{bmatrix},
\tag{10.11}
$$

where h_U1_{ij} signifies the channel gain from the ith antenna of CU-1 to the jth antenna of the BS. Like H_U1, the matrix H_U4 and H_UP are similar to the H_D4 and H_DP. The σ_U1_i and σ_U4_i represent the parallel channel gain coefficients from the ith antenna of the CU-1 and CU-4 to the ith antenna of BS, respectively. Similarly, σ_UP_i denotes the parallel channel gain coefficients from the ith antenna of the CU-4 to the ith antenna of PU. In addition to this, the R_{H_U1}, R_{H_U4}, and R_{H_UP} are the number of nonzero singular values of the channel matrix H_U1, H_U4, and H_UP, respectively. Moreover, the input vectors at CU-1 and CU-4 are X_U1 and X_U4; however, the output vectors at the BS and PU are Y_UBS and Y_UPU. The noise vector at the BS and PU are denoted as:

$$
N_UBS = [n_UBS_1\; n_UBS_2\; n_UBS_3.....n_UBS_{N_{BS}}]',
\tag{10.12}
$$

$$
N_UPU = [n_UPU_1\; n_UPU_2\; n_UPU_3.....n_UPU_{N_{PU}}]',
\tag{10.13}
$$

where n_UBS_i represents the noise power at the ith antenna of BS and n_UPU_i represents the noise power at the ith antenna of PU. Moreover, the input covariance matrix of input vector X_U1 and X_U4, is denoted as Q_U1 and Q_U4, respectively.

10.4 Performance Analysis

This section deals with the sum-rate computation of CUs for CR-orthogonal multiple access (CR-OMA), CR-NOMA, CR-MIMO, and CR-MIMO-NOMA frameworks by using the data rate/throughput achieved at each antenna for the downlink as well as uplink scenarios.

10.4.1 Downlink Scenario

As X_D1 and X_D4 are the input signal vectors transmitted from the BS and the data vector received at CU-1, that is Y_D1, is the sum of data received due to X_D1 and X_D4

transmission via H_D1 and noise. Similarly, the data received at CU-4, that is Y_D4, is the sum of data received due to X_D1 and X_D4 transmission via H_D4 and noise. Therefore, the Y_D1 and Y_D4 are mathematically derived as:

$$Y_D1 = H_D1X_D1 + H_D1X_D4 + N_D1, \tag{10.14}$$

$$Y_D4 = H_D4X_D1 + H_D4X_D4 + N_D4. \tag{10.15}$$

Now, the throughput at each user in case of CR-NOMA, CR-MIMO, and CR-NOMA-MIMO is presented.

10.4.1.1 Throughput Computation for MIMO-CR-NOMA

There are N_T and N_R number of antennas at the BS and each user, respectively; therefore, MIMO appears in addition to the CR-NOMA which is named as MIMO-CR-NOMA. This section comprises the throughput analysis of CU-1 and CU-4 when there is significant interference at CU-4 (far user/cell-edge user) due to PU transmission from another nearby cell. As the CU-1 and CU-4 are the near and far users, respectively, which means the channel gains from the BS to the CU-4 are very small as compared to the channel gains from the BS to the CU-1 that formulates a relation between H_D1 and H_D4, i.e. $H_D1 > H_D4$. Therefore, the total power assigned to the CU-1 is less when compared with the CU-4 as per the concept of NOMA. Moreover, the CU-1 first decodes the signal of CU-4 and then subtracts it from the total received signal in order to yield the signal intended for CU-1. Consequently, the CU-1 does not experience any interference from the signal intended for CU-4. Therefore, the throughput achieved at CU-1 is derived as follows using the MIMO channel capacity expression [3]:

$$RD_{CU1\_MCN} = B\log_2 \left| I + H_D1Q_D1H_D1^H \right|, \tag{10.16}$$

where A^H represents the Hermitian of matrix A. On the other hand, the CU-4 decodes its intended signal X_D4 by considering the signal received at CU-4 due to X_D1 transmission via the channel H_D4, as interference. In addition to this, the CU-4 is at the edge and gets interference due to PU transmitted signal via channel matrix H_DP. Therefore, the throughput achieved at the CU-4 is mathematically represented as:

$$RD_{CU4\_MCN} = B\log_2 \left| I + \left(I + H_D1Q_D4H_D1^H \right)^{-1} \right.$$

$$\left. H_D4Q_D4H_D4^H \left(I + H_DPQ_DPH_DP^H \right)^{-1} \right|. \tag{10.17}$$

Now, if we break the MIMO channel matrix H_D1 into the individual channels using the singular value decomposition, we can take out the parallel channel coefficients [3]. The MIMO user throughput is computed as the sum of throughput achieved through individual antennas which further is the function of power assigned to each antenna and channel coefficient offered to each antenna by the channel. Thus, eqs. (10.16) and (10.17) are rewritten as:

$$RD_{CU1\_MCN} = \sum_{i=1}^{R_{H_D1}} B \log_2 \left(1 + \frac{P_D1_i \sigma_D1_i^2}{N_D1_i}\right), \tag{10.18}$$

$$RD_{CU4\_MCN} = \sum_{i=1}^{R_{H_D4}} B \log_2 \left(1 + \frac{P_D4_i \sigma_D4_i^2}{N_D4_i + P_D1_i \sigma_D4_i^2 + P_D P_i \sigma_D P_i^2}\right), \tag{10.19}$$

where $P_D P_i$ denotes the power assigned for the ith antenna of PU transmitter, where R_{H1} and R_{H4} is the number of nonzero singular values of the channel matrix H_D1 and H_D4, respectively. $P_D P_i$ denotes the power assigned for the ith antenna of PU transmitter: however, σ_D1_i, σ_D4_i, and $\sigma_D P_i$ denotes the ith singular values of the channel matrix H_D1, $\cdot H_D4$, and, $H_D P$, respectively. Further, the throughput of CU-4 is analyzed when the interference at CU-4 is assumed to be zero due to either the absence of PU or very small channel gain from PU transmitter to CU-4. The throughput at CU-4 remains same as in eq. (10.19) except the interference term due to PU transmission, that is $P_D P_i \sigma_D P_i^2$. Therefore, the data throughput at the CU-4 is mathematically represented as:

$$R_{CU4\_MNO} = \sum_{i=1}^{R_{H4}} B \log_2 \left(1 + \frac{P_D4_i \sigma_D4_i^2}{N_D4_i + P_D1_i \sigma_D4_i^2}\right). \tag{10.20}$$

10.4.1.2 Throughput Computation for CR-NOMA Systems

The MIMO-CR-NOMA is a generalized framework from where we can achieve the CR-NOMA systems as its special case when the BS and CUs are equipped with single antenna. For the throughput calculation, the generalized eqs. (10.18) and (10.19) are used when we consider $R_{H1} = 1$ and $R_{H4} = 1$, which means only single channel. Therefore, the throughput at CU-1 and CU-4 are derived as:

$$RD_{CU1\_CN} = B \log_2 \left(1 + \frac{P_D1 \sigma_D1^2}{N_D1}\right), \tag{10.21}$$

$$RD_{CU4\_CN} = B \log_2 \left(1 + \frac{P_D4 \sigma_D4^2}{N_D4 + P_D1 \sigma_D4^2 + P_D P \sigma_D P^2}\right). \tag{10.22}$$

10.4.1.3 Sum Throughput for CR-OMA, CR-NOMA, CR-MIMO, and CR-NOMA-MIMO Frameworks

In case of CR-OMA, the Sector-1 supports only single-user CU-1 with single antenna; therefore, the sum throughput of the CR-OMA framework is as follows:

$$SRD_{CR-OMA} = RD_{CU1\_CN}. \tag{10.23}$$

However, the CR-NOMA supports two users that are CU-1 and CU-4; therefore, the sum throughput of the CR-NOMA system is as follows:

$$SRD_{CR-NOMA} = RD_{CU1\_CN} + RD_{CU4\_CN}. \tag{10.24}$$

Further, in case of CR-MIMO, the Sector-1 supports single-user CU-1 but with multiple antennas. Therefore, the sum throughput of the CR-MIMO systems is derived as follows:

$$SRD_{CR-MIMO} = RD_{CU1\_MCN}. \tag{10.25}$$

Moreover, in case of the MIMO-based CR-NOMA, the Sector-1 supports two users that are CU-1 and CU-4 with multiple antennas; therefore, the total throughput in this case is derived as follows:

$$SRD_{MIMO-CR-NOMA} = RD_{CU1\_MCN} + RD_{CU4\_MCN}. \tag{10.26}$$

10.4.2 Uplink Scenario

During the uplink scenarios, the CU-1 and CU-4 transmit the signals toward the BS and the signal received at the BS is the superposition-coded signal due to CU-1 and CU-4 transmission. Therefore, the BS needs to separate the CU-1 and CU-4 signals using the SIC techniques. The main difference between the downlink and uplink scenarios is that the interference at PU limits the power transmission by the CU-4. Therefore, in order to consider the interference at PU, we have exploited an important metric which is known as interference efficiency [45, 46].

10.4.2.1 Throughput Computation for MIMO-CR-NOMA

The BS initially decodes the signal of CU-4 by considering the signal due to CU-1 transmission as interference; however, the throughput of CU-1 is computed by subtracting the CU-4 signal from the total received signal. Therefore, the throughput of the CU-1 and CU-4 is computed as follows:

$$RU_{CU1\_NCM} = B \log_2 \left| I + H_U 1 Q_U 1 H_U 1^H \right|, \tag{10.27}$$

$$RU_{CU4\_NCM} = B \log_2 \left| I + \left(I + H_U 1 Q_U 1 H_U 1^H \right)^{-1} H_U 4 Q_U 4 H_U 4^H \right|. \tag{10.28}$$

In terms of power, eqs. (10.27) and (10.28) are rewritten as follows:

$$RU_{CU1\_NCM} = \sum_{i=1}^{R_{H_U 1}} B \log_2 \left(1 + \frac{P_U 1_i \sigma_U 1_i^2}{N_U B S_i} \right), \tag{10.29}$$

$$RU_{CU4\_NCM} = \sum_{i=1}^{R_{H_U 4}} B \log_2 \left(1 + \frac{P_U 4_i \sigma_U 4_i{}^2}{N_U BS_i + P_U 1_i \sigma_U 1_i{}^2} \right). \qquad (10.30)$$

The sum throughput for different cases of CR-OMA, CR-NOMA, CR-MIMO, and MIMO-CR-NOMA frameworks in the downlink scenarios can be computed similarly as computed for the downlink scenarios.

10.4.2.2 Throughput Calculation for CR-NOMA Systems

Similar to the downlink scenario, the MIMO-CR-NOMA is a generalized framework from where we can achieve the CR-NOMA systems as its special case when the BS and CUs are equipped with single antenna. For the throughput calculation, the generalized eqs. (10.29) and (10.30) are used when we consider $R_{H1} = 1$ and $R_{H4} = 1$, which means only single channel. Therefore, the throughput at CU-1 and CU-4 are derived as:

$$RU_{CU1\_CN} = B \log_2 \left(1 + \frac{P_U 1 \sigma_U 1^2}{N_D BS} \right), \qquad (10.31)$$

$$RU_{CU4\_CN} = B \log_2 \left(1 + \frac{P_U 4 \sigma_U 4^2}{N_U BS + P_U 1 \sigma_U 1^2} \right). \qquad (10.32)$$

10.4.2.3 Sum Throughput for CR-OMA, CR-NOMA, CR-MIMO, and CR-NOMA-MIMO Frameworks

In case of CR-OMA, the Sector-1 supports only single-user CU-1 with single antenna; therefore, the sum throughput of the CR-OMA framework is as follows:

$$SRU_{CR-OMA} = RU_{CU1\_CN}. \qquad (10.33)$$

However, the CR-NOMA supports two users that are CU-1 and CU-4; therefore, the sum throughput of the CR-NOMA system is as follows:

$$SRU_{CR-NOMA} = RU_{CU1\_CN} + RU_{CU4\_CN}. \qquad (10.34)$$

Further, in case of CR-MIMO, the Sector-1 supports single-user CU-1 but with multiple antennas. Therefore, the sum throughput of the CR-MIMO systems is derived as follows:

$$SRU_{CR-MIMO} = RU_{CU1\_MCN}. \qquad (10.35)$$

Moreover, in case of the MIMO-based CR-NOMA, the Sector-1 supports two users that are CU-1 and CU-4 with multiple antennas; therefore, the total throughput in this case is derived as follows:

$$SRU_{MIMO-CR-NOMA} = RU_{CU1\_MCN} + RU_{CU4\_MCN}. \qquad (10.36)$$

10.4.2.4 Computation of Interference Efficiency of CU-4 In Case of CR-MIMO-NOMA

The interference efficiency for the CU-4 is defined as the data rate achieved by the CU-4 per unit of interference power introduced at the PU [45, 46]. The interference power (IP) introduced at PU is the power received at PU due to CU-4 transmission via the $H_U P$ is

$$IPU_{CU4\_NCM} = \sum_{i=1}^{R_{H_U P}} \left(P_U 4_i \sigma_U P_i^2 \right). \tag{10.37}$$

Therefore, the interference efficiency of CU-4 is defined as the throughput achieved per unit of interference power introduced at PU which is computed as follows:

$$IEU_{CU4\_NCM} = \frac{RU_{CU4\_NCM}}{IPU_{CU4\_NCM}}. \tag{10.38}$$

10.5 Simulation and Results Analysis

This section presents the simulation details with results and discussion obtained using the simulation tool MATLAB-2010 [47], for the downlink as well as uplink scenarios. We have considered a frequency-selective fading channel with different numbers of transmitting and receiving antennas for downlink and uplink scenarios where the Doppler shift and sampling frequency of input signal is considered as 0 Hz and 1 MHz, respectively. Further, we have considered N_{BS}, N_{CU1}, N_{CU4}, and N_{PUT} number of antennas at the BS, CU-1, CU-4, and PU, respectively.

10.5.1 Simulation Results for Downlink Scenario

The MIMO channel is considered with $N_{BS} = N_{PUT} = N_T$, $N_{CU1} = N_{CU4} = N_R$ number of transmitting and receiving antennas, respectively. The channel gain matrices $H_D 1$ and $H_D 4$ are assumed to be related via a relation $H_D 4 = H_D 1/5$. Uniform power allocation method is used to allocate the power to each antenna of the CU-1 and CU-4. Moreover, it is worth mentioning that the sum of powers assigned to each antenna of CU-1 or CU-4 is predefined which is considered in terms of the SNR per unit channel gain $SNR_D 1 = \frac{P_D 1}{N_D 1}$ or $SNR_D 4 = \frac{P_D 4}{N_D 4}$. On the other hand, as per NOMA principle, the total power is the sum of powers assigned to each user, i.e. $P_D 14 = P_D 1 + P_D 4$ or $SNR_D 14 = SNR_D 1 + SNR_D 4$. The simulation value of $SNR_D 14$ is assumed as 10 dB. The $SNR_D P = P_D P/N_D 4$ is the SNR per unit channel gain from the PU transmitter to the CU-4 receiver. As the PU transmitter is considered far from the CU-4, therefore $SNR_DP = 1$ dB. The channel H_D1 is simulated using the MIMO Communication toolbox for N_T and N_R number of transmitting and receiving antennas, respectively.

Further, the singular values of the channel matrix are computed by using the singular value decomposition in order to obtain the parallel channels from the ith antenna of BS to the ith antenna of the CU-1 or CU-4. The variations of total throughput of CU-1 with the throughput of CU-4 for different numbers of transmitting and receiving antennas at BS and each user, respectively, is shown in Figure 10.2. The $N_T = N_R = 1$ is a special case of MIMO which makes it equal to the SISO. The increase in throughput of CU-1 results in the decrease in throughput of the CU-4 for all the considered cases as shown in Figure 10.2, which fully agrees with the NOMA concept. In addition to this, the increase in number of antennas results in the increase in throughput for the CU-1 and CU-4. The relation between the throughput of the CU-1 and CU-4 with the SNR per unit channel gain at CU-4 is depicted in Figure 10.3. It is perceived that the increase in SNR results in the increase in throughput of CU-1 as well as CU-4. Moreover, the throughput of CU-1 is significantly more as compared to that of the CU-4 and it is because of the less channel gains as well as interference due to CU-1 and PU transmission at the CU-4. In order to consider the scenarios where BS and users have different number

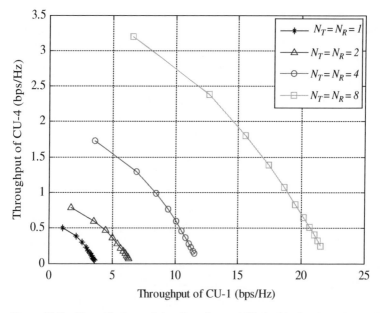

Figure 10.2 The variations of the throughput of CU-1 with the throughput of CU-4 for $N_T = N_R$ in the downlink scenario.

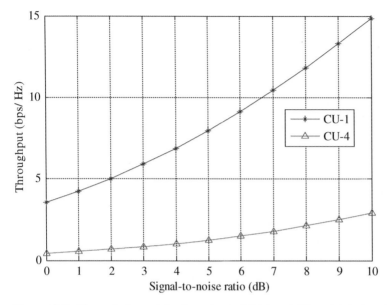

Figure 10.3 The variations of the throughput of CU-1 and CU-4 with SNR at CU-4 in the downlink scenario.

of antennas, we have simulated the results as depicted in Figures 10.4 and 10.5. In Figure 10.4, the number of antennas at the transmitter is 8; however, the number of antennas at each user varies from 2 to 8. The increase in the number of receiving antenna at each user results in the increase in the throughput of CU-1 as well as CU-4. Similarly, in Figure 10.5, the number of receiving antennas at each user is 8; however, the number of transmitting antennas varies from 2 to 8. As the number of receiving antennas at the CU-1 and CU-4 increases, there is a significant increase in the throughput of the CU-1 and CU-4. Moreover, the variations of the throughput of CU-1 with the SNR at CU-1 for CR-OMA, CR-NOMA, CR-MIMO, and CR-NOMA-MIMO systems are illustrated in Figure 10.6. The throughput in every case is increasing with increase in the SNR per unit channel gain at the CU-1. It is clear that the CR-NOMA-MIMO outperforms when compared to all other systems. It is worth mentioning that, in this graph MIMO-NOMA is missing because without CR, no channel is available and even after the use of MIMO-NOMA, the throughput of the system will be zero.

10.5.2 Simulation Results for Uplink Scenario

The MIMO channel is considered with $N_{CU1} = N_{CU4} = N_T$ and $N_{BS} = N_{PUT} = N_R$ number of transmitting and receiving antennas, respectively. The channel gain

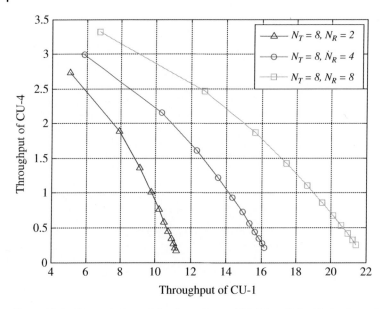

Figure 10.4 The variations of the throughput of CU-1 (bps/Hz) with the throughput of CU-4 (bps/Hz) for $N_T = 8$ and different values of N_R.

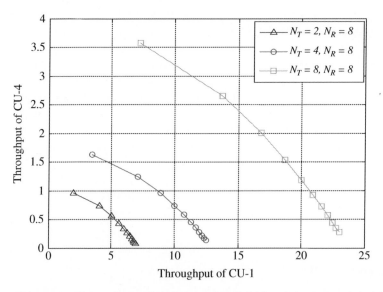

Figure 10.5 The variations of the throughput of CU-1 (bps/Hz) with the throughput of CU-4 (bps/Hz) for $N_R = 8$ and different values of N_T.

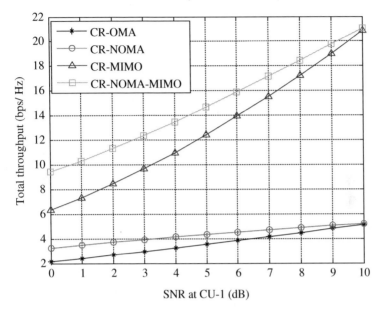

Figure 10.6 The variations of the throughput of CU-1 with the SNR at CU-1 for CR-OMA, CR-NOMA, CR-MIMO, and CR-NOMA-MIMO in the downlink scenario.

matrices H_U1 and H_U4 are assumed to be related via a relation $H_U4{=}H_U1/5$. Uniform power allocation method is used to allocate the power to each antenna of the CU-1 and CU-4. Moreover, it is worth mentioning that the sum of powers assigned to each antenna of CU-1 or CU-4 is predefined which is considered in terms of the SNR per unit channel gain $SNR_U1 = \frac{P_U1}{N_U1}$ or $SNR_U4 = \frac{P_U4}{N_U4}$. On the other hand, as per NOMA principle, the total power is the sum of powers assigned to each user, i.e. $P_U14 = P_U1 + P_U4$ or $SNR_U14 = SNR_U1 + SNR_U4$. The simulation value of SNR_U14 is assumed as 10 dB. The $SNR_UP = P_UP/N_U4$ is the SNR per unit channel gain from the PU transmitter to the CU-4 receiver. As the PU transmitter is considered far from the CU-4, therefore $SNR_UP = 1$ dB. The channel H_U1 is simulated using the MIMO Communication toolbox for N_T and N_R number of transmitting and receiving antennas, respectively. Further, the singular values of the channel matrix are computed using the singular value decomposition in order to obtain the parallel channels from the ith antenna of the CU-1 or CU-4 to the ith antenna of BS and PU. The variations of the throughput of CU-1 with that of the CU-4 for different numbers of transmitting and receiving antennas are shown in Figure 10.7. Similar to the downlink scenario, the throughput of the CU-4 and CU-1 has inverse relation; however, higher number of antennas results in the increased throughput region. Figure 10.8 depicts the relation between the total/sum throughput with the

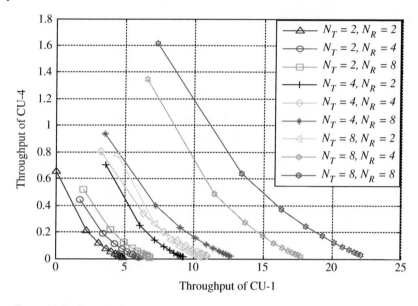

Figure 10.7 The relationship between the throughput of CU-4 (bps/Hz) and throughput of CU-1 (bps/Hz) for different values of N_T and N_R in uplink scenario.

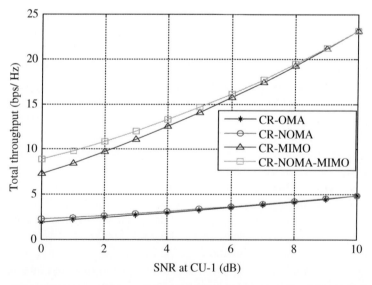

Figure 10.8 The variations of total throughput with the SNR at BS due to CU-1 transmission for CR-OMA, CR-NOMA, CR-MIMO, and CR-NOMA-MIMO in the uplink scenario.

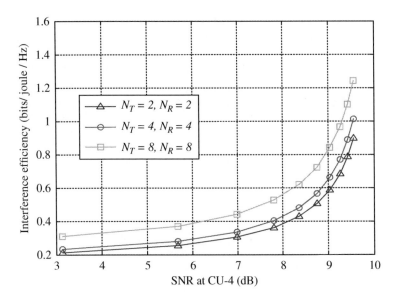

Figure 10.9 The relationship between the interference efficiency and SNR at BS due to CU-4 transmission in the uplink scenario.

SNR at the BS due to CU-1 transmission for different frameworks such as the CR-OMA, CR-NOMA, CR-MIMO, and CR-MIMO-NOMA/MIMO-CR-NOMA.

It is clear that the MIMO-CR-NOMA performs well as compared to that of the other frameworks. The relationship between the interference efficiency and SNR at BS due to CU-4 transmission for different numbers of transmitting and receiving antennas in the uplink scenario is presented in Figure 10.9. The interference efficiency increases with increase in the SNR at BS due to CU-4 transmission; however, more number of antennas in the MIMO systems results in the higher interference efficiency.

10.6 Summary

This chapter explores the simultaneous use of various technologies of the next-generation communication systems that are CR, MOMA, and MIMO for efficient spectral utilization. Since all of these three technologies provide different perspectives of the effective spectral utilization, simultaneous exploitation of these technologies results in the improved spectral efficiency for both the downlink and uplink scenarios; however, improved interference efficiency in the uplink scenario. The total throughput at each user is computed and it is perceived that use of multiple antennas at the BS and each user results in the significant improvement in the throughput. Moreover, in the downlink scenarios, the throughput of CU-4 is reduced due to interference from the PU transmission; however, in the uplink scenario, the same is limited in order to maintain the interference-free PU communication. In this chapter, the power allocation to the antenna is performed using uniform power allocation; however, power allocation optimization is a significant way to achieve higher throughput, which will be explored in the future communication.

References

1 Internet of Things (IoT) connected devices installed base worldwide from 2015 to 2025 (in billions). https://www.statista.com/statistics/471264/iot-number-of-connected-devices-worldwide/ (accessed 29 June 2019).

2 Haykin, S. (2005). Cognitive radio: Brain-empowered wireless communications. *IEEE Journal on Selected Areas in Communications* 23 (2): 201–220.

3 Mitola, J. and Maguire, G.Q. (1999). Cognitive radio: making software radio more personal. *IEEE Personal Communications* 6 (4): 13–18.

4 Thakur, P., Singh, G., and Satashia, S.N. (2016). Spectrum sharing in cognitive radio communication system using power constraints: a technical review. *Perspectives in Science* 8: 651–653.

5 Walko, J. (2005). Cognitive radio. *IET Review* 51 (5): 34–37.

6 Liu, Y., Qin, Z., Elkashlan, M. et al. (2017). Nonorthogonal multiple access for 5G and beyond. *Proceedings of the IEEE* 105 (12): 2347–2381.

7 Thakur, P., Kumar, A., Pandit, S. et al. (2019). Frameworks of non-orthogonal multiple access techniques in cognitive radio communication systems. *China Communications* 16 (6): 129–149.

8 Islam, S.M.R., Avazov, N., Dobre, O.A., and Kwak, K.S. (2017). Power domain non-orthogonal multiple access (NOMA) in 5G systems: potential and challenges. *IEEE Communications Surveys & Tutorials* 19 (2): 721–742.

9 Dai, L., Wang, B., Ding, Z. et al. (2017). A survey of non-orthogonal multiple access for 5G. *IEEE Communications Surveys & Tutorials* 19 (2): 721–742.

10 Yin, Y., Liu, M., Gui, G. et al. (2020). QoS-oriented dynamic power allocation in NOMA-based wireless caching networks. *IEEE Wireless Communications Letters* https://doi.org/10.1109/LWC.2020.3021204.

11 Abbasi, O., Yanikomeroglu, H., Ebrahimi, A., and Mokari, N. (2020). Trajectory design and power allocation for drone-assisted NR-V2X network with dynamic NOMA/OMA. *IEEE Transactions on Wireless Communications* https://doi.org/10.1109/TWC.2020.3008568.

12 Lee, I.H. and Jung, H. (2020). User selection and power allocation for downlink NOMA systems with quality-based feedback in Rayleigh fading channels. *IEEE Wireless Communications Letters* https://doi.org/10.1109/LWC.2020.3008174.

13 Zhang, H., Zhang, H., Long, K., and Karagiannidis, G. (2020). Deep Learning based radio resource management in NOMA networks: user association, subchannel and power allocation. *IEEE Transactions on Network Science and Engineering* https://doi.org/10.1109/TNSE.2020.3004333.

14 Jensen, M.A. and Wallace, J.W. (2004). A review of antennas and propagation for MIMO wireless communications. *IEEE Transactions on Antennas and Propagation* 52 (11): 2810–2824.

15 Paulraj, A.J., Gore, D.A., Nabar, R.U., and Bolcskei, H. (2004). An overview of MIMO communications – a key to Gigabit wireless. *Journals & Magazines* 92 (2): 198–218.

16 Yang, S. and Hanzo, L. (2015). Fifty years of MIMO detection: the road to large-scale MIMOs. *IEEE Communications Surveys & Tutorials* 17 (4): 1941–1988.

17 Thakur, P. and Tiwari, K. (2018). Error rate analysis of precoded-OSTBC MIMO system over generalized-K fading channel. In: *Advances in Systems, Control and Automation, Lecture Notes in Electrical Engineering*, vol. 442. Singapore: Springer.

18 Federal Communications Commission. (2002). Notice of proposed rule making and order: facilitating opportunities for flexible, efficient, and reliable spectrum use employing cognitive radio technologies. ET Docket No. 03-108.

19 Akyildiz, I.F., Lee, W.Y., Varun, M.C., and Mohanty, S. (2008). A survey on spectrum management in cognitive radio networks. *IEEE Communications Magazine* 46 (4): 40–48.

20 Thakur, P., Kumar, A., Pandit, S. et al. (2017). Spectrum mobility in cognitive radio network using spectrum prediction and monitoring techniques. *Physical Communication* 24: 1–8.

21 Higuchim, K. and Benjebbour, A. (2015). Non-orthogonal multiple access (NOMA) with successive interference cancellation for future radio access. *IEICE Transactions on Communications* E99-B (3): 403–414.

22 Ding, Z., Lei, X., Karagiannidis, G.K. et al. (2017). A survey on non-orthogonal multiple access for 5G networks: research challenges and future trends. *IEEE Journal on Selected Areas in Communications* 35 (10): 2181–2195.

23 Mietzner, J., Schober, R., Lampe, L. et al. (2009). Multiple-antenna techniques for wireless communications—a comprehensive literature survey. *IEEE Communications Surveys & Tutorials* 11 (2): 87–105.

24 Amin, M.R. and Tarapasiya, S.D. (2012). Space-time coding scheme for MIMO systems—literature survey. *Process Engineering* 38: 3509–3517.

25 Alamouti, S.M. (2018). A simple transmit diversity technique for wireless communications. *IEEE Journal on Selected Areas in Communications* 16 (2): 1451–1458.

26 Vouyioukas, D. (2013). A survey on beamforming techniques for wireless MIMO relay networks. *International Journal of Antennas and Propagation* 2013: 1–22.

27 Alexiou, A. and Hardt, M. (2004). Smart antenna technologies for future wireless systems: trends and challenges. *IEEE Communications Magazine* 42 (9): 90–97.

28 Wornell, G. and Trott, M. (1997). Efficient signal processing techniques for exploiting transmit antenna diversity on fading channels. *IEEE Transactions on Signal Processing* 45 (1): 191–205.

29 Faissal, E.B., and Hussain, B.A. (2015). Efficient performance evaluation for EGC, MRC, and SC receivers over Weibull multipath fading channel. *Proceedings of International Conference on Cognitive Radio Oriented Wireless Networks (CROWNCOM)*, Doha, Qatar, 346–357.

30 Lv, L., Chen, J., Ni, Q. et al. (2018). Cognitive non-orthogonal multiple access with cooperative relaying: a new wireless frontier for 5G spectrum sharing. *IEEE Communications Magazine* 56 (4): 188–195.

31 Shokair, M., Saad, W., and Ibraheem, S.M. (2018). On the performance of downlink multiuser cognitive radio inspired cooperative NOMA. *Wireless Personal Communications* 101 (2): 875–895.

32 Liu, X., Wang, Y., Liu, S., and Meng, J. (2018). Spectrum resource optimization for NOMA-based cognitive radio in 5G communications. *IEEE Access* 6: 24904–24911.

33 Miridakis, N.I., Tsiftsis, T.A., and Alexandropoulos, G.C. (2018). MIMO underlay cognitive radio: optimized power allocation, effective number of transmit antennas and harvest-transmit tradeoff. *IEEE Transactions on Green Communications and Networking* 2 (4): 1101–1114.

34 Hao, W., Muta, O., Gacanin, H., and Furukawa, H. (2018). Power allocation for massive MIMO cognitive radio networks with pilot sharing under SINR requirements of primary users. *IEEE Transactions on Vehicular Technology* 67 (2): 1174–1186.

35 Satheesan, U. and Sudha, T. (2016). Achievable data rate in hybrid MIMO cognitive radio networks. *Processing Technology* 24: 873–879.

36 Kumar, V., Malarvizhi, S., and Hariprasath, M. (2015). Low cost reception scheme for MIMO cognitive radio. *Proceedings of 2nd International Conference on Electronics and Communication Systems (ICECS)*, Coimbatore, India (February 2015), 43–46.

37 Iwata, R., Va, V., Sakaguchi, K., and Araki, K. (2013). Experiment on MIMO cognitive radio using Tx/Rx beamforming. *Proceedings 2013 IEEE 24th Annual International Symposium on Personal, Indoor, and Mobile Radio Communications (PIMRC)*, London, UK (September 2013), 2871–2875.

38 Liu, Y. and Dong, L. (2014). Spectrum sharing in MIMO cognitive radio networks based on cooperative game theory. *IEEE Transactions on Wireless Communications* 13 (9): 4807–4820.

39 Choi, J. (2016). On the power allocation for MIMO-NOMA systems with layered transmissions. *IEEE Transactions on Wireless Communications* 15 (5): 3226–3237.

40 Senel, K., Cheng, H.V., Bjornson, E., and Larsson, E.G. (2019). What role can NOMA play in massive MIMO? *IEEE Journal on Selected Topics in Signal Processing* 13 (3): 597–611.

41 Lin, C., Chang, Q., and Li, X. (2019). A deep learning approach for MIMO-NOMA downlink signal detection. *Sensors* 19 (11): 1–22.

42 Yu, Y., Chen, H., Li, Y. et al. (2017). Antenna selection in MIMO cognitive radio-inspired NOMA systems. *IEEE Communications Letters* 21 (12): 2658–2266.

43 Givi, S.S., Shayesteh, M.G., and Kalbkhani, H. (2020). Energy-efficient power allocation and user selection for mmWave-NOMA transmission in M2M communications underlaying cellular heterogeneous networks. *IEEE Transactions on Vehicular Technology* https://doi.org/10.1109/TVT.2020.3003062.

44 Goldsmith, A. (2012). *Wireless Communications*. Cambridge, United Kingdom: 4.

45 Mili, M.R. and Musavian, L. (2017). Interference efficiency: a new metric to analyze the performance of cognitive radio networks. *IEEE Transactions on Wireless Communications* 16 (4): 2123–2138.

46 Thakur, P., Kumar, A., Pandit, S. et al. (2018). Spectrum monitoring in heterogeneous cognitive radio network: how to cooperate? *IET Communications* 12 (17): 2110–2118.

47 (2010a). *MATLAB and Statistics Toolbox Release*, 2010. Natick, Massachusetts: The Math Works, Inc.

11

Interference Management in Cognitive Radio Networks

11.1 Introduction

Interference is the disruption in the existing signal operating at one particular frequency channel by the other signal that also starts operating on the same frequency channel either deliberately or by misapprehension. Therefore, it is very popular and worth discussing about the interfering channel and as per Carleial [1–3], the interference channel is a communication medium shared by the M number of sender–receiver pair, and more than one pair try to start communication on that channel or more than two transmitters transmit the data at the same time. The general situation for interference channel is depicted in Figure 11.1 which is defined as follows: M number of transmitters transmit the data toward M number of intended receivers. It is worth and interesting to mention that even if individual non-sharing channels are available between transmitter and receiver pair, the interference is still a problem at the receiver end and cannot be avoided. The terminology of the links between the transmitter and receivers with their links is defined as follows. The link between the transmitter and its intended receivers is known as communication link; however, the link between the transmitter and its unintended receivers is known as interference link.

The signal received at the unintended receiver via the interference link is known as interference. For the M number of direct communication links, there are $M\times(M-1)$ number of interference links. For example, $M = 3$ which means there are three transmitters and three receivers, then the direct communication links will be three; however, $3\times2 = 6$ number of interference links will be there. As there are M direct communication links, therefore, the M number of independent information sources can be transmitted from the transmitter toward the receiver. In general, the number of transmitters and receivers are different; therefore, it is worth analyzing the direct communication and interference links for this generalized case. Assume that the number of transmitters are M; however, the number of

Spectrum Sharing in Cognitive Radio Networks: Towards Highly Connected Environments, First Edition. Prabhat Thakur and Ghanshyam Singh.
© 2021 John Wiley & Sons, Inc. Published 2021 by John Wiley & Sons, Inc.

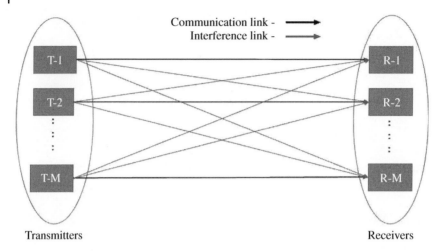

Figure 11.1 A scenario of communication and interference link from transmitter end to the receiver end.

receivers are N. Thus, in an $M \times N$ network, each transmitter is permitted to establish an individual communication with 2^{N-1} combinations of the receivers. Therefore, the capacity regions for an $M \times N$ network is $M \times 2^{N-1}$ dimensional [2, 3].

The interference at the receiver is assumed to be an undesired entity because it shows inverse relation with the capacity regions of the users which can be easily perceived by analyzing the Shannon's channel capacity formula which is defined as $C = B \log_2 \left(1 + \frac{Signal\ received\ power}{Noise\ power\ +\ Interfernce\ Power}\right)$, where C and B denotes the channel capacity and bandwidth available, respectively. The increase in interference power results in the reduction in the channel capacity. Therefore, it is worth analyzing the prominent scenarios of the interference introduction either at the networks' receiving end.

We have seen in Chapters 1 and 3 that the cognitive radio is a potential framework which transforms the concept of dynamic spectrum access into reality, and is going to play a vital role in the next-generation communication systems. Therefore, it is worth and compulsory to investigate the interference scenarios in the CR networks for successful implementation in the real world. The potential scenarios of interference in the CR networks are as follows:

- The use of licensed spectrum by the CU imposes an inherent challenge of interference avoidance at the PU network.
- The multiple CUs in CRN try to access the same group of channels which also lead to the data collision among the CUs and results in the multiple access interference.

- The interference can be either at the CRN or PU networks due to PU or CU transmission.

It is worth discussing the potential terminologies of CR and interference to real-ize the effect and to find the probable ways of interference mitigation in the CRNs. Those potential terminologies are white space, grey space, black space, and inter-ference temperature, which are illustrated as follows.

11.1.1 White space

White space is defined as the geographical region where a particular channel or a set of channels remains unoccupied by the PU communication or if that band of spectrum channels is not allocated for certain kind of services. One prominent example of white spaces appears in terms of television white spaces (TVWS), when as per IEEE 802.22 standards have permitted the use of analog TV spectrum bands by the wireless regional area networks (WRANs) [4].

11.1.2 Grey Spaces

The geographical location where the PUs use the spectrum channel or set of chan-nels on the non-regular basis which means for a particular time, the channel can be free; however, for other time slot, it is used by the PU.

11.1.3 Black Spaces

The geographical location where the spectrum channel or set of channels are occu-pied by the PU communication for most of the time and have very rare chances to be free from the PUs' communication.

11.1.4 Interference Temperature

As per FCC, the interference temperature metric is defined as the maximum received strength of noise plus undesired signals at particular radio frequency bands which is acceptable for successful decoding of the received desired signal [5]. For the first time the FCC has proposed to exploit the concept of interference temperature to allow unlicensed operation in the 6525–6700 MHz band that was allocated for satellite uplinks and terrestrial fixed microwave links and in the 12.75–13.25 GHz band, assigned for the broadcast auxiliary stations, cable relay stations, and other fixed microwave services [6].

It is important to calculate the interference temperature mathematically, if the f_c and B_c denotes the central frequency (hertz) and bandwidth (hertz) of the channel C, respectively. P signifies the power received at the receiving end in Watt and k is the Boltzmann's constant having mathematical value 1.38×10^{-23} Joule/Kelvin. Then, the interference temperature for the channel C is defined as [7]:

$$IT_C = \frac{P(f_c, B_c)}{kB_c}. \tag{11.1}$$

The FCC also mentions the meaning of interference temperature in terms of the power flux density (e.g. $\mu W/m^2$ over a bandwidth) multiplied by the antenna capture area in meters square divided by both the associated RF bandwidth in Hertz and Boltzmann's constant. Further, the interference temperature density can also be defined as the interference temperature per unit area, expressed in units of Kelvin per meter square, and calculated as the interference temperature divided by the effective capture area (aperture) of the receiving antenna. This quantity could be measured for particular frequencies using a reference antenna and, thereafter, would be independent of receiving antenna characteristics.

Further, this chapter is structured as follows. The next section comprises the interfering and non-interfering CRN. In Section 11.3, various interference avoidance and mitigation techniques are well explored. Section 11.4 presents the cross-layer perspective of interference. Section 11.5 illustrates the interference management and potential challenges for different constituents of the cognitive cycle and finally Section 11.6 summarizes the chapter.

11.2 Interfering and Non-interfering CRN

The CRNs on the basis of interference are categorized into two prime types that are the interfering CRN and non-interfering CRN as shown in Figure 11.2, which are further illustrated in detail.

11.2.1 Interfering CRN

The interference in the interfering CRN is classified into following two prime types of networks, which are inter-network and intra-network. The inter-network interference is defined as the interference between two networks and that two networks can be two CUs networks, two PUs networks, or one CU and one PU network. The inter-network interference between two PU-networks is much popular in the conventional cellular communication, satellite, radar, etc., and various mitigation techniques are well explored in the literature such as co-channel and adjacent channel interference (ACI) mitigation techniques [8–14] and its discussion is away from CR interference management; however, the interference between two CUs, one PU, and one CU is of prime importance. The intra-network interference is

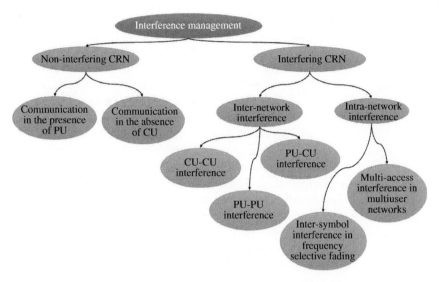

Figure 11.2 Interference classification in the cognitive radio networks.

defined as the possible distortions at any CU receivers due to other CUs in the networks.

11.2.2 Non-Interfering CRN

The non-interfering CRN is the idle and most desired scenario or framework to establish the CU's communication with zero or minimum interference. The prime idea behind the non-interfering CRN is that the CUs are allocated with a spectrum on which the probability of PU communication is very low such as TVWS. With the introduction and compulsion of digital TV over analog TV, most of the TV bands are unutilized for further allocation of these bands to new services. Therefore, the CUs using those spectrum bands have very less probability of interference from the PU network; however, interference among the CUs during multiple access remains the challenge. Thus, the CRN without interference comprises two scenarios that are CRN in the presence of PU and CRN in the absence of PU. The CRN in the presence of PU signifies the scenarios where PU still uses the TVWS even with lower probability; conversely the scenarios where PU uses the TVWS with zero probability signifies the communication in the absence of PU.

In the non-interfering CRN, the CUs are assumed to be establishing their communication on the channels which are assumed to be completely unoccupied or in other sense we can say that the CU establishes its communication only on the white spaces.

A) **CU-to-PU Interference:** The most popular and undesired interference of the CRN is the interference introduced at the PU receiver because of the CU transmission via the interference link. The entire idea and backbone of the CRN relies on the fact that the CU-to-PU should be less than the tolerable limit of interference which is already defined by the PU network and known as interference temperature as defined in eq. (11.1) [7]. The idea of interference temperature provides the flexibility to follow the receiver-centric approach; however, in the conventional interference networks, it follows the transmitter-centric approach. In the transmitter-centric approach, the interference is managed by controlling the transmission power with reference to time and location of the transmitter. Conversely, in the receiver-centric approach, the interference is managed at the PU receiver end by using the interference temperature limit.

B) **PU-to-CU Interference:** The PU-to-CU interference is defined as the amount of received signal at the CU receiver due to PU transmission while PU is assumed to be operating at its particular assigned channel at a given time and in geographical region. The PU-to-CU interference is an integral part of the CRN and computing this interference is the first step of the CR cycle, that is, the spectrum sensing. Once the interference due to PU-to-CU transmission is available at the CRN, only then other decisions of spectrum accessing and sharing are performed. To classify the level of interference at the CU due to PU transmission, the particular geographical location at particular time and frequency channel is divided as per their received signal strength, which is illustrated as follows.

The report of FCC revealed that there are a number of white spaces as per the allocated radio spectrum [15] and a theoretical model is proposed by Chen et al. [16] to characterize the spatial distribution of white, grey, and black spaces in the presence of a random primary network with homogeneous nodes. The authors have presented a good work with the mathematical modeling of the interference temperature. In [17], the authors have considered MIMO-based cognitive radio environment and have presented the interference mitigation techniques for the considered environment. The authors have exploited the subspace-projection-based precoding schemes that are full-projection and partial projection based in order to manage the interference from the CU to PU receivers and vice-versa. The performance of the proposed schemes is measured by the fact the full-projection-based precoding scheme improves the interference at PU significantly with increased CR throughput. Conversely, the partial projection-based scheme enhances the CR throughput by partially projecting its transmission onto the null space. Further, the authors have considered a multiuser MIMO interference system and have formulated a multicriteria optimization problem [18]. The authors have exploited Pareto region criteria by defining sufficient conditions for the convexity of the region. Two prominent convexification approaches are used that are multistage interference cancellation and a full-projection-based interference avoidance scheme. Further, the Nash equilibrium is used to transform the multi-objective problem into the single objective optimization problem. The authors have also investigated the convexity of the rate region and the existence of the FP-based NB solution for MIMO interference systems, numerically. Chen et al. [19, 20] have extended their work for the aggregate interference modeling in cognitive radio networks with power and contention control.

C) **CU-CU interference:** The CU-CU interference is defined as the undesired received power at the particular CU due to the transmission of the other CUs in the same CRN. Thus, we can say that the CU-CU interference is very similar to the kind of any other intra-network interference in the conventional networks; however, one additional difference is that CU also needs to address the interference introduction/reception to/from the PU network.

11.3 Interference Cancellation Techniques in the CRN

In the previous section, we have seen that for the successful implementation of the CRN, the interference management needs to be performed. Therefore, in this section, we have classified the interference cancellation techniques into two parts that are at the transmitter end and the receiver end as shown in Figure 11.3.

11.3.1 At the CU Transmitter

The key idea behind the CRN is to use the PU channel and exploit it for CUs' communication while avoiding the interference at the PU receiver. The interference due to CU transmission at the PU receiver can be avoided by controlling or managing the signal at the CU transmitter and this process in known as interference cancellation at the CU transmitter. The potential phenomenon of signal processing is applied on signal before transmission in order to mitigate the effect of interference at the PU receiver. The potential techniques illustrated in the literature for interference cancellation at the CU transmitter are spectrum shaping, predistortion filtering, spread spectrum, and transmit beamforming, which are elaborated as follows.

A) **Spectrum Shaping:** The spectrum shaping is defined as a potential technique of waveform generation for the transmitted signal in order to minimize the power leakage in the PU bands either in the form of adjacent channel or co-channel interference [21]. There are some well-explored spectrum shaping

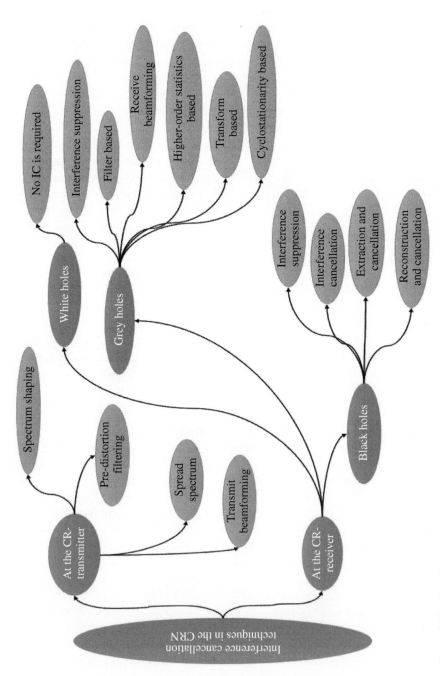

Figure 11.3 Interference cancellation techniques in the CRN at the CR transmitter and receiver.

techniques for the ultra-wideband (UWB) and software-defined radio where the key emphasis is to design a pulse waveform that can dynamically regulate as per the spectral environment and produce desired spectral shapes/notches [21–26]. Preferably, the signal waveforms should be constructed as the linear combination of a limited number of orthogonal basis functions, also known as the core pulse wavelets. The basic functions should be bandwidth limited, time limited, orthogonal to each other, and flexible enough to form any desired shape of the power spectrum. The orthogonal frequency division multiple access (OFDMA) is a potential and very popular technique of the spectrum shaping and active interference cancellation technique is used to suppress the large side-lobe in the OFDM signal. The authors have exploited a network coding for spectrum shaping where each CU carries out the adaptive channel sensing by dynamically updating the list of the (predicted) idle PU channels and providing priority to these channels for spectrum sensing. The proposed phenomenon of network coding improves the throughput of the PUs and CUs. Zhang and Mitra [23] have presented a potential approach to establish the CU communication where the probability to get the PU channel is very low; however, it become possible with the help of spectrum shaping of the CU signal. Lee et al. [25] have explored an interference management problem in the multi-carrier CRN where a potential framework is proposed for protecting the PU by reshaping the envelope of interference on each subcarrier. Afterward, the authors have coined the concept of minimum individual interference budget that the PU needs to permit on each subcarrier in order to minimum quality of service at the CUs. Further, a joint optimal transmit power and interference budget allocation algorithm for the PU to achieve improved throughput is proposed, which confirms its effectiveness via simulation results when comparing with the pre-existing approaches.

B) **Pre-distortion Filtering:** The ACI is a prominent interference for the inter- as well as intra-network. The key source of the ACI is the transmission nonlinearity due to cascaded nonlinear components in the radio-frequency chain. The less interference introducing CR transmitters demand high linearity; however, this requires the high price and results in less power-efficient system. The potential way to reduce the need of linearity is to use pre-distortion technique. A pre-distortion module pre-compensates the signal entering a nonlinear device for anticipated distortion so that the output from the combined pre-distortion module and nonlinear device is undistorted [27]. The pre-distortion filtering is exploited to suppress the ACI.

C) **Spread Spectrum:** The spread spectrum is a key technology of the wireless communication specifically saying for code division multiple access (CDMA), which is used for low power transmission in the UWB communication [28–31]. The originated wideband CU signals have low power spectral density and thus, the interference to a particular narrowband PU network is reduced. However, the cost we pay for the spread spectrum wideband is interference introduction to the multiple PUs. In the context of CR, the spread spectrum reduces the co-channel interference at the expense of increasing the interference in adjacent channels.

D) **Transmit Beamforming:** Similar to the receive beamforming, transmit beamforming [32] and transmit precoding [33] can be applied to CR networks for mitigating interference to the primary systems by adaptively choosing weights on the transmit antenna elements to form an emission pattern with nulls toward the directions of primary receivers. It is an effective and flexible approach to balance between the interference minimization for the primary users and the signal-to-interference-plus-noise ratio (SINR) maximization for the secondary users. The implementations of transmit beamforming are more complicated than the receive beamforming since a feedback mechanism is required to inform CR transmitters about the instantaneous channel-state information (CSI). The transmit beamforming is effective in suppressing both the co-channel and ACI at the expense of high hardware costs.

11.3.2 At the CR-Receiver

Geographical regions on the basis of the frequency channel and at particular time are categorized into three spaces which are white-space, grey-space, and black-space and the interference mitigation at the CR-receiver is further divided as follows. As we have seen in the previous discussion that the white spaces are the geographical regions where the probability of PU communication is very low, which means there is no question of interference at the PU network due to CRN; therefore, there is no need of interference cancellation techniques in the white spaces. On the other hand, in the grey spaces, the probability of the PU communication is significant and therefore, there is need of interference cancellation techniques. Some popular interference cancellation techniques are the interference suppression, filter based, transform based, and receive beamforming. Prominent interference approaches for black holes are interference suppression, interference cancellation, reconstruction and cancellation, and extraction and reconstruction [27]. Further, the interference cancellation techniques for the grey hole and black hole regions are classified as follows.

A) **Grey Hole IC techniques:** The interference amount at the CU receivers due to PU communication lies between low to medium in case of grey spaces; therefore, the interference suppression is a suitable technique which suppresses the strength of PU signal. The received signal at the CU receiver is suppressed [27].

i) **Filter-based approach:** The key principle of the filter-based approach is to differentiate the signal of interest (SOI) for CU and interference signals based on their power-spectrum properties which is achieved by designing a filter that can provide a desired frequency–response function and improves the regions of spectrum with high SINR in addition to suppressing the same frequencies with low SINR [20, 34, 35]. The Weiner filter is very popular for this purpose which works with prerequisites of the power spectrum (covariance) of SOI and interference. If the covariance is unavailable, then adaptive filter is used to regulate the weights of the filter. The filter-based approach is a well-explored technique having low complexity; however, as it exploits the only power spectrum, the suppression of co-channel interference is a difficult task. This limits the use of filter-based approach in the interference cancellation technique for CR networks.

ii) **Signal Transform Based:** The signal transform is a very important tool in the signal processing to retain/remove some of the desired/undesired properties of the signal and that is the key of the signal transform method [36]. Initially, the received signal is transformed into a particular domain such as frequency domain using Fourier transform, then remove/retain certain transform components, and then finally, the inverse transform is used to synthesize the SOI. One popular example is the OFDMA, which exploits the frequency domain to remove narrowband interference by eliminating the interfered sub-bands. It is worth mentioning that achieving the complete interference cancellation in the SOI using either time or frequency domain processing is difficult. Therefore, the simultaneous exploitation of both means that time-frequency analysis proves to be a potential candidate for signal separation and classification [37]. The popular time frequency analysis tools are the short-time Fourier transform (STFT), wavelet, and chirplet. The selection of the particular transform domain relies on the fact that the signal and interference should have different and separate components in that particular domain. The CU signal/waveform can be designed by keeping in mind all the properties of PU signal and make sure that some separate components are added during waveform design, thus the signal transform-based approach is much suitable for the CU networks.

iii) **Receive Beamforming:** Similar to the transform beamforming, the receive beamforming also plays an important role for the interference management. The beamforming is a technique that exploits the spatial domain to separate the SOI and interference by using the different spatial signatures. This put a prerequisite on the receive hardware that multiple antennas must be mounted on the CR receiver which applies weights on the antenna array to form a desirable reception pattern. If the SOI and interference arrive from different directions, a multi antenna CR receiver can adaptively form different beam patterns to enhance reception in the direction of the SOI and put nulls toward the directions of the interference. In the complex propagation environments, a trade-off is often needed between SOI signal enhancement and interference suppression. The beamforming suppresses the co-channel interference and number of antenna elements improves the suppression gain. However, the hardware cost is high because of the need of using multiple antennas and RF chains. Moreover, the achievable gain is opportunistic since it relies on a favorable propagation condition. Overall, the beamforming approach is a promising candidate for CR base stations and access points. Even for single-antenna CR users, the collaborative beamforming can potentially be used to obtain a high-interference suppression gain.

iv) **Cyclostationarity-Based Approach:** Cyclostationarity is the property of the signal according to which certain statistical parameters of the signal, such as frequency, data rate, phase offset, etc., repeat their self periodically and potential signal processing tool that is used to detect this repetition of the statistical parameters is known as Cyclostationarity-based approach [38, 39]. This signal processing tool exploits the spectral correlation density function and is a much powerful method when compared with only data rate calculation for signal analysis. The signals overlapping in the power spectrum and transform domain can be separable in the cyclic spectrum; therefore, the Cyclostationarity-based approach is very popular for the spectrum sensing techniques in the CR networks and can provide the 10dB interference separation. The Cyclostationarity-based approach is more suitable for the interference separation and cancellation; however, it requires training to compute optimal weights of the filter and that the SOI should have a large excess bandwidth. These two requirements, however, can be satisfied in a CR design and 17dB interference separation can be achieved [26].

v) **Higher-order statistics:** The first- and second-order statistics calculation and use of these statistics for signal processing such as the interference cancellation, signal detection, and separation is very common. However, if we could not separate the interference and signal at the second order statistics, it is probable that the high-order statistics can provide separate signature for the interference and SOI [40]. One potential application of higher order statistics plays an important role when the multiple diversity copies of the received signal are available which can be obtained from the antenna arrays. The higher order statistics are able to suppress both the co-channel as well as ACI, like the beamforming and interference cancellation technique and it is reported that 17dB interference separation can be achieved. The key issue with the higher order statistics is the computation and processing time.

B) **Interference Cancellation for Black Spaces:** The black spaces are the geographical regions where the probability of the PU communication is very high with sufficient power transmission. Therefore, it is very difficult to establish the CU communication in the black spaces as per the fundamental principle or idea of the CR network. However, if we explore the CU communication, then two kinds of interference scenarios have originated which are: (i) the CU communication introduces an interference to the PU communication; however, if the interference introduced to the PU receiver is under the tolerable limits, that is interference temperature, then CU communication can be established by controlling the CU transmission power. Thus, the CU transmission should be regulated in such a way that the interference introduced is less than the interference temperature. (ii) The PU communication introduces the interference to the CU receiver; however, the PU transmission power is not a controlling factor in the CR network, therefore, the CU should be able to avoid those interferences from the PU communication by using the interference separation or cancellation techniques. The successive interference cancellation (SIC) is a potential technique of the interference avoidance according to which the interference signal is estimated from the received signal and then subtracted from the complete received signal in order to yield the SOI [41]. There are two approaches for estimating interference: interference extraction and interference reconstruction. The interference avoidance can be achieved in two ways: the interference extraction and interference separation.

i) **Interference Extraction:** Extraction uses the suppression of the SOI and already explored interference suppression techniques can be exploited for suppressing the SOI and extracting PU interfering signal.

ii) **Interference Reconstruction:** In the case of digitally modulated primary signals, if a CR receiver receives a strong primary signal and knows its transmission structure (e.g. its coding and modulation schemes), it can first demodulate and decode the primary signal to recover the original primary information bits. Then, the CR receiver can reconstruct the corresponding primary signal based on the knowledge of its transmission structure and channel information. To further explain the concept of interference reconstruction and cancellation, a simple simulation model has built. In this model, a CR receiver operates in the black space of a terrestrial digital-video broadcasting (DVB-T) system (the primary system). The CR transmission is assumed to be synchronized to an 8-MHz DVB-T channel and applies quadrate phase-shift keying (QPSK) and OFDM for signal modulation.

11.4 Cross-Layer Interference Mitigation in Cognitive Radio Networks

In the previous sections, the interference mitigation techniques in the various scenarios such as at the transmitter end, receiver end, etc., specifically saying the layer-wise interference avoidance, are discussed. The transmit precoding appears to be a potential candidate for the interference management in the multiple-input-multiple-output (MIMO) systems; however, available with a prerequisite of the known information about the channel between the CU and PU. The availability of this channel information is a questionable thing sometimes in CRN; therefore, the authors have exploited the subspace technology to estimate the channel between CU to PU before coding [17, 42, 43]. The authors in [42, 43] have proposed the precoding schemes by exploiting the multiple signal classification (MUSIC) algorithm for estimating the channel information from CU to PU. The point to notice in these techniques is that the precoding at CU is performed by considering the effects of the interference that causes the data loss of the CU. Therefore, in [43], the authors have estimated the interference as well as exploited it for the precoding technique at the CR receiver which results in the improved throughput of CU. This entire process is a physical layer phenomenon; on the other hand, the selection of the channel for communication of CU among a set of PU channels is known as medium access control (MAC) layer phenomenon. This sharing of the channel between PU and CU leads to another kind of interference assumed to be a MAC layer phenomenon and is well explored for the cellular communication as well as cognitive radio networks [44–46]. In CRN, the channel allocation is studied by the authors in [47], where the channel allocation problem is illustrated by using a game-theoretic approach. Thus, for feasible and practical systems, the physical and MAC layer challenges need to be addressed simultaneously. By keeping this in mind, the authors have presented an idea of interference mitigation using the beamforming algorithm at the physical layer and iterative channel allocation at the MAC layer [48, 49]. The authors have proposed two algorithms in two particular scenarios: (i) when CU-CU interference channel information is known to the system and (ii) when the CU-CU interference channel information is unknown to the system. The authors have confirmed the effectiveness of proposed cross-layer mechanism using simulation results. Both of the proposed cross-layer algorithms outperform the without cross-layer approach in terms of the CR-primary interference and the CR throughput; (iii) the non-iterative algorithm achieves similar performance to its iterative counterpart without incurring much communication overhead.

11.5 Interference Management in Cognitive Radio Networks via Cognitive Cycle Constituents

The interference management is an important aspect in the conventional wireless networks; however, it becomes of a prime importance in the future applications such as vehicular networks, healthcare industry, fourth industrial revolution, etc., because even a small amount of interference can affect the communication which may directly cause the accident. Therefore, in this section, we have investigated the various aspects of interference management in the cognitive-inspired various networks, namely (i) spectrum sensing, (ii) spectrum prediction, (iii) transmission below PU tolerable interference, (iv) using advanced encoding techniques, and (v) spectrum monitoring as shown in Figure 11.4.

11.5.1 Spectrum Sensing

The spectrum sensing is the prime element of the CR-inspired wireless networks for interference avoidance where the channel is sensed before the data transmission and data are only transmitted on the idle sensed channel. On the emergence of PU, the CU needs to switch the communication on another idle sensed channel in order to avoid the interference with the PU. Various spectrum sensing techniques are well illustrated in literature and discussed in Chapter 1. The key issue with the spectrum sensing technique is that it needs to provide dedicated time slot or extra sensing unit [50, 51]. In addition, the complexity as well as prerequisites for the spectrum sensing technique used needs to be considered during the design of a particular wireless network. Therefore, the selection of a particular spectrum accessing technique for the particular wireless network application scenarios (inside-body, on-body, slow and high-speed moving vehicles, etc.) is a potential research issue.

11.5.2 Spectrum Prediction

The spectrum prediction is a technique which exploits the historical information about the PU channel states in order to predict the future states of the channel [51–53]. The spectrum prediction is used for interference management in two ways which are: (i) assist the spectrum sensing technique and (ii) spectrum mobility by predicting the emergence time of PU on the current communication channel. In the first approach, the CU predicts the states of channels in the high traffic environments and among which highly idle predicted channel is selected for the spectrum sensing. If that channel is sensed as idle, then data transmission is started,

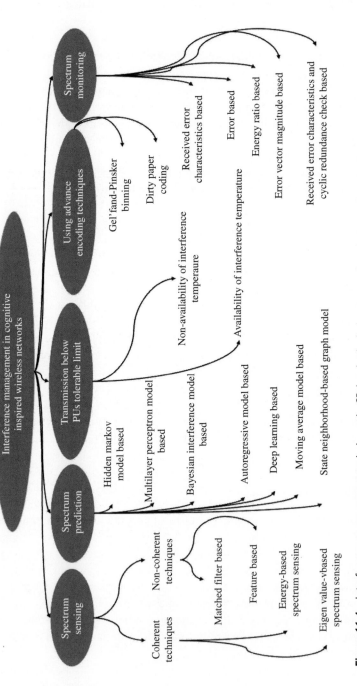

Figure 11.4 Interference management techniques in CR-inspired wireless networks using constituents of cognitive cycle.

otherwise second highest idle sensed channel is sensed, and so on [51]. In the second approach, the CU predicts the emergence time of PU during the CUs' transmission and enables the CU to switch its communication before PUs' emergence and results in the proactive spectrum mobility [52]. Thus, it is clear that spectrum prediction enables the CU to avoid the interference with PU that occurred during the emergence and detection time of PU. Various spectrum prediction techniques are reported in the literature, that are Hidden Markov Model Based, Multilayer Perceptron Neural-network-based Prediction, Bayesian-inference-based Prediction, Moving-average-based Prediction, Autoregressive-model-based Prediction, Static-neighbor-graph-based Prediction, and Deep-learning-based Prediction [53–56].

The key issues with the spectrum prediction technique are as follows: (i) the need of the historical information which is not so easy to get available, (ii) the storage unit as well as power requirements for the predictive analysis because for predictive analyses huge data need to be analyzed which introduces the challenges of Big-Data analytics, and (iii) moreover, the complexity of the prediction technique is also a prominent point that needs to be considered during the design of particular wireless network because more complex nature of the technique may affect the power consumption as well as size of the·unit.

11.5.3 Transmission Below PUs' Interference Tolerable Limit

This is a very effective way to avoid the interference at PU if we have information of the PUs interference tolerable limit and the high channel capacity per unit bandwidth is optional. In this technique, the CU controls its power in such a way that the power received at the PU receiver is below the already defined interference threshold which is much popular with the name of interference temperature as we have discussed in the previous sections of this chapter. This technology is achieved by using UWB technology, where the entire information is spread over a large bandwidth with low power similar to noise. This enables the CU to avoid the interference at PU by exploiting the well-explored spread spectrum technique [57].

The key issue with this technique is the availability of the interference threshold level that can be tolerable by the PU. In addition to this, the low data rate per unit bandwidth restricts the application of this technique only at the scenarios where the generated data rate is less than that of the achieved data rate and make unsuitable for the high data generated scenarios.

11.5.4 Using Advanced Encoding Techniques

This technique allows the CU to transmit data with full power even in the presence of PU on the channel; however, for this the CU should have the ability of advanced encoding techniques such as dirty paper coding [58] and Gelfand-Pinsker bining

[59]. The use of advanced encoding technique makes a system more complex and needs more power for processing. This technique is unsuitable for the inside-body sensing devices in the medical networks because of the high complex nature and demand for high-power; however, suitable for the on-body or surrounding medical devices where the high data rates are required as well as cost and complexity can be compromised with better medical facilities. The key issues that need to be addressed by the researchers for this technique are complexity and power consumption. The sufficient power availability in the vehicular networks reduces the demands of power efficiency; however, complexity is still an issue. Moreover, the practical implementation of this technique is yet a major milestone.

11.5.5 Spectrum Monitoring

The spectrum monitoring is a very prominent and recently explored technique which is used to detect the emergence of PU during the CUs' communication so that the CU can stop/switch the communication on that channel. This technique serves the same purpose as the spectrum sensing; however, does not require extra hardware unit and time which makes it more suitable for spectrum mobility/hand-off. The CU only exploits the characteristics of the received signal for processing in order to know the deviation from the normal received signal that indicates the emergence of PU. On detection of the emergence of PU, the CU stops its communication immediately, and switches its communication on another available channel. In literature, various spectrum monitoring techniques are reported such as received error count-based spectrum monitoring, received error count and cyclic redundancy check-based spectrum monitoring, energy-based spectrum monitoring, energy ratio-based spectrum monitoring, and error vector magnitude-based spectrum monitoring [60–62]. The challenges for spectrum monitoring technique is to analyze the processing power that requires a particular monitoring technique because an increase in the processing power affects the total power consumption of the medical unit that makes it unsuitable for inside-body units. For vehicular networks, the detection delay of the monitoring techniques is a potential metric.

Some potential recent researches for the interference management in the CRNs and cellular/5G networks are illustrated as follows [63–70]. Hossain et al. [63] have presented a well-illustrative qualitative comparison of the existing cell association and power control schemes to demonstrate their limitations for interference management in 5G networks. The authors have discussed the vision and requirements for 5G multitier cellular networks which are data rate and latency, machine-type communication, millimeter wave communication, multiple radio access technologies, base station densifications, prioritized spectrum access, network-assisted communication, and energy harvestings. Further, the interference management challenges in the 5G multitier networks are explored, which are "designing optimized cell association and power control" (Capc) methods for multitier network, "designing efficient methods to support simultaneous association to multiple BSs," "designing efficient methods for cooperation and coordination among multiple tier." Moreover, the authors have illustrated the distributed power control schemes, joint cell association and power control schemes, and prioritized power control schemes. In [64], Wang et al. have illustrated an interference management and resource allocation in multitier adhoc CRNs where a joint interference management and resource allocation problem in a multichannel ad hoc CRN is considered. Further, the authors have formulated a problem as an overlapping coalition formation game to maximize the sum rate of CU links while guaranteeing the QoS of PU links. It is revealed by using the simulation results that the proposed scheme achieves appreciable performance improvement in terms of the sum rate of SU links. Bakhsh et al. have examined a cooperative D2D communication in CRNs that comprises two secondary links (one primary link and a relay network) [65]. For the relay networks, the authors have proposed a clustering relay selection method to enable simultaneous transmission of primary and secondary links. Further, the authors have proposed an interference management scheme to improve the system performance whose effectiveness is claimed using the simulation results. Ranjan et al. [66] have presented an approach to minimize the interference among the secondary nodes by exploiting the interference index as the interference minimization key that maximizes the system capacity. The authors have thoroughly exploited an existing distributed greedy algorithm, which, on the introduction of interference index, furnished a gain of 60% in the CR network capacity. Further, a trade-off analysis between the interference index and channel leakage ratio is presented with an interference bound of 10 dBm, which may form the basis of interference management in CRNs. In [67], the authors have presented the potential frameworks which exploit the cognitive-inspired IoMT so that the issue of spectrum scarcity in wireless body area network is resolved. The potential frameworks with their pros and cons as well as major research challenges with probable solutions are discussed. As "security and interference with human body organs" is of primary concern due to direct human body involvement, further security concerns and interference management techniques are described. Haroon et al. [68] have examined the bottleneck uplink converge performance of the micro base station in the presence of intercell interference (ICI) and drones interference (DI). Further, the authors have exploited the efficient resource allocation technique that is reverse frequency allocation in order to mitigate the ICI and DI. Moreover, the authors have explored the decoupled association in place of coupled association for improvement in the uplink signal-to-noise ratio. In [69], Martinez et al. have examined the coexistence of CRNs on TVWS for rural and suburban connectivity. Qamar et al. [70] have explored the interference management issues for 5G heterogeneous networks and D2D communication by using different potential techniques such as intercell interference coordination, coordinated multipoint, and coordinated scheduling. Moreover, the authors have critically reviewed the methodologies, advantages, and limitations along with the future work. Further, the authors have outlined the interference mitigation proposal in the 3GPP releases.

11.6 Summary

This chapter primarily emphasizes over the concept of interference in conventional wireless networks and further the concept is extended for the cognitive radio networks, where the CRN with and without interference scenarios are described. Further, the inter- and intra-network interference scenarios such as CU-to-CU, CU-to-PU, and PU-to-CU interference are illustrated. Afterward, the interference mitigation techniques in CRN for different levels of interference in the white, grey and black spaces such as spectrum shaping, filter based, transform based, transmit beamforming, etc., are presented in detail. For the complete implementation of CRN with interference, cross-layer perspective of interference management is also explored and finally the interference management and potential challenges for different constituents of the cognitive cycle are illustrated.

References

1 Carleial, A.B. (1978). Interference channels. *IEEE Transactions on Information Theory* 24 (1): 60–70.
2 Carleial, A.B. (1975). On the capacity of multiple-terminal communication network. Stanford Electronics Lab. Report, 6603-1, Stanford University, Stanford, CA.
3 Zahir, T., Arshad, K., Nakata, A., and Moessner, K. (2013). Interference management in femtocells. *IEEE Communications Surveys & Tutorials* 15 (1).
4 IEEE802.22 (2020). Working group on wireless regional area networks enabling broadband wireless access using cognitive radio technology and spectrum sharing in white spaces. https://www.ieee802.org/22/ (accessed 1–8 August 2020).
5 NTIA (2004). NTIA Comments on the Establishment of an Interference Temperature Metric to Quantify and Manage Interference and to Expand Available Unlicensed Operation in Certain Frequency Bands ET Docket No. 03-237 13 August 2004. https://www.ntia.doc.gov/fcc-filing/2004/ntia-comments-establishment-interference-temperature-metric-quantify-and-manage-inte.

6 FCC Reveals Plans for 'Interference Temperature' Implementation. https://www. tvtechnology.com/news/fcc-reveals-plans-for-interference-temperature-implementation (accessed 20 July 2020).

7 Sharma, I.M., Sahoo, A., and Nayak, K.D. (2018). Channel modeling based on interference temperature in underlay cognitive wireless networks. *Proceedings of IEEE International Symposium on Wireless Communication Systems*, Reykjavik, Iceland (October 2018), 224–228.

8 Usha, S.M. and Nataraj, K.R. (2016). Analysis and mitigation of adjacent and co-channel interference on AWGN channel using 8-PSK modulation for data communication. *Proceedings of International Conference on Wireless Communications, Signal Processing and Networking (WiSPNET)*, Chennai, India, (March 2016), 922–926.

9 Hu, D. and Mao, S. (2011). Co-channel and adjacent channel interference mitigation in cognitive radio networks. *Proceedings of Military Communications Conference (MILCOM 2011)*, Baltimore, MD, USA (November 2011), 13–18.

10 Hu, D. and Mao, S. (2013). On co-channel and adjacent channel interference mitigation in cognitive radio networks. *Ad Hoc Networks* 11 (5): 1629–1640.

11 Almeida, J., Alam, M., Arnaldo, J.F., and Oliveira, S.R. (2016). Mitigating adjacent channel interference in vehicular communication systems. *Digital Communication Network* 2 (2): 57–64.

12 Lasisi, H. and Okedere, O.B. (2017). Interference management techniques in cellular networks: a review. *Cogent Engineering* 4 (1): 1–10.

13 Song, Y.S., Lee, S.K., Lee, J.W. et al. (2019). Analysis of adjacent channel interference using distribution function for V2X communication systems in the 5.9-GHz band for ITS. *ETRI Journal* 41 (6): 703–714.

14 Ali, M.J. (2017). Wireless body area networks: Co-channel interference mitigation & avoidance. Ph. D Thesis, University Paris Descartes, Paris, France. https://tel.archives-ouvertes.fr/tel-02109264/document (accessed 30 July 2020).

15 FCC-03-322A1.pdf. http://web.cs.ucdavis.edu/~liu/289I/Material/FCC-03-322A1.pdf (accessed 18 February 2018).

16 Chen, Z., Wang, C.-X., Hong, X., Thompson, J., Vorobyov, S.A. and Ge, X. (2010). Interference modeling for cognitive radio networks with power or contention control. *Proceedings of IEEE Wireless Communication and Networking Conference*, Sydney, Australia (April 2010), 1–6.

17 Chen, Z., Wang, C.-X., Hong, X. et al. (2013). Interference mitigation for cognitive radio MIMO systems. *Phy. Commun.* 9: 308–315.

18 Chen, Z., Vorobyov, S.A., Wang, C.-X., and Thompson, J. (2012). Pareto region characterization for rate control in MIMO interference systems and Nash bargaining. *IEEE Transactions on Automatic Control* 57 (12): 3203–3208.

19 Chen, Z., Wang, C.-X., Hong, X. et al. (2013). Aggregate interference modeling in cognitive radio networks with power and contention control. *IEEE Transactions on Communications* 60 (2): 456–468.

20 Hong, X., Chen, Z., Wang, C.-X., and Vorobyov, S.A. (2009). Cognitive radio networks. *IEEE Vehicular Technology Magazine* 4 (4): 76–84.

21 Wang, S., Sagduyu, Y.E., Zhang, J., and Li, J.H. (2011). Spectrum shaping via network coding in cognitive radio networks. *Proceedings 30th IEEE International Conference on Computer Communications (IEEE INFOCOM)*, Shanghai, China, (April 2011), 396–400.

22 Clancy, T.C. and Walker, D. (2020). Spectrum shaping for interference management in cognitive radio networks. *Proceeding of the SDR 06 Technical Conference and Product Exposition*, Orlando, FL (16 November 2006), 1–6. https://www.wirelessinnovation.org/assets/Proceedings/2006/sdr06-3.6-4-clancy.pdf (accessed 22 July 2020).

23 Zhang, W. and Mitra, U. (2010). Spectrum shaping: A new perspective on cognitive radio-part I: coexistence with coded legacy transmission. *IEEE Transactions on Communications* 58 (6): 1857–1867.

24 Jorswieck, E.A., and Lv, J. (2011). Spatial shaping in cognitive MIMO MAC with coded legacy transmission. *Proceedings of 12th IEEE International Workshop on Signal Processing Advances in Wireless Communications* (SPAWC), San Francisco, CA, USA (June 2011), 451–455.

25 Lee, H.-W., Chang, W., and Jung, B.C. (2018). Optimal power allocation and allowable interference shaping in cognitive radio networks. *Computers and Electrical Engineering* 71: 265.

26 Akhtar, M. and Nakhai, M.R. (2011). Transmit beamforming and interference shaping in cellular cognitive radio networks. *IET Communications* 5 (14): 2052–2058.

27 Ofcom (2020). A study into the application of interference cancellation techniques. Rep. 72/06/R/037/U. http://www.ofcom.org.uk/research/technology/research/emer_tech/intcx/summary.pdf (accessed 30 July 2020).

28 Scholtz, R. (1977). The Spread Spectrum Concept. *IEEE Transactions on Communications* 25 (8): 748–755.

29 Flikkema, P.G. (1997). Spread-spectrum techniques for wireless communication. *IEEE Signal Processing Magazine* 14 (3): 25–36.

30 Pickholtz, R., Schilling, D., and Milstein, L. (2020). Theory of Spread-Spectrum Communications - A Tutorial. *IEEE Transactions on Communications* 30 (5): 855–884.

31 Cook, C. and Marsh, H. (1983). An introduction to spread spectrum. *IEEE Communications Magazine* 21 (2): 8–16.

32 Phan, T.K., Vorobyov, S.A., Sidiropoulos, N.D., and Tellambura, C. (2009). Spectrum sharing in wireless networks via QoS-aware secondary multicast beamforming. *IEEE Transactions on Signal Processing* 57 (6): 2323–2335.

33 Zhou, J. and Thompson, J.S. (2008). Linear precoding for the downlink of multiple input single output coexisting wireless systems. *IET Communications* 2 (6): 742–752.

34 Ambede, A., Sirigina, R.P., and Madhukumar, A.S. (2018). Adaptive filter based blind residual self-interference cancellation in full-duplex MIMO transceivers. *Proceedings of IEEE Wireless Communications and Networking Conference* (WCNC), Barcelona, Spain, (15–18 April 2018).

35 Kim, B., Yu, H., and Noh, S. (2020). Cognitive interference cancellation with digital channelizer for satellite communication. *Sensors* 20 (2): 1–15.

36 Huang, D. and Letaief, K.B. (2005). An interference-cancellation scheme for carrier frequency offsets correction in OFDMA systems. *IEEE Transactions on Communications* 53 (7): 1155–1165.

37 Chen, V.C. and Ling, H. (1991). Joint time-frequency analysis for radar signal and image processing. *IEEE Signal Processing Magazine* 16: 12.

38 Zali, A., and Hendessi, F. (2002). Cyclostationary based interference cancellation in long code DS/CDMA systems. *Proceedings of 4th International Workshop on Mobile and Wireless Communications Network*, Stockholm, Sweden (9–11 September 2002), 115–118.

39 Gelli, G., Paura, L., and Tulino, A.M. (1998). Cyclostationarity-based filtering for narrowband interference suppression in direct-sequence spread spectrum systems. *IEEE Journal on Selected Areas in Communications* 16 (9): 1747–1755.

40 Kostanic, I. and Mikhael, W. (2002). Blind source separation technique for reduction of co-channel interference. *Electronics Letters* 38 (20): 1210–1211.

41 Sheng, M., Li, X., Wang, X., and Xu, C. (2017). Topology control with successive interference cancellation in cognitive radio networks. *IEEE Transactions on Communications* 65 (1): 37–48.

42 Yi, H., Hu, H., Rui, Y., Guo, K., and Zhang, J. (2009). Null space-based precoding scheme for secondary transmission in a cognitive radio MIMO system using second-order statistics. *Proceedings of IEEE International Conference on Communication*, Dresden, Germany (14–18 June 2009), 1–5.

43 Zhang, R., Gao, F., and Liang, Y.-C. (2010). Cognitive beamforming made practical: Effective interference channel and learning-throughput tradeoff. *IEEE Transactions on Communications* 58 (2): 706–718.

44 Hoang, A.T., Liang, Y.-C., and Islam, M.H. (2010). Power control and channel allocation in cognitive radio networks with primary users' cooperation. *IEEE Transactions on Mobile Computing* 9 (3): 348–360.

45 Cao, L., Zhao, H., Li, X., and Zhang, J. (2016). Matching theory for channel allocation in cognitive radio networks. *Proceedings of 83rd IEEE Vehicular Technology Conference* (VTC Spring), Nanjing, China (15–18 May 2016), 1–5.

46 Thakur, P. and Singh, G. (2019). Energy and spectral efficient SMC-MAC protocol in distributed cognitive radio networks. *IET Communications* 13 (17): 2705–2713.

47 Shrivastav, V., Dhurandher, S.K., Woungang, I., and Rodrigues, J.J.P.C. (2016). Game theory-based channel allocation in cognitive radio networks. *Proceedings of IEEE Global Communications Conference* (GLOBECOM), Washington, DC, USA (4–8 December 2016), 1–5.

48 Zeydan, E., Kivanc-Tureli, D., and Tureli, U. (2007). Joint iterative channel allocation and beamforming algorithm for interference avoidance in multiple-antenna Ad Hoc networks. *Proceedings of IEEE Military Communication Conference*, Orlando, USA (October 2007), 1–7.

49 Zeydan, E., Kivanc-Tureli, D., and Tureli, U. (2008). Cross layer interference mitigation using a convergent two-stage game for ad hoc networks. *Proceedings of 42nd IEEE Annual Conference on Information Sciences and Systems*, Princeton, USA (19–21 March 2008), 671–675.

50 Thakur, P., Kumar, A., Pandit, S. et al. (2017). Advanced frame structures for hybrid spectrum accessing strategy in cognitive radio communication system. *IEEE Communications Letters* 21 (2): 410–413.

51 Thakur, P., Kumar, A., Pandit, S. et al. (2018). Performance analysis of high-traffic cognitive radio communication system using hybrid spectrum access, prediction and monitoring techniques. *Wireless Networks* 24 (6): 2005–2015.

52 Thakur, P., Kumar, A., Pandit, S. et al. (2017). Spectrum mobility in cognitive radio network using spectrum prediction and monitoring techniques. *Physical Communication* 24 (2): 1–8.

53 Xing, X., Jing, T., Cheng, W. et al. (2013). Spectrum prediction in cognitve radio networks. *IEEE Wireless Communications* 20 (2): 90–96.

54 Cristian, I. and Moh, S. (2015). A low-interference channel states prediction algorithm for instantaneous spectrum accessin cognitive radio networks. *Wireless Personal Communications* 84 (4): 2599–2610.

55 Barnes, S.D., Maharaj, B.T., and Alfa, A.S. (2016). Cooperative prediction for cognitive radio networks. *Wireless Personal Communications* 89 (4): 1177–1202.

56 Mannes, R., Clayes, M., Figueiredo, F.A.P.D. et al. (2019). Deep Learning based spectrum prediction collision avoidance for hybrid wireless environments. *IEEE Access* 7: 45818–45830.

57 Pereira, M., Postolache, O., and Girao, P. (2009). Spread spectrum techniques in wireless communication. *IEEE Instrumentation and Measurement Magazine* 12 (6): 21–25.

58 Saradka, B., Bhashyam, S., and Thangaraj, A. (2012). A dirty paper coding scheme for the Multiple Input Multiple output broadcast channel. *Proceedings of National Conference on Communications (NCC)*, Kharagpur, India (3–5 February 2012), 1–5.

59 Tyagi, H., and Narayan, P. (2009). The Gelfand-Pinsker channel: Strong converse and upper bound for the reliability function. *Proceedings of IEEE International Symposium on the Information Theory*, Seoul, South Korea (28 June to 3 July 2009), 1954–1957.

60 P Thakur, and G Singh, "Spectrum monitoring techniques for spectrum mobility in cognitive radio networks: a technical reviews," *Journal of Electromagnetic Waves and Applications*, 2020 (Under Review).

61 Thakur, P., Kumar, A., Pandit, S. et al. (2019). Performance analysis of cooperative spectrum monitoring in cognitive radio network. *Wireless Networks* 25 (17): 989–997.

62 Thakur, P., Kumar, A., Pandit, S. et al. (2018). Spectrum monitoring in heterogeneous cognitive radio network: how to cooperate? *IET Communications* 12 (17): 2110–2118.

63 Hossain, E., Rasti, M., Tabassum, H., and Abdelnasser, A. (2014). Evolution toward 5G multi-tier cellular wireless networks: an interference management perspective. *IEEE Wireless Communications* 21 (3): 118–127.

64 Wang, K., Heng, W., Li, X., and Wu, J. (2020). Interference management and resource allocation in multi-channel ad hoc cognitive radio network. *IEICE Transactions on Communications* https://doi.org/10.1587/transcom.2020EBP3103.

65 Bakhsh, Z.M., Moghaddam, J.Z., and Ardebilipour, M. (2020). An interference management approach for CR-assisted cooperative D2D communication. *AEU International Journal of Electronics and Communications* 115 https://doi.org/10.1016/j.aeue.2019.153026.

66 Ranjan, R., Agrawal, N., and Joshi, S. (2020). Interference mitigation and capacity enhancement of cognitive radio networks using modified greedy algorithm/channel assignment and power allocation techniques. *IET Communications* 14 (9): 1502–1509.

67 Thakur, P. and Singh, G. (2020). Security and interference management in the cognitive-inspired internet of medical things. In: *Intelligent Data Security Solutions for e-Health Applications*, 131–149. Academic Press.

68 Haroon, M.S., Muhammad, F., Abbas, G. et al. (2020). Interference management in ultra-dense 5g networks with excessive drone usage. *IEEE Access* 8: 102155–102164.

69 Martinez, A.R., Plets, D., Deruyck, M., Martens, L., Nieto, G.G., and Joseph, W. (2020). Dynamic interference optimization in cognitive radio networks for rural and suburban areas," *Wireless Communications and Mobile Computing.*, vol. 2020, May 2020.

70 Qamar, F., Hindia, M.H.D.N., Dimyati, K. et al. (2019). Interference management issues for the future 5G network: A review. *Telecommunication Systems* 71 (4): 627–643.

12

Simulation Frameworks and Potential Research Challenges for Internet-of-Vehicles Networks

12.1 Introduction

Due to rapid and significant development in the various communication and computational technologies, the industry is moving toward the fourth Industrial revolution, that is Industry 4.0 (I4.0). In order to clear the perspective of Industry 4.0, it is worth having briefs about previous industrial revolutions [1]. The first industrial revolution gets registered in the history around the mid-seventeenth century with the development of steam engine. However, it further revolutionized in the late-eighteenth century with the emergence of new sources of energy that are electricity, oil, and gas. Further, the twentieth century, with the development of nuclear energy, progresses in the electronics due to transistor and microcontroller, as well as rise in the telecommunication and computer, has witnessed the third industrial revolution. The key element of third industrial revolution is high-level automation in the production and it became possible due to automatons-programmable logic control (PLCs) and robots. Now, we are on the way to evolve the concept of third industrial revolution by using the digital developments and Internet in the late and early stages of twentieth and twenty-first centuries, respectively. In other words, the key emphasis of Industry 4.0 is the exploration of digital developments and Internet for the drastic evolvement of the industry such as physically, digitally, and biologically by keeping in view the transformations to entire production, management, and governance systems. Today, the significant digital developments are internet-of-things (IoT), big data analytics, cyber-security, augmented reality, autonomous-robot, 3D-printing/additive manufacturing, system integration, cloud as well as fog computing, artificial intelligence, machine learning, deep learning, neural networking, 5G, etc., as shown in Figure 12.1 [2–9]. All the mentioned techniques are evolving in theirself and contributing for the Industry 4.0; however, it is worth mentioning that the IoT is progressing immensely.

Spectrum Sharing in Cognitive Radio Networks: Towards Highly Connected Environments,
First Edition. Prabhat Thakur and Ghanshyam Singh.
© 2021 John Wiley & Sons, Inc. Published 2021 by John Wiley & Sons, Inc.

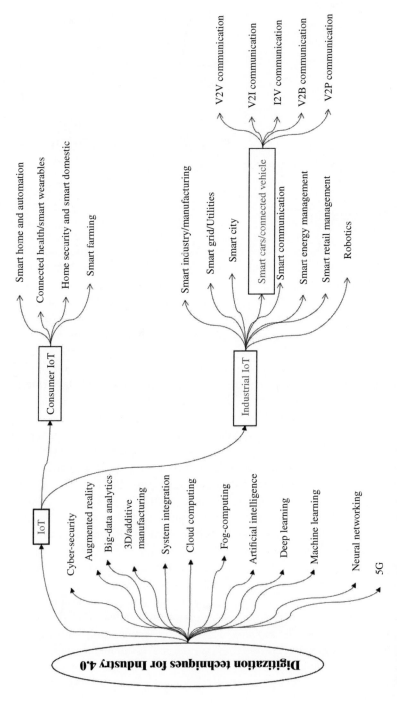

Figure 12.1 The digitization techniques for Industry 4.0.

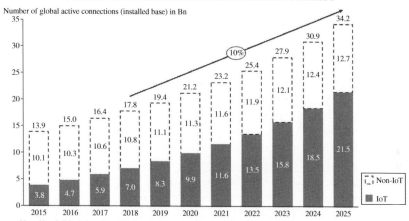

Figure 12.2 The number of active connections for IoT and non-IoT devices worldwide. *Source:* Lueth [10]. © 2018, IoT Analytics GmbH.

As per a report by the IoT Analytics, the number of active device connections worldwide for non-IoT (devices such as mobile phones, PCs, tablets, laptops, fixed line phones, etc.) and IoT devices will reach 34.2 and 21.5 Billion [10] as shown in Figure 12.2. On the other hand, as per statista, the number of wireless connected devices in the year 2025 will be 75 Bn, which is almost five times when compared with 15 Bn in the year 2015. The IoT is showing its applications in various consumer and industrial areas; thus, IoT is further subdivided into two parts that are Consumer IoT (CIoT) and Industrial IoT (IIoT) [11].

The consumer applications of the IoT comprise smart home, connected wearables/smart health, smart farming, etc. On the other hand, smart industry/manufacturing, smart grid/utilities, smart communication, smart city, smart cars/connected vehicles, smart energy management, smart retail, robotics, etc., are the major applications of industrial IoT. The key emphasis of this chapter is to discuss about the smart cars/connected vehicles because high-speed movement of nodes/vehicles and predicted paths make it a challenging as well as interesting ad-hoc networking problem. For communication and networking in the connected vehicles, the five ways are presented namely, the in-vehicle communication (IV), vehicle-to-vehicle communication (V2V), vehicle-to-infrastructure (V2I), vehicle-to-broadband (V2B), and vehicle-to-pedestrians (V2P).

Further, in this section, we have emphasized to perceive the difference between consumer and industrial IoT, which is presented as follows.

12.1.1 Consumer IoT

The IoT which is human centered or connected to human awareness for the ease of human life in any perspective, such as safety, time, cost, life management, etc., is classified as CIoT. The CIoT generally comprises two classes of IoT that are personal assistant devices and home assistant devices. The devices which assist the human being directly for its daily routine works such as in office, during traveling, in playground, etc., are called as personal assistant devices and key examples are smartphones, wearables, voice assistant, smart fashion, hearables, etc. On the other hand, the devices which are used to manage the home automatically are called as home assistant devices such as hubs and controllers, home appliances, smart plugs, meter and charging, lightning and climate, and entertainment devices.

12.1.2 Industrial IoT

The IoT which is used to gear up the revolution in the industrial manufacturing, design, automation, etc., is called as IIoT by exploiting the various digitization techniques such as big data analytics, cloud and fog computing, machine learning, etc. The prime aims of the IIoT are quality product, reduced cost, revenue growth, and reduced time conversely; the use cases of the IIoT are product as a service, remote condition monitoring, digital work instructions, and real-time visibility. The key difference between the CIoT and IIoT is due to non-intervention of human beings in the IIoT because here major communication is among the machines, that is machine-to-machine communication.

IIoT is classified as the smart industry/manufacturing, smart grid/utilities, smart communication, smart city, smart retail, robotics, smart cars/connected vehicles, etc. Further, this chapter is structured as follows. The next section comprises the applications of CIoT and IIoT.

12.2 Applications of CIoT

The key applications of the CIoT are smart home, smart wearables, home security and smart domestics, personal healthcare, healthcare carrier, etc., which are illustrated as follows.

12.2.1 Smart Home and Automation

The smart home is defined as a home where most of the devices such as lighting, heating, air-conditioning, etc., are operated by human beings remotely through mobile or computer [12–16]. This provides comfort, security, cost-effective, energy efficiency (low operating costs), and convenience at all times, regardless of whether anyone is at home or outside. The key elements which can be controlled via remotely and smart way are locking system, Television/speakers, security cameras, thermostats, lighting, other appliances, etc. The key advantages of the smart home are that it saves time, cost, increases the comfort as well as reduces the thought process. On the other hand, as everything connected through technology may be annoying for someone who is unaware about the technology. Moreover, the small failure in technology may result in the nonfunctioning of maximum devices.

12.2.2 Smart Wearables

The smart wearables refer to the wearables, which are equipped with the medical sensing devices such as pressure, temperature, pulse measure, blood levels and cardiac conditions, etc., as well as communication devices in order to connect this information with another unit so that necessary actions can be taken [17, 18]. The smart wearables are suitable for both the healthy and unhealthy persons. For the healthy persons, we use smart wearables to monitor the effect due to daily routine works such as walking, jogging, games, gym, etc.; however, in case of unhealthy persons, the doctors can use the desired critical health sensing devices such as heart beat variations, etc., in order to monitor the patients' health conditions so that necessary measures can be taken. The smart wearables are used on the human body directly; however, the power required to operate these devices and the electromagnetic waves used for communication may affect the health and these effects need to be studied.

12.2.3 Home Security and Smart Domestics

A smart-home device is a thing whose main functionality is extended with networking abilities to create a new one. The additional infrastructure for those devices, like a base or control station, falls also in the smart home. The smart home systems are installed by considering most of the security concerns; however, these are vulnerable to some of the threats and the authors have presented a classification of these threats namely, the internal system threats and external system threats [16, 19]. The interior threats are further classified as the failure of home devices, malfunction of power and Internet, software failure, and confidential data leakage. The external threats are reported as the denial of services, malicious codes injection, eavesdropping attack, and man-in-the-middle attack.

12.2.4 Smart Farming

The smart farming is an advanced way of farming in order to improve the production per hectare [20–22]. These days, following technologies are available for the farmers: the sensing technologies (including soil scanning, water, light, humidity, temperature management), software technologies (specialized software solutions that target specific farm types), and communication technologies (such as cellular communication, positioning technologies, including GPS, hardware and software systems) which enable the IoT-based solutions, robotics and automation, and data analytics. Near future will witness the fully automation in the farming and will reduce the human intervention to a very small extent.

12.3 Applications of Industrial IoT

The key applications of the IIoT which will be discussed in this section are smart industry/manufacturing smart grid/utilities, smart communications, smart city, smart cars/connected vehicles, smart energy management, smart retail, and robotics [11].

12.3.1 Smart Industry

The smart manufacturing/industry is the acronym for the next-generation industries which comprises fully integrated, collaborative, manufacturing systems that respond in real time to meet changing demands and conditions in the factory, in the supply network, and in the customer needs. The smart manufacturing adopts the combination of the physical technology and cyber technology with the existing discrete system for particular technology [23]. In [8], the authors have presented a hierarchical architecture of the smart factory and have analyzed various technologies for smart industry for the physical (PHY), medium access control (MAC), network, cloud application as well as terminal layers' perspectives. Further, the potential issues regarding the prime emerging technologies such as IoT, big data, and cloud computing are investigated. In addition to this, a candy packing line was used to verify the key technologies of the smart factory and claims that overall equipment effectiveness of the equipment is significantly improved.

In [9], Wang et al. have proposed a smart industry framework that comprises industrial network, cloud, and supervisory control terminals with smart shop-floor objects where a classification of the smart objects into various types of agents with a defined coordinator is presented. The proposed framework relies on the distributed cooperation and autonomous decision that leads to high flexibility of the network. On the basis of these, the smart factory is defined as a self-organized multi-agent system assisted with big data-based feedback and coordination.

12.3.2 Smart Grid/Utilities

A grid is defined as a network of transmission lines, substations, transformers, and more that deliver electricity from the power plant to your home or business places. The conventional grid/electric-power systems are facing a radical transformation worldwide with the decarbonized electricity supply to replace aging assets and control the natural resources with new information and communication technology (ICT). This leads to the origination of smart grid technology which is a self-sufficient electricity network that relies on the digital automation technology for monitoring, control, and analysis within the supply chain. The key aim of smart grid is to detect the malfunctions in the network quickly as well as to repair if possible so that the target sustainable, reliable, safe, and quality electricity is delivered to the consumers [24, 25].

12.3.3 Smart Communication

The communication is a method to convey information from the source to the destination via a channel that can be RF cable, optical fiber cable, air, light, etc. These days the wireless communication systems are in mature stage where communication across the world can be set up as well as maintained for number of hours in the form of voice, media, video calls, conference call, etc. However, still we are moving toward the next generation of communication systems where key emphasis is on the improved data rates, reduced latency, as well as high reliable and secure communication. The various smart technologies make it possible to achieve the complete smart communication framework such as smart antenna, smart sensor systems, advanced multiple accessing and modulation techniques, smart metering, etc. [26]. The rapid proliferation of wireless connected devices due to IoT increases the demand of spectrum; however, its physical availability originates the problem of efficient use of spectrum. The machine-to-machine (M2M) communication is an important part of communication for Industry 4.0 where human intervention is very less and machines have to deal with each other; therefore, the reliability of communication is of prime importance in order to avoid the delay and misbehavior of the machines [27, 28]. Thus, the smart communication is backbone of almost all the upcoming next-generation technologies.

12.3.4 Smart City

The smart city is defined as an urban area which exploits various IoT sensors to collect the data so that the resources can be used efficiently [29, 30]. The collected data are used to monitor as well as manage the city traffic and transport systems, power plants, water supply networks, waste management, school management, smart hospitals, libraries, offices as well as other services of the society. In addition to this, the smart city is equipped with advanced monitoring systems such as CCTV cameras, and immediate action systems such as drones or artificial birds that can stop the accused/intruders. Ang et al. [31] have presented a comprehensive review of the current and emerging internet-of-vehicles (IoVs) paradigms and communication models with an emphasis on deployment in smart cities. The key focus is on the applications, architecture, and challenges in the IoVs for smart cities.

12.3.5 Smart Energy Management

The energy and power are the integral and undetachable entities of the human beings after the second industrial revolution that was started with the electricity, oil, and gas. Every revolution after that relies on the electricity, either electronics or computer or cyber physical systems. However, this prime need of electricity increases its demand which motivates us to explore various possible ways of electricity production and we did well in this direction. With the increase in the population and introduction of various technical services of society such as home devices, public devices, IoT, etc., the production of electricity is still insufficient. Therefore, in order to manage the production, transfer, as well as distribution of electricity, various conventional methods are present in the literature; however, the smart energy management systems are desired that can reduce the losses during transfer, production, and distribution. Therefore, various researchers have presented the smart energy management frameworks which are responsible for the fault detection during the generation, smart grids for efficient transfer, etc. [32–35]. Du and Lu [32] have presented a load-scheduling model in order to manage the power of the household devices. A linear-sequential-optimization-enhanced, multi-loop algorithm is used to solve the appliance commitment problem and through simulation results, and it is proved that the proposed algorithm is fast, robust, and flexible. The authors in [33] have started with a fact that 20–30% of building energy consumption can be saved through optimized operation and management without changing the building structure and the hardware configuration of the energy supply system. On the basis of these, the authors have exploited the micro-grid for the energy-efficient design for the building.

12.3.6 Smart Retail Management

Retail was an important as well as integral part of the market and customers after the formation of the market long back ago which advances day-by-day with every industrial revolution; however, third revolution affects it drastically on the emergence of computer technologies which assemble the entire retail in a computer from the long files. However, further advancement in the technology made retail smarter as compared to its past with different kinds of facilities such as touch and non-touch cashier screen, stock control (pack size and recipes), blockmans test – meat cutting schedule, advanced customer analysis, advanced supplier control, purchase orders, online customer ordering, user friendly interface, online payment options, cashless options (prepaid cards) – canteens and clubs, powerful report search engine, custom reporting, track purchase orders, track customer/supplier payments, track goods received documents, track invoices/purchase orders, multiuser interface, and multilevel security levels [36]. A survey is presented by the KPMG in the 10 cities across the Greater Bay Area where 14 000 people were under consideration [37]. The key emphasis of the survey was to understand the self-identification level of the smart-customers and their attitude to the smart platforms and payment methods in retail. Moreover, in the report, a survey of 286 CEO to understand the challenges and opportunities after opting the smart platforms in Hong Kong is presented.

12.3.7 Robotics

The robotics is an interesting word which becomes possible due to different engineering aspects which are mechanical engineering, electronic engineering, information engineering, computer science, and others. It deals with the design, construction, operation, and use of robots as well as computers for their control, sensory feedback, and information processing. The applications of robots are increasing in the industry day-by-day in order to reduce the manpower and increase the speed of production. Some interesting applications of robots are in the remote places such as glaciers, space, desserts, mountains, etc. Some well-known names among these applications are ROBONAUT, LEONARDO, S-BOTS, Mars Rovers, etc. [38]. Further, the industrial robots are generally classified into five types namely, the Cartesian, Cylindrical, SCARA, 6 Axis, and Delta [39]. The most commonly used robots are Cartesian types because of ease of programming and use which is generally used for pick and place application. The cylindrical type of robots are equally popular to the Cartesian robots, however, used for linear and rotary motions; therefore, the popular application is pick, rotate, and then place. The SCARA robots can be considered as the combination of Cartesian and cylindrical types since it comprises the X, Y, Z as well as rotary motions. In addition to this, the SCARA-type robots are faster when compared with that of the Cartesian and cylindrical robots. The 6-axis-type robot is more advanced compared to the SCARA and all other discussed types and is considered as all in one robot; however, its programming is more complex. The last type robots are delta robots which are very fast but expensive.

12.3.8 Smart Cars/Connected Vehicles

Today, we are moving toward the connected static as well as moving world via Internet and the smart cars or connected vehicle is an integral element of the moving world. The smart cars are considered as the computer systems on wheels and multiple cars connected to each other are named as smart cars' network or connected vehicles [40, 41]. The smart features in a car allows for human-less driving, autonomous, parking, data collection, fuel efficient nature, minimum distance/time path selection, etc. Moreover, some advanced features can be added for the ease and amusement of vehicles. The smart cars or connected vehicles on or nearby the roads comprises following things such as vehicles, infrastructure units, humans, clouds, etc., which originates the network connected via Internet and known as IoVs.

12.4 Communication Frameworks for IoVs

The US Department of Transportation (DoT) has been established in 1966, whose one wing, National Highway Traffic Safety Administration (NHTSA), is responsible for the safe and reliable road transportation. Motivated by the fact that 94% of the road fatalities happened because of the human errors, the NHTSA has revealed that automated driving system (ADS) has potentials to reduce the highway fatalities by addressing their root cause [42, 43].

The journey of driving system starts with zero stage which means no automation and progress step-by-step toward its final stage, that is full automation. The entire journey is classified into different stages as follows:

Zeroth Stage (No Automation): Zero autonomy; the driver performs all driving tasks.

First Stage (Driver Assistance): Vehicle is controlled by the driver, but some driving-assist features may be included in the vehicle design.

Second Stage (Partial Automation): Vehicle has combined automated functions, like acceleration and steering, but the driver must remain engaged with the driving task and monitor the environment at all times.

Third Stage (Conditional Automation): Driver is a necessity, but is not required to monitor the environment. The driver must be ready to take control of the vehicle at all times with notice.

Fourth Stage (High Automation): The vehicle is capable of performing all driving functions under certain conditions. The driver may have the option to control the vehicle.

Fifth Stage (Full Automation): The vehicle is capable of performing all driving functions under all conditions. The driver may have the option to control the vehicle.

The NHTSA had responsibilities to evolve the ADS from the zero stage to the fifth stage by addressing all the challenges regarding the safe and reliable road transportation. By keeping this in mind, the NHTSA has launched their programs as per the available industrial technologies, namely (i) USDOT Automated Vehicles 2.0 (AV 2.0) Activities [44], (ii) preparing for the Future of Transportation: Automated Vehicles 3.0 [45], and (iii) Ensuring American Leadership in Automated Vehicle Technologies: Automated Vehicles 4.0 [46].

1) AV 2.0: (USDOT Automated Vehicles 2.0) was launched by the USDOT in September 2017, where key emphasis is on the vision for safety 2.0. As per USDOT, "*A Vision for Safety 2.0* calls for industry, state and local governments, safety and mobility advocates and the public to lay the path for the deployment of automated vehicles and technologies."

2) AV 3.0 (*Preparing for the Future of Transportation: Automated Vehicles 3.0*) builds upon *Automated Driving Systems 2.0: A Vision for Safety* (ADS 2.0) and was launched in October 2018. AV 3.0 expands the scope to all surface on-road transportation systems, and was developed through the input from a diverse set of stakeholder engagements, throughout the Nation. AV 3.0 is structured around three key areas:

   a) *Advancing multimodal safety,*
   b) *Reducing policy uncertainty, and*
   c) *Outlining a process for working with U.S. DOT.*

The U.S. DOT sees AV 3.0 as the beginning of a national discussion about the future of our on-road surface transportation system.

3) AV 4.0 (Ensuring American Leadership in Automated Vehicle Technologies: Automated Vehicles 4.0) was launched in January 2020 which is built upon AV3.0 and is structured around three main areas as follows: USG AV Principles, Administration Efforts Supporting AV Technology Growth and Leadership, and USG Activities and Opportunities for Collaboration.

AV 4.0 and full automation lead to driverless activities for which the vehicles need to be aware about their immediate surrounding environment that comprises nearby vehicles, their speed, direction of arrival, road directions and conditions, etc. In order to achieve this, the vehicle needs to communicate with the surrounding environment by using suitable sensing and actuating devices. Thus, the communication is an integral and prominent part of IoVs which are exploited by various researchers. Wan et al. [47] have presented a 5G-enabled edge computing-IoV system framework in order to enhance the performance of the existing edge computing-IoV systems. The authors have illustrated a specific computation offloading in 5G-enabled EC-IoV systems under different cases. Duan et al. [48] have presented an architecture of 5G-V2X communication networks by exploiting the technologies of 5G new radio (NR), network slicing, and device-to-device communications. Further, authors have emphasized on the principles and key features of the IoVs as well as investigated the challenges, opportunities, and future research directions. The potential applications of the SDN and multi-access edge computing technology are illustrated.

The popular communication frameworks in the IoT are V2V, V2I, infrastructure-to-vehicle (I2V), infrastructure-to-infrastructure (I2I), vehicle-to-human (pedestrians), infrastructure to cloud, etc., which are further discussed as follows.

12.4.1 Vehicle-to-Vehicle (V2V) Communication

The vehicles have to communicate with each other in order to share and know about the speed, location, braking, direction of travel, loss of stability, etc. The V2V communication forms a random wireless network with mobile nodes (vehicles) having different speed, direction of travel, etc. [40, 49–53]. The popular technology for V2V communication is dedicated short-range communication (DSRC), as well as standardized by the different communication bodies such as Federal Communication Commission (FCC) and International Standard Organization (ISO) [54, 55]. Thus, it is perceived that the V2V communication comprises a wireless network where automobiles send the messages to each other in order to share the information about theirself. Ideally, the range of the network is up to 300 meters or about 10 seconds at highway speed. V2V would be a mesh network as every vehicle/node transmits, receives, captures, and retransmits the signals. The prime requirement of the V2V communication is exchanging the safety-related messages among neighboring vehicles quickly and reliably. The single hope broadcast is a fundamental mechanism to handle the delay requirements; however, it fails to achieve reliability constraints because of the continuously changing V2V network topology and high density of vehicle at highway.

The rate adaptation is an effective way to achieve better system performance in dynamic mobile network according to which vehicle/nodes estimates the channel condition and regulates the data rate accordingly [56]. However, the conventional RA is not suitable for direct implementation for the highway V2V environments because it is difficult to estimate the channel conditions in such an open loop environment (Sender node receives no feedback from the receiver) and rapid change in topology, position, speed, and direction of arrival makes it more challenging. At highway, the communication suffers at PHY layer due to channel losses such as fading and shadowing; however, at MAC layer, communication suffers several nontrivial collisions such as channel conditions and hidden terminal. Generally, the low data rates can tolerate poor channel conditions; however, they introduce interference. On the other hand, the high data rate reduces the collision chances; however, it needs better channel conditions to endure the effective communication. Thus, it is perceived that the vehicle needs to adapt its data rates as per the difference between the interference loss and fading losses. By keeping this in mind, the authors have presented a LOss differentiation rate adaptation (LORA) scheme for V2V communication in highway environment [57]. In this method, the vehicle perceives the traffic intensity, distance from the neighboring vehicles and on the basis of which, it estimates and differentiates packet loss due to several reasons. Further, the effective data rate is selected on the basis of fading and interference losses.

The potential technology that exactly fits the V2V communication is M2M communication for moving nodes.

12.4.2 Vehicle to Infrastructure (V2I) Communication

The infrastructure in the connected vehicle environments defined as a roadside unit is responsible to work as an intermediator between vehicles and clouds, several decision-making processes such as traffic management, spectrum management, power management, gateway between vehicles and human, etc. Therefore, the V2I communication is used to store the information from vehicles to the infrastructure unit so that the database can be ready for further use such as prediction [40]. Another key infrastructure unit comprises: (i) advance traffic management system that shares the routing and pricing information, (ii) autonomous vehicles maintenance, storage, and charging points, (iii) reinvention of curb by rezoning on street parking into flexible spaces pickup/drop-off and transit stops during rush hours, commercial use in the evening, and freight use overweight, (iv) mobility hubs to bring together multiple modes of transportation, and (v) smart parking meters or market price parking permits for remaining on street parking. Moreover, the vehicles and nodes form a wireless network where the vehicles are mobile nodes; however, the infrastructure units are assumed as stationary. Thus, in V2I communication, the transmitting nodes are mobile; however, the receiving nodes are static which means reception is easier when compared to that of the V2V communication. However, in some cases the infrastructure units are also of mobile nature that further originates the network similar to V2V environments. The authors have discussed about power management for the different frameworks such as vehicle-to-home (V2H), vehicle-to-grid (V2G), V2V, etc.

12.4.3 Infrastructure to Vehicles (I2V) Communication

The wireless network in the I2V is very much similar to that of the V2I [40]. The key difference is the nature of transmitting and receiving nodes, which means the transmitting nodes are static; however, the receiving nodes are mobile with high speed (50–150 kmph). In addition to this, the mobile nature of infrastructure units leads I2V to the V2V communication.

12.4.4 Vehicle-to-Broadband (V2B) Communication

The vehicles need to be manage their communication among nearby vehicles as well as other remote controlling units via a broadband service such as 3G/4G/ 5G [58]. The broadband cloud can play an important role for driver assistance and vehicle tracking as the information at the cloud about the network such as traffic information, monitoring data, road directions, etc., is available.

12.4.5 Vehicle-to-Pedestrians (V2P) Communication

In the conventional transport system, the interaction between vehicle and human/ pedestrians relies on the hand/eye expression of the pedestrian that needs to be understand by the driver. However, in the driverless or semi/fully autonomous vehicles, there is need of a system that can communicate with the pedestrians [59]. The vehicle needs to be aware about the pedestrian's present state as well communicate its current and next move with him so that he can manage accordingly. Thus, it is perceived that the most prominent way for V2P communication is machine-to-human (M2H) which is explored by various researchers in the literature [60–63]. The authors have proposed a M2H interface, which utilizes the external surface and surrounding of vehicle to communicate with the road users/ humans. This type of interface is responsible to communicate the information concerning vehicle driving mode, imminent vehicle maneuvers, perception of vehicle surrounding, as well as cooperation capabilities.

The driving modes are generally of three types, which are manual, semi-, or fully autonomous vehicles, where fully autonomous vehicles are driverless and actually originate the need of full interaction between human and pedestrian. The cooperation in V2I environments refers to the ability of vehicle/human to communicate their intension with other vehicles. Further, it is avoided to provide explicit advice or instructions to the pedestrians due to diversity in its population, which limits us to work on the vehicle side for most of the communication and perceiving phenomenon [59, 64]. Several authors have introduced the communication by using color such as turquoise would be the most appropriate color to utilize in light-based autonomous vehicle-to-pedestrian communication, due to its saliency, discriminability, attractiveness, and uniqueness in the traffic system [65]. Ackermann et al. [66] have presented an experimental study to investigate design and assessment criteria for communication between pedestrians and vehicles. The experimental results revealed that direct instructions to cross the street are preferred when compared with the information exchange between the vehicle and pedestrian. Several prominent proposed frameworks where autonomous vehicle is equipped with technologies to communicate its intension or next move with pedestrians so that they can behave accordingly as follows. In [67], Lagstrom has proposed a prototype for HMI communication, which uses light emitting diodes (LED) strip in the upper windscreen that conveys the vehicle mode and intension to the pedestrian. The messages, which are conveyed to the pedestrians, are "automated driving mode," "about to yield," "about to start," "about to stop" which must be easily understood by the human beings. It is claimed, that the pedestrians like this framework as it can replace the driver activities in the autonomous vehicles. A study by Clamann et al. [68] has also presented a prototype where a display is mounted on the radiator grill that displayed two messages on the black screen namely "Walk" or "Do not Walk" in addition to the speed of vehicle. It is reported that "Forty-six percent of participants stated that using the display would make their crossing decision easier, but only 4% used the display as primary piece of information when deciding to cross the road in front of the vehicle." The authors have investigated a study regarding the various prototypes and experimental studies which revealed that pedestrians like mostly direct messages displayed on the screen of the vehicle that are "Walk" or "Do not Walk" [69, 70]. There are further proposals made by research teams and vehicular industries without delivering empirical data, but those proposals should also be taken into account here. Visual information is also displayed by using projections on the street surface [71]. Mercedes-Benz showed a zebra crossing, which can be projected on the road surface. The question remains, which design criteria is meaningfully applied in the context of the interaction of pedestrians and automated vehicles [72]? Current DIN EN ISO 9241-110 guidelines are not applicable as there is no active interaction of the pedestrian with the HMI; therefore, the criteria of individualism and controllability do not apply.

12.5 Simulation Environments for Internet-of-Vehicles

The IoVs is going to become an integral part of our personal, social, as well as professional life, that increases its need to be more reliable, robust, and always working. The human life directly relies on the functionality of the IoVs networks and emulation and simulation environments need to be explored which needs to be very precise and reliable. The simulation and emulation are the prominent ways to recreate a computer- or hardware-based environment so that similar environment scenarios and its challenges can be explored as well as analyzed. The key difference between the emulation and simulation needs to be explained clearly.

As per the Cambridge dictionary, the **simulation** is defined as "A model of a set of problems or events that can be used to teach someone how to do something, or the process of making such a model." Banks et al. [73] have defined the simulation as "an approximate imitation of the operation of a process or system that represents its operation over time." The simulation is used in many contexts, such as simulation of technology for performance tuning or optimizing, safety engineering, testing, training, education, and video games. Frequently, the computer experiments are used to study simulation models. The interesting side of the simulation is inclusion of computer which is used for scientific modeling of natural systems or human systems to gain insight into their functioning as in economics and is used to show the eventual real effects of alternative conditions and course of action [74]. The simulation is also used when the real system cannot be engaged, because it may not be accessible, or it may be dangerous or unacceptable to engage, or it is being designed but not yet built, or it may simply not exist and this reason is more suitable for IoVs.

Emulation: However, as per the Cambridge dictionary, the emulation is defined as: "to copy something achieved by someone else and try to do it as well as they have." The key difference between the simulation and emulation is that, simulation is software implementation of a model where the internal functions of the original systems are not taken into consideration (for example, a "flight simulator" does not have any "component" of an actual aircraft). However, the Emulation is a replica of the internal system functions on a different host (for example, on a Mac OS X I can write a software emulator of a Windows OS where a Win program can run). The emulators can also be strictly hardware-based and are normally based on a partial or complete "reverse engineering" phase.

The prime software tools that offer the simulation facility for the IoVs networks are Network Simulator (NetSim), Simulation Environment for Mobility (SUMO), Network Simulator-2 (NS-2), NS-3, MATLAB, and OmNet++.

12.5.1 SUMO

SUMO is a free, open source, highly portable, microscopic and continuous traffic simulation package which is designed to handle large road networks and is mainly developed by employees of the Institute of Transportation Systems at the German Aerospace Center. It is open source, licensed under the EPLv2 and permits modeling of intermodal traffic systems including road vehicles, roadside infrastructure units, public transport, and pedestrians [75, 76]. The SUMO includes a wealth of supporting tools, which handle tasks such as route finding, visualization, network import, and emission calculation. It can be enhanced with custom models and provides various APIs to remotely control the simulation. The SUMO offers various features, namely (i) online interaction, (ii) multimodal simulation including vehicles, public transport, pedestrians, (iii) automatic generation of time schedules of traffic lights, (iv) no limitations in network size and number of simulated vehicles, (v) evaluation of eco-aware routing based on pollutant emission and investigations of autonomous route choice on the overall network, and (vi) supports the import formats Open Street Map, VISUM, VISSIM, and Nav Teq. SUMO is implemented in C++ and uses only portable libraries.

12.5.2 Network Simulator (NetSim)

NetSim is software which provides the facility of simulation and emulation for a wide range environment such as wireless networks, wireless ad-hoc networks, mobile ad-hoc network, vehicular ad-hoc networks, cognitive radio networks, wireless sensor networks, etc. Specifically saying, the NetSim modeling and simulation are supported for Aloha, Slotted Aloha, Token Ring/Bus, Ethernet CSMA/CD, Fast and Gigabit Ethernet, WLAN – IEEE 802.11 a/b/g/n and e, X.25, Frame Relay, TCP, UDP, IPv4 and IPv6, Routing – RIP, OSPF, BGP, MPLS, Wi-Max, MANET, GSM, CDMA, Wireless Sensor Network, ZigBee, Cognitive radio, etc. NetSim is classified into three categories known as NetSim Professional [77], NetSim Standard [78], and NetSim Academic [79] which are further illustrated as follows.

NetSim Professional is an end-to-end, full stack, packet-level network simulator and emulator. It provides network engineers with a technology development environment for protocol modeling, network R&D, and military communications. The behavior and performance of new protocols and devices can be investigated in a virtual network within NetSim at significantly lower cost and in less time than with hardware prototypes. The key feature defined for the NetSim Pro are as follows: (i) Network Simulator and Emulator, (ii) Network capacity and growth studies, (iii) New technology and protocol evaluation, and (iv) used by military-/defense/space organizations, utilities distribution companies, network equipment manufacturers, and services providers. The key features and use direction for NetSim standard are as follows: (i) network simulator/emulator for R & D in universities, (ii) comes with protocol source C Code, (iii) develop and simulate your own protocols and algorithms, (iv) debug in runtime and "watch" all variables, (v) interface with external tools like MATLAB, SUMO, Wireshark, etc., and (vi) discounted educational pricing. NetSim Academic is defined as a Network simulator for teaching/network lab experimentation at universities and has the following features that are defined as follows: (i) TCP/IP, Ethernet, Routing, Wi-Fi, MANET, IoT, LTE, WSN ... and more, (ii) well-designed experiment manual with 45+ labs covering the latest technologies, (iii) packet animation for visual understanding, and (iv) deeply discounted pricing.

Netsim Academic is also called as Network Lab on your Desktop. Netsim Pro version is suited for commercial (enterprise/defense) customers while NetSim standard and NetSim Academic are targeted at education customers. The version comparison table shows the features available in the different versions. NetSim Academic is an economical option for educational customers intending to use NetSim for lab experimentation and teaching. The key difference of features can be cleared in the from the comparison table [80].

12.5.3 Ns-2

Ns-2 is a discrete event simulator, which is designed for the research of networking phenomenon and protocols which substantially support the simulation of HTTP, TCP, UDP, SRM, RTP, routing, and multicast protocols over wired as well as wireless (local and satellite) networks [81, 82]. The Ns-2 has emerged as the variant of REAL network simulator in 1989 and was supported by DARPA in 1995 through VINT project at LBL, Xerox, PARC, UCB, and USC/ISI. The Ns-2 has evolved significantly in the last couple of years due to dramatic change of the wireless technologies and networking. The interesting point about the Ns-2 is that it is free software, licensed under the GNU GPLv2 license, and is publicly available for research, development, and use. The Ns-2 is used to simulate and model several ad-hoc routing protocols and propagation models. It provides simulation of a variety of protocols HTTP, TCP, UDP, SRM, RTP, and some routing algorithms.

12.5.4 Ns-3

Similar to the Ns-2, the Ns-3 is also a discrete-event network simulator for Internet systems, targeted primarily for research and educational use. Ns-3 is a new software development effort focused on improving upon the core architecture, software integration, models, and educational components of Ns-2. The Ns-3 is also free software, licensed under the GNU GPLv2 license, and is publicly available for research, development, and use. The Ns-3 also works as emulator; however, it is not possible in case of Ns-2. There are certain advances and differences when we compare the Ns-2 with Ns-3 which are as follows and discussed in detail [83, 84]. It also provides a graphical user interface which makes it easier to use and more popular. The coding in Ns-2 is supported with the C++ and OTCL languages; however, in Ns-3 coding is supported with C++ and Python languages.

12.5.5 OMNeT++

OMNeT++ is an extensible, modular, component-based C++ simulation library and framework which also includes an integrated development and a graphical

Table 12.1 Comparison of various IoVs simulation tool.

Name of the software	Ns-2	Ns-3	NetSim	MATLAB	OMNET++
License type	Free Open Source	Free Open Source	Commercial	Commercial	Commercial
Language	C++ and OTCL	C++ and Python	C and JAVA	MATLAB	C++
GUI support	Limited	Yes	Yes	Yes	Yes
Simulation event type	Discrete event	Discrete event	Stochastic discrete event	Discrete event	Discrete event
Available module	Wired, Wireless, Ad-hoc, and Wireless sensor networks	Wired, Wireless, Ad-hoc, and Wireless sensor networks	Wired and Wireless sensor network (wireless LAN, WiMAX), VANET, SDR, CRN	Wired and Wireless sensor network (wireless LAN, WiMAX), VANET, SDR, CRN	Wired, Wireless, Ad-hoc and Wireless Sensor Networks
Scalability	Limited	Limited	Large enough	Large enough	Enough

runtime environment [85]. Domain-specific functionality (support for simulation of communication networks, queuing networks, performance evaluation, etc.) is provided by model frameworks, developed as independent projects. There are extensions for real-time simulation, network emulation, and support for alternative programming languages (Java, C#), database integration, System C integration, HLA, and several other functions. Therefore, it is worth exploring the various simulation and emulation environments which are tabulated in Table 12.1.

12.6 Potential Research Challenges

A lot of milestones in the field of IoVs are achieved; however, the IoVs is in the emerging state and need to be explored a lot to achieve its complete working model on the roads. For this, we are presenting some potential issue that needs to be addressed very carefully and categorized as Social and Technical issues as shown in Figure 12.3.

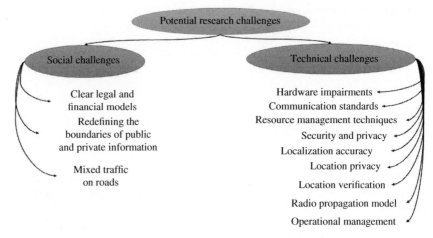

Figure 12.3 Potential research challenges in IoVs networks.

12.6.1 Social Challenges

The arena of the vehicles is well defined on the roads; however, the challenges are introduced due to need of its interaction with the other vehicles, roadside units, cloud, and passengers/human/pedestrians. The dispersion of driver/human means the dispersion of responsibilities and the vehicles misbehaving due to any cause may directly result in the property and life loss/injury. Therefore, clear legal and financial models are required to assign responsibilities to various stakeholders of IoVs. The complete autonomous nature of vehicle demands a lot of information about the pedestrians and road-traffic, which introduce a need of redefining the boundaries of public and private information. The mixed traffic on roads due to some of the semi-/fully autonomous and non-autonomous vehicles makes the IoVs networks more complex and challenging. The protocols for the interaction between the driver and autonomous vehicles needs to be defined very clearly and preferences of the decision-making either by driver or autonomous vehicle must be clear.

12.6.2 Technical Challenges

A) **Hardware impairments**
The hardware impairments have a long history of introducing challenges for the physical implementation of the various technical frameworks and the case is same of IoVs. It is worth mentioning that, even a small or single hardware

impairment restricts the working or implementation of IoVs due to direct involvement of human life and property.

B) **The communication standard**

Various communication standards are defined for the vehicular communication such as IEEE 822.11 P for Wireless Access for Vehicular Environments (WAVE), IEEE 1609, LTE, LTE-Advanced, etc. [86–88]. However, the existing cellular or WAVE lacks many features for the implementation of the large-scale IoVs. The rules and protocols defined in the cellular or WAVE standards are defined only for fully autonomous environments (where no driverless car on the road); however, this scenario is very away from the current time. Therefore, potential communication protocols and standards need to be developed for the mixed traffic environments (with fully/semi-autonomous vehicles and vehicles with drivers). This demand leads to the use of the machine learning, artificial intelligence, and decision-making algorithms for IoVs.

C) **Resource management techniques**

Most of the IoVs' components on the roads comprise resource constraint devices such as energy, spectrum, power, etc. The failure of the IoVs network due to lack of any resource directly leads to loss of human life. Therefore, resource optimization protocols for various scenarios of the IoVs needs to be explored. The information exchange and sharing is an integral part of the IoVs or any other network; however, IoVs faces major challenges due to rapid change of topology on the network because of high speed of vehicles and unpredicted road density. Thus, advance and rapid routing algorithms needs to be designed [89–92].

D) **Security and privacy**

The IoVs suffers from all the security and privacy effects which affect the conventional mobile ad-hoc network and IoT [93]. However, the autonomous and driverless nature makes it more prone to the intruder and misbehaving or denial of service by the vehicles. The small security or attacking activities can lead to the failure of network and loss of human and property. Therefore, the security and privacy are of prime importance in the IoVs and the authors have explored various security attacks and privacy issues in the vehicular networks and have presented a state-of-the-art of existing vehicular attacks. The manipulation of the safety critical systems such as compromising the cruise control, steering, and braking systems needs to avoid for working of the IoVs. The companies such as Hyundai, Volkswagen, etc., have proposed different digital key mobile-based applications and claimed its sufficient security and privacy.

E) **Localization accuracy**

The entire IoVs network model is function of the location of the users and distances among them; therefore, the exact localization of the vehicle is very important. Conversely, the high moving speed of vehicles on the roads makes it a challenging problem. One potential method for vehicle localization is GPS based [94–96]; however, the issue with this technique is that it does not consider the speed of vehicles. Therefore, more efforts to improve the accuracy of localization are required.

F) **Location privacy**

The continuous need of the localization of vehicle risks the privacy of location of the vehicles which is also of prime importance in the IoVs. The key techniques in literature that addresses the location privacy are pseudonym switching [97], Silent period [98], and Mix zone [99]; however, still we are looking for another appropriate technique because of the following issues in the existing approaches. Pseudonym switching is suitable for higher vehicle density and is easily detectable in lower vehicle density environments. On the other hand, silent period is applicable for non-real-time ITS applications and unsuitable for the real-time applications. Moreover, the Mix zone is useful on multilane roads with larger zone area and ineffective in the single and street roads.

G) **Location verification**

The location verification is a very prominent aspect in the vehicular communication as the error in location detection of other vehicle can cause the accident. Some malicious users/vehicles or intruders can alter the actual location detection and shows false location of vehicle and disturbs the communication of IoVs network. Therefore, exploring the location verification technique is of prime importance in the IoVs networks and some efforts in this direction are mentioned in the literature that are directional antenna [100], beaconing-based belief [101], and cooperative approach [96]. However, still a lot need to be explored for successful implementation in terms of cost of infrastructure in antenna, overload issue in the beaconing approach, and untrustworthiness in the cooperative approach.

H) **Radio propagation models**

The road environments vary in a number of ways as we move on the street roads, single roads, city roads, highways, forest, flyovers, etc., and channel also varies similarly and shows different channel models. Therefore, designing a single-channel propagation model for such number of different scenarios is a very challenging task; however, some efforts are mentioned in the literature [102–104]. Additional prime research issues that are yet to be explored are to realize the effect of moving and static vehicle at the same time, maintaining line of light for long time, etc.

I) **Operational Management**

The number of vehicles on the roads, their high speed movement, and need of coordination with the surrounding vehicles have increased the volume of data drastically. The capturing, storing, and processing of such huge data demands highly reliable and secure operational management efforts. The potential areas where IoVs needs effective operation management are telecom services, sensing services, actuating services, computational services, etc.

12.7 Summary

In this chapter, the frameworks and challenges for the implementation of IoVs networks are explored. It is perceived that the semi-full-autonomous nature of the vehicles and their need to interact with drivers and pedestrians makes IoVs network of heterogeneous nature. These IoVs network scenario are simulated by using various simulation tools so that performance and challenges of the network can be realized before implementation. The heterogeneous nature and high mobility of vehicles introduces potential social and technical challenges for the implementation of the IoVs network, which are elaborated in detail.

References

1 Desoutter. Industrial revolution: from Industry 1.0 to Industry 4.0. https://www.desouttertools.com/industry-4-0/news/503/industrial-revolution-from-industry-1-0-to-industry-4-0 (accessed 18 February 2020).

2 i-SCOOP. Industry 4.0: the fourth industrial revolution – guide to Industrie 4.0. https://www.i-scoop.eu/industry-4-0/#Industry_40_the_essence_explained_in_a_nutshell (accessed 18 February 2020).

3 Villagran, N.V., Estevez, E., Pesado, P., and Marquez, J.D.J. (2019). Standardization: A Key Factor of Industry 4.0. *Proceedings of 6th International Conference on eDemocracy & eGovernment* (ICEDEG), Quito, Ecuador, Ecuador (24–26 April 2019), 1–6.

4 Brettel, M., Friederichsen, N., Keller, M., and Rosenberg, M. (2014). How virtualization, decentralization and network building change the manufacturing landscape: an industry 4.0 perspective. *International Journal of Mechanical, Aerospace, Industrial, Mechatronic and Manufacturing Engineering* 8 (1): 37–44.

5 Li, F., Wan, J., Zhang, P. et al. (2016). Usage-specific semantic integration for cyber-physical robot systems. *ACM Transactions on Embedded Computing Systems* 15 (3): 50.

6 Zhou, J. (2015). Intelligent manufacturing-main direction of 'Made in China 2025'. *China Mechanical Engineering* 26 (17): 2273–2284.

7 Wan, J., Yi, M., Li, D. et al. (2016). Mobile services for customization manufacturing systems: an example of industry 4.0. *IEEE Access* 4: 8977–8986.

8 Chen, B., Wan, J., Li, P. et al. (2018). Smart factory of industry 4.0: key technologies, application case, and challenges. *IEEE Access* 6: 6505–6519.

9 Wang, S., Wan, J., Zhang, D. et al. (2016). Towards smart factory for industry 4.0: a self-organized multi-agent system with big data based feedback and coordination. *Computer Networks* 101: 158–168.

10 Lueth, K.L. State of the IoT 2018: number of IoT devices now at 7B – market accelerating. https://iot-analytics.com/state-of-the-iot-update-q1-q2-2018-number-of-iot-devices-now-7b/ (accessed 18 February 2020).

11 Liao, Y., Loures, E.F.R., and Deschamps, F. (2018). Industrial internet of things: a systematic literature review and insights. *IEEE Internet of Things Journal* 5 (6): 4515–4525.

12 Erfani, S., Ahmadi, M., and Chen, L. (2017). The Internet of things for smart homes: An example. *Proceedings of 8th Annual Industrial Automation and Electromechanical Engineering Conference (IEMECON)*, Bankok, Thailand (16–18 August 2017), 153–157,

13 Ningqing, L., Haiyang, Y., and Chunmeng, G. (2013). Design and implementation of a smart home control system. *Proceedings of 3rd International Conference on Instrumentation, Measurement, Computer, Communication and Control*, Shenyang, China (September 2013), 1535–1538.

14 Khan, A., A-Zahrani, A., A-Harbi, S., A-Nashri, S., and Khan, I.A. (2018). Design of an IoT smart home system. *Proceedings of 15th Learning and Technology Conference* (L&T), Jeddah, Saudi Arabia (February 2018), 1–5.

15 Zungeru, M., Gaboitaolelwe, J., Diarra, B. et al. (2019). A secured smart home switching system based on wireless communications and self-energy harvesting. *IEEE Access* 7: 25063–25085.

16 Kumar, P., Braeken, A., Gurtov, A. et al. (2017). Anonymous secure framework in connected smart home environments. *IEEE Transactions on Information Forensics and Security* 12 (4): 968–979.

17 Godfrey, A., Hetherington, V., Shum, H. et al. (2018). From A to Z: Wearable technology explained. *Maturitas* 113: 40–47.

18 Venugopal, K. and Heath, R.W. (2016). Millimeter wave networked wearables in dense indoor environments. *IEEE Access* 4: 1205–1221.

19 Shin, D., Yun, K., Kim, J. et al. (2019). A security protocol for route optimization in DMM-based smart home IoTs networks. *IEEE Access* 7: 142531–142550.

20 Braun, S., Carbon, R., and Naab, M. (2016). Piloting a mobile-app ecosystem for smart farming. *IEEE Software* 33 (4): 9–14.

21 Taneja, M., Jalodia, N., Malone, P. et al. (2019). Connected cows: Utilizing fog and cloud analytics toward data-driven decisions for smart dairy farming. *IEEE Internet of Things Magazine* 2 (4): 32–37.

22 Jukan, A., Carpio, F., Masip, X. et al. (2019). Fog-to-cloud computing for farming: Low-cost technologies, data exchange, and animal welfare. *Computer* 52 (19): 41–51.

23 Ramadhan, A. (2019). Industry 4.0: Development of smart sunroof ambient light manufacturing system for automotive industry. *Proceedings of Advances in Science and Engineering Technology International Conferences* (ASET), Dubai, United Arab Emirates (26 March to 10 April 2019), 1–4.

24 Singhal, A. and Saxena, R.P. (2018). Data storage in smart grid systems. *Proceedings of IEEE 6th International Istanbul Smart Grids and Cities Congress and Fair (ICSG)*, Istanbul, Turkey (25–26 April 2018), 110–113.

25 Yilmaz, E.N., Polat, H., Oyucu, S., Aksoz, A., and Saygin, A. (2012). Software models for Smart Grid. *Proceedings of 1st International Workshop on Software Engineering Challenges for the Smart Grid (SE-SmartGrids)*, Zurich, Switzerland (3 June 2012), 42–45.

26 Whelan, B.-M., Angus, D., Wiles, J. et al. (2018). Toward the development of smart communication technology: automating the analysis of communicative trouble and repair in dementia. *Innovation in Aging* 2 (3): 1–15.

27 Yuan, J., Shan, H., Huang, A. et al. (2017). Massive machine-to-machine communications in cellular network: distributed queueing random access meets MIMO. *IEEE Access* 5: 2981–2993.

28 Khoueiry, B.W. and Soleymani, M.R. (2016). A novel machine-to-machine communication strategy using rateless coding for the internet of things. *IEEE Internet of Things Journal* 3 (6): 937–950.

29 Priby, O., Priby, P., Lom, M., and Svitek, M. (2019). Modeling of smart cities based on its architecture. *IEEE Intelligent Transportation Systems Magazine* 11 (4): 28–36.

30 Ahlgren, B., Hidell, M., and Ngai, E.C.-H. (2016). Internet of things for smart cities: interoperability and open data. *IEEE Internet Computing* 20 (6): 52–56.

31 Ang, L.-M., Seng, K.P., Ijemaru, G.K., and Zungeru, A.M. (2018). Deployment of IoVs for smart cities: applications, architecture, and challenges. *IEEE Access* 7: 6473–6492.

32 Du, P. and Lu, N. (2011). Appliance commitment for household load scheduling. *IEEE Transactions on Smart Grid* 2 (7): 411–419.

33 Guan, X., Xu, Z., and Jia, Q.S. (2010). Energy-efficient buildings facilitated by microgrid. *IEEE Transactions on Smart Grid* 1 (3): 243–252.

34 Nunna, H.S.V.S.K., Aziz, N.A.B., and Srinivasan, D. (2018). A smart energy management framework for distribution systems with perceptive residential consumers. *Proceedings of IEEE PES Asia-Pacific Power and Energy Engineering Conference (APPEEC)*, Kota Kinabalu, Malaysia (7–10 October 2018), 434–438.

35 Singh, S., Roy, A., and Selvan, M.P. (2019). Smart load node for nonsmart load under smart grid paradigm: a new home energy management system. *IEEE Consumer Electronics Magazine* 8 (2): 22–27.

36 Pantanoa, E. and Timmermansa, H. (2014). What is smart for retailing? *Procedia Environmental Sciences* 22: 101–107.

37 KPMG. Tapping into smart retail: a survey of CEOs and Consumers in the Greater Bay Area. https://assets.kpmg/content/dam/kpmg/cn/pdf/en/2018/11/tapping-into-smart-retail.pdf (accessed 21 February 2019).

38 The 50 robots best ever. https://www.wired.com/2006/01/robots-3/ (accessed 14 October 2019).

39 Marquiss, R. (2018). Five types of industrial robots and how to choose the best fit. Valin Corporation. https://www.valin.com/resources/articles/five-types-of-industrial-robots-and-how-to-choose-the-best-fit (accessed 14 October 2019).

40 Liu, C., Chau, K.T., Wu, D., and Gao, S. (2013). Opportunities and challenges of vehicle-to-home, vehicle-to-vehicle, and vehicle-to-grid technologies. *Proceedings of the IEEE* 101 (11): 2409–2427.

41 Uhlemann, E. (2018). Time for autonomous vehicles to connect [Connected Vehicles]. *IEEE Vehicular Technology Magazine* 13 (3): 10–13.

42 NHTSA. Automated vehicles for safety. https://www.nhtsa.gov/technology-innovation/automated-vehicles-safety.

43 Thakur, P. and Singh, G. (2020). Potential simulation frameworks and challenges for internet of vehicles networks. *2020 International Conference on Artificial Intelligence, Big Data, Computing and Data Communication Systems (icABCD)*, Durban, South Africa (6–7 August 2020), 1–6, IEEE, 2020.

44 U.S. Department of Transportation. Automated driving systems: a vision for safety 2.0. https://www.transportation.gov/av/2.0 (accessed 18 February 2020).

45 U.S. Department of Transportation. Preparing for the future of transportation: automated vehicles 3.0. https://www.transportation.gov/av/3 (accessed 18 February 2020).

46 U.S. Department of Transportation. Ensuring american leadership in automated vehicle technologies: automated vehicles 4.0 https://www.transportation.gov/av/4 (accessed 18 February 2020).

47 Wan, S., Renhao, G., Umer, T. et al. (2020). Toward offloading internet of vehicles applications in 5G networks. *IEEE Transactions on Intelligent Transportation Systems*: 1–9. https://doi.org/10.1109/TITS.2020.3017596.

48 Duan, W., Gu, J., Wen, M. et al. (2020). Emerging technologies for 5G-IoV networks: applications, trends and opportunities. *IEEE Network* 34 (5): 283–289.

49 Aoki, M. and Fujii, H. (1996). Inter-vehicle communication: technical issues on vehicle control application. *IEEE Communications Magazine* 34 (10): 90–93.

50 Lu, N., Cheng, N., Zhang, N. et al. (2014). Connected vehicles: Solutions and challenges. *IEEE Internet of Things Journal* 1 (4): 289–299.

51 Madabushi, R. (2019). Connected cars: evolvement, solution and challenges. https://labs.sogeti.com/connected-cars/ (accessed 14 October 2019).

52 Masini, B.M., Ferrari, G., Silva, C., and Thibault, I. (2017). Connected vehicles: applications and communication challenges. *Mobile Information Systems* 2017: 1–2.

53 Ericson Connected Vehicles. https://www.ericsson.com/en/internet-of-things/ automotive (accessed 14 October 2019).

54 Ghafoor, K.Z., Guizani, M., Kong, L. et al. (2020). Enabling efficient coexistence of DSRC and C-V2X in vehicular networks. *IEEE Wireless Communications* 27 (2): 134–140. https://doi.org/10.1109/MWC.001.1900219.

55 Naik, G., Choudhury, B., and Park, J.-M. (2019). IEEE 802.11bd & 5G NR V2X: evolution of radio access technologies for V2X communications. *IEEE Access* 7: 70169–70184.

56 Xie, H.F., Huang, D.W., Long, Y., and Wang, Q.H. (2019). A rate adaptation scheme with enhanced loss differentiation for LoRaWAN communications. *Journal of Physics Conference Series* 1237 (3): 1–7.

57 Yao, Y., Chen, X., Rao, L. et al. (2017). LORA: Loss differentiation rate adaptation scheme for vehicle-to-vehicle safety communications. *IEEE Transactions on Vehicular Technology* 66 (3): 2499–2512.

58 Liang, W., Li, Z., Zhang, H. et al. (2015). Vehicular ad hoc networks: architectures, research issues, methodologies, challenges, and trends. *International Journal of Distributed Sensor Networks* 11 (8): 1–11.

59 Habibovic, A. (2018). Communicating intent of automated vehicles to pedestrians. *Frontiers in Psychology* 9 (1336): 1–17.

60 Haeuslschmid, R., Pfleging, B., and Alt, F. (2016). A design space to support the development of windshield applications for the car. *Proceedings of CHI Conference on Human Factors in Computing Systems*, New York, USA, 2016, 5076–5091.

61 Colley, A., Hakilla, J., Pfleging, B., and Alt, F. (2017). A design space for external displays on cars. *Proceedings of 9th International Conference on Automotive User Interfaces and Interactive Vehicular Applications Adjunct*, New York, USA, 146–151.

62 Mirnig, A.G., Wintersberger, P., Meschtscherjakov, A., Riener, A., and Boll, S. (2018). Workshop on communication between automated vehicles and vulnerable road users. *Proceedings of 10th International Conference on Automotive User Interfaces and Interactive Vehicular Applications*, New York, USA, 65–71.

63 Schieben, A., Wilbrink, M., Kettwich, C. et al. (2018). *"Designing the interaction of automated vehicles with other traffic participants: A design framework based on human needs and expectations" Cognition, Technology and work*. Berlin: Springer.

64 Owensby, C., Tomitsch, M., and Parker, C. (2018). A framework for designing interactions between pedestrians and driverless cars: insights from a ridesharing design study. *Proceedings of 30th Australian Conference on Computer-Human Interaction*, New York, USA, 359–363.

65 Dey, D., and Terken, J. (2017). Pedestrian interaction with vehicles: roles of explicit and implicit communication. *Proceedings of 9th International*

Conference on Automotive User Interfaces and Interactive Vehicular Applications, NewYork, USA, 109–113.

66 Ackermann, C., Beggiato, M., Schubert, S., and Krems, J.F. (2019). An experimental study to investigate design and assessment criteria: What is important for communication between pedestrians and automated vehicles? *Applied Ergonomics* 75: 272–282.

67 Lagstrom, T. (2015). Autonomous Vehicles 'interaction with Pedestrians. An Investigation of Pedestrian-driver Communication and Development of a Vehicle Eternal Interface. Master Thesis. Chalmers University of Technology, Gothenborg. Sweden.

68 Clamann, M., Aubert, M., and Cummings, M.L. (2017). Evaluation of vehicle-to-pedestrian communication displays for autonomous vehicles. *Proceedings of Transportation Research Board 96th Annual Meeting*, 2017 (No. 17-02119), Washington, DC (8–12 January 2017), 1–12

69 Yang, S. (2017). Driver behavior impact on pedestrians' crossing experience in the conditionally autonomous driving context. Master of Science Dissertation, KTH Royal Institute of Technology, Stockholm, Sweden.

70 Fridman, L., Mehler, B., Xia, L. et al. (2017). To walk or not to walk: Crowdsourced assessment of external vehicle-to-pedestrian displays. arXiv preprint. arXiv:1707.02698 (accessed 12 August 2020).

71 Hillis, W.D., Williams, K.I., Tombrello, T.A. et al. (2016). Communication between autonomous vehicle and external observers. U.S. Patent No. 9,475,422. Washington, DC: U.S. Patent and Trademark Office.

72 Deb, S., Rahman, M.M., Strawderman, L.J., and Garrison, T.M. (2018). Pedestrians' receptivity toward fully automated vehicles: research review and roadmap for future research. *IEEE Transactions on Human-Machine Systems* 48 (3): 279–290.

73 Banks, J., Carson, J., Nelson, B., and Nicol, D. (2011). *Discrete-Event System Simulation*, 3. Prentice Hall. ISBN: 978-0-13-088702-3.

74 Sokolowski, J.A. and Banks, C.M. (2009). *Principles of modeling and simulation*, 6. Wiley. ISBN: 978-0-470-28943-3.

75 CIVITAS. SUMO: Simulation of Urban Mobility. https://civitas.eu/tool-inventory/sumo-simulation-urban-mobility. https://www.ericsson.com/en/internet-of-things/automotive (accessed 14 October 2019).

76 SUMO: Simulation of Urban Mobility. http://sumo.sourceforge.net/. https://www.ericsson.com/en/internet-of-things/automotive (accessed 2 July 2020).

77 TETCOS. NetSim professional. https://www.tetcos.com/netsim-pro.html. https://www.ericsson.com/en/internet-of-things/automotive (accessed 02 July 2020).

78 TETCOS. NetSim standard. https://www.tetcos.com/netsim-std.html. https://www.ericsson.com/en/internet-of-things/automotive (accessed 02 July 2020).

79 TETCOS. NetSim academic. https://www.tetcos.com/netsim-acad.html. https://www.ericsson.com/en/internet-of-things/automotive (accessed on 02 July 2020).

80 TETCOS. How do the different versions of NetSim compare. https://www.tetcos.com/version-comparison.html. https://www.ericsson.com/en/internet-of-things/automotive (accessed 02 July 2020).

81 ISI. The network simulator – ns-2. https://www.isi.edu/nsnam/ns/. https://www.ericsson.com/en/internet-of-things/automotive (accessed 02 July 2020).

82 https://www.tutorialsweb.com/ns2/NS2-1.html. https://www.ericsson.com/en/internet-of-things/automotive (accessed 02July 2020).

83 Difference between NS-2 and NS-3. https://rishikeshteke.wordpress.com/tag/difference-between-ns2-and-ns3/. https://www.ericsson.com/en/internet-of-things/automotive (accessed 02July 2020).

84 Difference between NS-2 and NS-3. http://www.startechnologychennai.com/Difference%20between%20ns2%20and%20ns3.html. https://www.ericsson.com/en/internet-of-things/automotive (accessed 02 July 2020).

85 Varga, A. (2010). OMNeT++. In: *Modeling and Tools for Network Simulation* (eds. K. Wehrle, M. Güneş and J. Gross). Berlin, Heidelberg: Springer.

86 IEEE 802.11p-2010 IEEE standard for information technology – local and metropolitan area networks – specific requirements – Part 11: Wireless LAN Medium Access Control (MAC) and physical layer (PHY) specifications amendment 6: wireless access in vehicular environments. https://standards.ieee.org/standard/802_11p-2010.html (accessed 27 February 2020).

87 Hameed Mir, Z. and Filali, F. (2014). LTE and IEEE 802.11p for vehicular networking: a performance evaluation. *EURASIP Journal on Wireless Communications and Networking* 89: 1–15.

88 Bilgin, B.E. and Gungor, V.C. (2013). Performance comparison of IEEE 802.11p and IEEE 802.11b for vehicle-to-vehicle communications in highway, rural, and urban areas. *International Journal of Vehicular Technology* 2013: 1–10.

89 Haitao, Z., Yuting, Z., Hongbo, Z., and Dapeng, L. (2018). Resource management in vehicular ad hoc networks: multi-parameter fuzzy optimization scheme. *Procedia Computer Science* 129: 443–448.

90 Cordeschi, N., Amendola, D., Shojafar, M., and Baccarelli, E. (2015). Distributed and adaptive resource management in cloud-assisted cognitive radio vehicular networks with hard reliability guarantees. *Vehicular communications* 2 (1): 1–12.

91 Liang, L., Xie, S., Li, G. Y., Ding, Z., and Yu, X. (2018). Graph-based radio resource management for vehicular networks. *Proceedings of IEEE International Conference on Communications (ICC)*, Kansas City, USA (20–24 May 2018), 1–6.

92 Mahmood, A., Butler, B., and Jennings, B. (2018). Towards efficient network resource management in SDN-based heterogeneous vehicular networks. *Proceedings of 42nd IEEE Annual Computer Software and Applications Conference (COMPSAC)*, Tokyo, Japan (23–27 July 2018), 813–814.

93 Khelifi, H., Luo, S., Nour, B., and Shah, S.C. (2018). Security and privacy issues in vehicular named data networks: An overview. *Mobile Information Systems* 2018: 1–18.

94 Kaiwartya, O.M., Hanan, A., Cao, Y. et al. (2016). Internet of vehicles: motivation, layered architecture, network model, challenges, and future aspects. *IEEE Access* 4: 5356–5373.

95 Yao, J., Balaei, A.T., Hassan, M. et al. (2011). Improving cooperative positioning for vehicular networks. *IEEE Transactions on Vehicular Technology* 60 (6): 2810–2823.

96 Fogue, M., Martinez, F.J., Garrido, P. et al. (2015). Securing warning message dissemination in VANETs using cooperative neighbor position verification. *IEEE Transactions on Vehicular Technology* 64 (6): 2538–2550.

97 Huang, X., Yu, R., Kang, J. et al. (2016). Software defined networking with pseudonym systems for secure vehicular clouds. *IEEE Access* 4 (1): 3522–3534.

98 Tyagi, A.K. and Sreenath, N. (2015). Location privacy preserving techniques for location based services over road networks. *Proceedings of IEEE ICCSP*, Melmaruvathur, India (2–4 April 2015), 1319–1326.

99 Ying, B., Makrakis, D., and Mouftah, H.T. (2013). Dynamic mix-zone for location privacy in vehicular networks. *IEEE Communications Letters* 17 (8): 1524–1527.

100 Monteiro, M.E.P., Rebelatto, J.L., and Souza, R.D. (2016). Information-theoretic location verification system with directional antennas for vehicular networks. *IEEE Transactions on Intelligent Transportation Systems* 17 (1): 93–103.

101 Malandrino, F., Borgiattino, C., Casetti, C. et al. (2014). Verification and inference of positions in vehicular networks through anonymous beaconing. *IEEE Transactions on Mobile Computing* 13 (10): 2415–2428.

102 Qureshi, M.A., Noor, R.M., Shamshirband, S. et al. (2015). A survey on obstacle modeling patterns in radio propagation models for vehicular ad hoc networks. *Arabian Journal for Science and Engineering* 40 (5): 1385–1407.

103 Dubey, B.B., Chauhan, N., Chand, N., and Awasthi, L.K. (2014). Analyzing and reducing impact of dynamic obstacles in vehicular ad-hoc networks. *Wireless Networks* 21 (5): 1631–1645.

104 Sommer, C., Joerer, S., Segata, M. et al. (2015). How shadowing hurts vehicular communications and how dynamic beaconing can help. *IEEE Transactions on Mobile Computing* 14 (7): 1411–1421.

13

Radio Resource Management in Internet-of-Vehicles

13.1 Introduction

We have seen in Chapter 12 that US DoT is consistently involved in the management of administrative and technical perspectives of the transportation since very long time and has defined various standards and protocols for the smooth implementation of current as well as next-generation intelligent transport systems (ITSs). The internet-of-vehicles (IoVs) appear to be a very prominent aspect of the future generation ITSs. The potential wireless access technologies for IoVs' applications are generally classified into the vehicular communication, cellular communication, and short-range static communication as shown in Figure 13.1. The vehicular communication comprises the dedicated short-range communication (DSRC) [1], wireless access for vehicular communication (WAVE) [2], and communication architecture for land mobile (CALM) [3] technologies. The cellular communication has potential technologies such as 4G/LTE/LTE-Advanced [4], 5G/New radio [5], Worldwide Interoperability for Microwave Access (WiMAX) [6], and Satellite communication. The short-range static communication also has a significant role for intra-vehicle communication applications and has following potential techniques such as Wireless Fidelity (Wi-Fi), Bluetooth, and ZigBee [7].

As per the ITS programs, the spectrum allocated for the connected vehicles are in the range of 5.9 GHz which was allocated in 1999 [8] and amended in 2004 and 2006 [9]. Recent amendments by the Federal Communication Commission (FCC) in the DSRC standards are presented in 2016 [10]. The spectrum band 5.9 GHz is defined for DSRC as well as non-DSRC technologies where the non-DSRC technologies are radio-frequency identification (RFID), WiMAX, Wi-Fi, Bluetooth, and cellular communication. The key aim of DSRC 5.9 GHz band which is a derivative technology of Wi-Fi is to fulfill the prime need for security, privacy, and

Spectrum Sharing in Cognitive Radio Networks: Towards Highly Connected Environments, First Edition. Prabhat Thakur and Ganshyam Singh.
© 2021 John Wiley & Sons, Inc. Published 2021 by John Wiley & Sons, Inc.

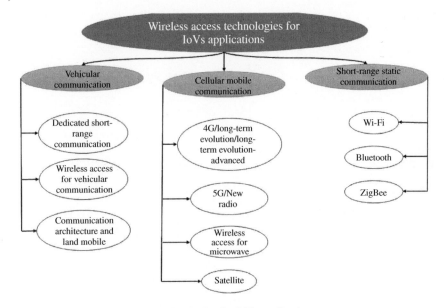

Figure 13.1 Wireless access technologies for IoVs applications.

low latency in the vehicular networks. Over the period of time, the cellular communication has performed well in terms of improved latency and high-speed communication. However, the DSRC has proved as a remarkable technology for vehicular networks in terms of security and safety critical nature. The first allocated DSRC channel at the band of 5.9 GHz is 75 MHz in the range of 5.850–5.925 GHz, especially for the connected vehicle technologies such as vehicle-to-vehicle (V2V), vehicle-to-infrastructure (V2I), vehicle-to-devices (V2D), and/or vehicle-to-anything (V2X).

DSRC is a very important and popular standard suitable for vehicular networks which is defined at the band of 5.9 GHz with 75 MHz band as per the FCC Order no. 99-305 with following key benefits in the mind: to improve traveller safety, manage the traffic congestion, aid in decreasing the air pollution, and help to conserve vital fossil fuels. DSRC is a two-way wireless communications protocol suite that integrates the IEEE 802.11 [10], IEEE 1609.x standards [11], SAE J2735 [12], and SAE J2945 [13].

The IEEE 802.11 is defined for the lower layers that are physical (PHY) and medium access control (MAC) of open system interconnection (OSI) model. The IEEE standard for WAVE is responsible for the architecture, security, and management standards for V2V and V2I frameworks. Further, IEEE 1609 family is responsible for the network layer functionalities and cryptographic processes for establishing the trust and protecting confidentiality. The payload definitions and performance constraints for the common data are established in the SAE standards. As per the definition of ITS program: The Intelligent Transportation Systems (ITS) program definition of connected vehicles includes both 5.9 Gigahertz (GHz) DSRC and non-DSRC technologies as means of facilitating communication for vehicle-to-vehicle (V2V) and vehicle-to-infrastructure (V2I) applications. Non-DSRC technologies (e.g. Radio Frequency Identification (RFID), Worldwide Interoperability for Microwave Access (WiMAX), Wi-Fi, Bluetooth, and cellular communication) enable use of existing commercial infrastructure for additional capacity support, but may not meet the low-latency needs of transmitting safety-critical information.

13.1.1 Dedicated Short-Range Communication

As per the ITS, the DSRC supports a centralized architecture which comprises an onboard vehicular radio system, a roadway backbone communications system also known as roadside units/infrastructure unit (IU), connectivity-internet-database networks [8]. The vehicular nodes can communicate with each other by broadcasting their information in the neighboring environment or the IU receives the information from all the vehicles about their location, moving speed, moving direction, and so on, and by processing that information the IU broadcasts in the environment so that all the vehicles know about their surrounding's neighboring vehicles and other information such as traffic flow (speed and volume), lane occupancy, priority signal pre-emption, toll collection, freight tracking, and roadway condition. The centralized and infrastructure-based architecture of the DSRC makes it unsuitable for the less dense roads such as remote and rural areas or city roads in the idle hours. The key feature of the roads in the rural and urban environments can be defined as follows. In the rural environments, there are lots of space when compared to the number of vehicles on the roads; in the urban areas that leads to sparse utilizations of the roads as well as roads are less travelled. Conversely, in the urban environments, the space is very limited when compared with the number of vehicles which is an advantageous feature with reference to the plenty utilization of the roads, however, resulting in the congestion of roads. Therefore, the installation cost of the DSRC is high and infrastructure-based architecture makes it costlier for the less dense roads. Since the per vehicle cost is the ratio of the total cost of the infrastructure and number of vehicles, it is defined as:

$$Per\,vehicle\,cost = \frac{Total\,cost\,of\,infrastructure}{Number\,of\,vehicles}.$$

This prerequisite of the infrastructure for the DSRC implementation motivates the researchers to look for an alternate of the DSRC technology which was assumed to be resolved by using the third generation partnership project (3GPP).

13.1.2 Wireless Access for Vehicular Environments

The WAVE is an integral standard of IEEE for the vehicular applications which defined the architecture and standardized the set of services and interfaces in order to achieve the secure V2V and V2X communication [2]. Specifically saying, the IEEE 1609 Family of Standards WAVE defines the architecture, communications model, management structure, security mechanisms, and physical access for high speed (up to 27 Mb/s), short range (up to 1000 m), low latency wireless communications in the vehicular environment. The primary architectural components defined by these standards are the on board unit (OBU), road side unit (RSU), and WAVE interface. The potential enabling applications in the ITS are vehicle safety, automated tolling, enhanced navigation, traffic management, and many more. The key standards for IEEE 1609 and its developments at the various stages are as follows [2]: (i) IEEE P1609.0 Draft Standard for Wireless Access in Vehicular Environments (WAVE)-Architecture, (ii) IEEE 1609.1-2006 – Trial Use Standard for Wireless Access in Vehicular Environments (WAVE)-Resource Manager, (iii) IEEE 1609.2-2006 – Trial Use Standard for Wireless Access in Vehicular Environments (WAVE) – Security Services for Applications and Management Messages, (iv) IEEE 1609.3-2007-Trial Use Standard for Wireless Access in Vehicular Environments (WAVE) – Networking Services, (v) IEEE 1609.4-2006-Trial Use Standard for Wireless Access in Vehicular Environments (WAVE) – Multi-Channel Operations, and (vi) IEEE P1609.11 Over-the-Air Data Exchange Protocol for Intelligent Transportation Systems (ITS).

13.1.3 Communication Access for Land Mobile (CALM)

The communication access for land mobiles is also a potential perspective by the ISO TC 204/ working group 16 in order of define the wireless communication protocols for the movement of the vehicles in ITS [14]. The CALM architecture relies on the IPv6 convergence layer and provides a standardized set of air interface protocols to achieve the effective resource utilization, safety critical communication, using one or more several media with multipoint transfer. The key communication links that are supported by the CALM are V2V, V2I, I2V, and I2I.

13.2   Cellular Communication

Before understanding the role of cellular communication for IoVs, it is worth see-ing how the 3GPP has opened the door of cellular communication in different fields such as IoVs. Therefore, in this section initially we have elaborated the 3GPP with its different releases and the potential applications of these releases. Further, we will see how 4G/LTE/LTE-Advanced and New Radio (NR) has poten-tial significance in the next-generation IoVs in order to fulfill the high throughout and reliability as well as low latency requirements. The evolvement of the 3GPP since its birth in 1998 is presented in the next section.

13.2.1   3GPP Releases

The 3GPP is a consortium of various international communication organizations such as European Telecommunication Standard Institute (ETSI) which was established in 1998 to develop the standards for the third generation (3G) mobile communication whose headquarter is located at the Sophia Antipolis Park in France. The key highlights of the various releases of 3GPP are presented in Figure 13.2 [15, 16]. The first release, that is Phase-1 of the 3GPP, was launched in 1998 for global system for mobile communication (GSM) and since then 15 releases have been launched including the Release 99 for the ultra mobile tel-ecommunication services (UMTS)/wideband code division multiple access (WCDMA), Release 8 for the first time introduction of LTE with consistent improvement toward the LTE-Advanced, Release 14 that promises the C-V2X communication, mission-critical enhancement, carrier aggregation, IoTs, voice and multimedia-related items, radio improvements and systems improvements, Release 15 for the 5G where the concept of NR is introduced, massive machine-type communication and IoTs, and mission-critical interworking with legacy net-works. The researchers/scientists/academicians are working continuously for the Release 16 and Release 17 where the key emphasis is on the "5G phase-2 imple-mentation, multimedia priority service, vehicle-to-everything (V2X) application layer services, 5G satellite access, LAN support in 5G, wireless and wireline con-vergence for 5G, terminal positioning and location, communications in vertical domains and network automation and novel radio techniques [17]. Further items being studied include security, codecs and streaming services, local area network (LAN) interworking, network slicing and the IoT".

It is clear from the afore-discussion that release 8 and release 14 open the door of the cellular communication for the vehicular networks with permissions to the LTE and its advances that are LTE-A. It is worth mentioning that initial genera-tions of the cellular communication which are 2G and 3G also played an important

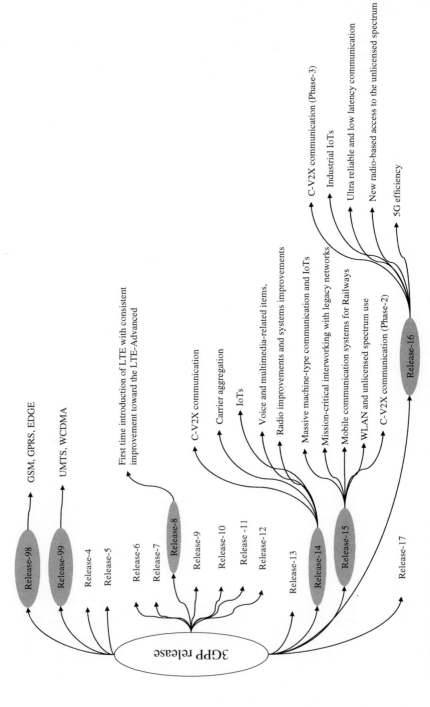

Figure 13.2 Highlights of the various releases of the 3GPP.

role to serve the communication requirements of vehicular networks for conventional services [18, 19]. However, with the introduction of LTE, direct implementation of cellular communication becomes possible as it proposes the sufficient speed and latency requirements to support the current demands of the vehicular networks. It is difficult to fulfill the future demands of the vehicular networks by using the LTE or DSRC only; therefore, the next generation that is 5G or NR are assumed to be potential technologies for the next 3GPP release. Briefs about the cellular communication technologies suitable for the current and future communication networks are presented as follows.

13.2.2 Long-Term Evolution

Conventionally, the long-term evolution (LTE) is a very prospective technology in the broadband wireless communication for mobile users after the GSM, code division multiple access (CDMA), General Packet Radio Service (GPRS), Enhanced Data rates for GSM Evolution (EDGE), High-Speed Packet Access, (HSPA) and UMTS. The key emphasis of LTE is to improve the capacity and speed of the network by exploiting various radio interfaces under the guidelines of the 3GPP [20–25]. The prime feature of the LTE is to support high-speed moving vehicles, with large capacity of 300 and 75 Mbps. The LTE-Advanced (LTE-A) is a progressed version of the LTE where we further achieve the improved performance metrics such as capacity, round trip latency, etc.

13.2.3 New Radio

Recently, the researchers have explored the 5G for vehicular communication in order to fulfill the demand of very high speed of internet and reduced latency [26–30]. Storck and Figueiredo [26] have proposed an ecosystem that relied on the software-defined networking (SDN) concept. The key idea is the assumption of vehicles as entertainment consumer point and to fulfill their demands, there should be guarantee of delivery and quality. For this, the authors have computed the internet-based video service traffic and V2V communication in urban and rural scenarios. The network simulator NS-3 is used while using the millimeter wave (mmWave) communication where simulation results are witness of the adequate performance to the IoV communications requirements when adopting the 5G network with V2X communications. Tayyaba [31] has explored the role of SDN and machine learning for the improvement in the vehicles' connectivity using 5G for communication. The vehicular networks achieve the excellent flexible platform because of SDN-based 5G connectivity. The authors have designed a resource allocation scheme relying on the potential deep learning algorithms, namely: (i) convolutional neural network (CNN), (ii) deep learning network (DNN), and long-short term memory (LSTM), on the basis of mininet-Wi-Fi-based setup with floodlight as SDN controller. The designed scheme framework optimizes the resource allocations according to the changing demands and network dynamics in vehicular networks. The simulation results validate the outperformance of the LSTM algorithm when compared with rest of the classification techniques. Ge et al. [32] have proposed a vehicular network architecture that comprises the 5G mobile communication, cloud computing, and SDN technologies. The interesting contribution is the exploitation of fog cells for flexible coverage of the vehicles and reduced number of handovers between vehicles and RSUs by using multi-hop communication. It is concluded through the simulation results that the 5G software defined networks outperform in terms of transmission delay and fog cell in 5G leading to improved throughput. Masini et al. [33] have investigated the next-generation radio access technique in the vehicular networks while exploring their potential limitations and potential challenges. Further, the authors have emphasized on the beaconing for cooperative awareness and compare the performance of LTE when adopted in different fashions, from the actual legacy solution in V2I mode to future potential alternatives, such as LTE-V2V, LTE-V2V with short subframe, and LTE-V2V with full duplex (FD) radios. The performance results are provided in terms of medium access efficiency and success probability of resource allocation, and show the potential advantages of future solutions, especially those provided by LTE-V2V with FD radios. As per scientists/researchers of the Nokia [34], the feasibility of automotive vehicles is the function of the ability of the researchers/engineers to exploit the automotive and telecommunication industry. The automotive industry comprises the car manufacturer as well as their suppliers, car spare-part manufacturer, map providers, and road operators. Conversely, the telecommunication industry consists of the telecom service providers and equipment manufacturers. The telecommunication industry is rapidly approaching toward its next generation, that is fifth generation (5G), and by keeping this in mind, Nokia has formulated a "5G automotive association (5GAA)" in 2016. 5GAA predicted almost 100 million vehicles on the roads in 2016 which is an increasing function of time. It is perceived that serving such large number of vehicles via centralized architecture while maintaining the capacity and latency constraints is a challenging task. The edge computing appears to be a promising fourth generation (4G) technology to handle the distributed architecture that leads to improved reliability and security besides reduced latency of the network. Ruan et al. [35] have investigated a trade-off problem between the energy- and spectral-efficiency in the cognitive satellite vehicular networks while exploiting the 5G.

13.2.4 Dynamic Spectrum Access

The dynamic spectrum access is a very prominent technique of spectrum allocation where the users/nodes/vehicles are permitted to access any available channel; however, the constraint is that it should be free from any kind of communication by the legacy/primary users. The cognitive radio is a potential technology that employs the concept of DSA using the cognitive engine having four technologies that are spectrum sensing, spectrum analysis and decision, spectrum sharing, and spectrum mobility which are discussed in detail in Chapter 1. The ultimate theme of the dynamic spectrum access is to manage and access the spectrum very efficiently so it is worth saying that efficient spectrum management is the prime objective behind the DSA mechanism. As per US DoT, the spectrum management is defined as an "art and science of managing the use of the radio spectrum in order to minimize interference as well as ensure that the radio spectrum is used to its most efficient extent so that it can beneficial for the public. For USDOT this is focused on supporting safe, efficient and economical transportation". The need for efficient spectrum management arises due to the fact that the electromagnetic spectrum is very limited and a fraction of the radio spectrum in the entire electromagnetic spectrum can support the communication in our surrounding environments such as buildings, trees, and city areas. This demand for commercial spectrum to support broadband wireless communications has led the government to consider repurposing various radio frequencies, including the satellite communications bands next to global positioning systems (GPS).

13.3 Role of Cognitive Radio for Spectrum Management

Spectrum is a valuable natural resource in this era of technology as well as communication and its significance reaches extreme when we think about the connected world comprising smart home, smart cities, ITSs, connected vehicles, etc. Thus, the spectrum has similar significance as the other resources in the environment such as water, air, etc., and needs the equal attention as they have. Therefore, there are dedicated organizations which are responsible for spectrum analysis, allocation, maintenance, and reuse of the spectrum on the world level as well as on the country level such as International Telecommunication Union (ITU), FCC, National Telecommunications and Information Administration (NTIA), Telecommunication Regulatory Authority of India (TRAI), and department of culture, media, and support [36–41]. The prime focus of the spectrum management is to manage the interference among the users which are spaced adjacently or nearby geographical areas. The importance of the spectrum management comes into picture especially in defense, security, and healthcare scenarios where zero interference tolerance policy is adopted. The role of the spectrum management agencies is defined very clearly and following are the prime responsibilities regarding the federal spectrum: (i) establishing and issuing the policies, (ii) development of plans and policies in the peace and war time, (iii) frequency/spectrum allocation, (iv) maintenance of spectrum-use database, (v) reviewing and certifying the federal agencies' new telecommunication systems, (vi) providing technical/engineering expertise for spectrum resource assessment, (vii) participating in all aspects of the federal government communication, and (viii) participating in federal government telecommunication and automated information as shown in Figure 13.3a. Further, one more classification for the role of spectrum management organizations is presented in Figure 13.3b, and according to this classification, the prime role of spectrum management organizations are: (i) to define the spectrum properly, (ii) prepare the spectrum policies and regulations, (iii) spectrum sharing and trading, (iv) decide about the spectrum pricing, and (v) establishing and monitoring the international co-ordinations.

The frequency/spectrum allocation is a very prominent aspect that directly affects the service providers as well as consumers in terms of the costs. The static spectrum allocation policy is very popular till the 15 release of 3GPP/4G. However, the key challenges with the static spectrum allocation policy are: (i) the static nature of spectrum allocation binds the use of spectrum by only the licensed/owner users; however, it is very impractical situation that the licensed user always needs the spectrum for communication. (ii) The spectrum allocation is performed by the standard organization either on the national or international level, and for a long period of time such as couple of years, this leads to centralized control for allocation and use of the spectrum, and (iii) the spectrum demand can vary during small span of time for particular place or state, this leads to wastage of that spectrum and provokes us for distributed nature of spectrum allocation policies.

However, more spectral efficient design and underutilization of the spectrum motivates the researchers and policy makers to exploit the dynamic spectrum access. The dynamic spectrum access will play a significant role as a major milestone to overcome the issue of its unutilization/under-utilization; however, cognitive radio is the prominent technology that transforms the idea of DSA into reality. The special cognitive engine provides the facility to the users to access the licensed spectrum for a fraction of time if it is free from any kind of communication without any major modifications in the spectrum allocation policies. The independent nature of each user to access the spectrum makes it more popular to the distributed architecture of the network.

13.4 Effect of Mobile Nature of Vehicles/Nodes on the Networking

On the consideration of mobile nature of nodes in the network, the ad-hoc networks are transformed into mobile-ad-hoc networks (MANETs), which address as well as manage the challenges introduced in the ad-hoc networks because of the mobile nature of nodes such as Doppler effect, weak signal reception, etc.

(a)

(b)

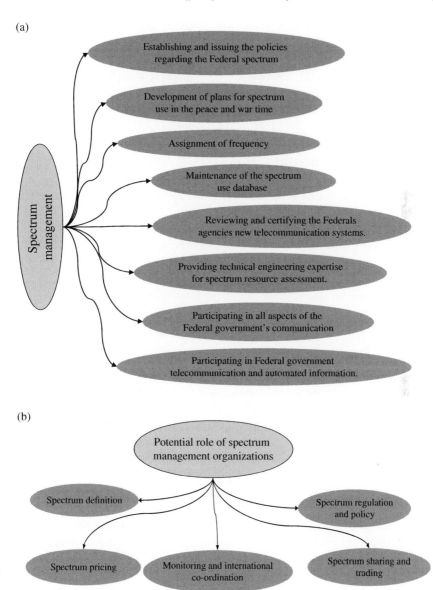

Figure 13.3 (a) The key responsibilities of the spectrum management agencies. (b) The role of spectrum management organizations.

The need of communication in the vehicles give birth to a sub-domain of MANET which is known as vehicular ad-hoc network (VANET) where additional challenges emerge because of high speed of vehicles/nodes. The major challenges appearing in the vehicular networks are: (i) frequent channel switching/mobility, (ii) frequent handovers as per the cognitive radio concept, and (iii) rapid topology change, Doppler effect, fast signal detection, processing, and reacting.

13.5 Spectrum Sharing in IoVs

We have perceived in Chapters 1 and 4 that the spectrum sharing plays an important role in the conventional networks to diminish the effect of spectrum underutilization. The IoVs has a significant role for the vehicular networks and spectrum sharing is highly desired to fulfill the demand of spectral efficiency [42, 43]. The high speed of vehicles and heterogeneous nature of nodes in vehicular networks such as vehicles, IUs, pedestrians, cloud communication, etc., introduced additional challenges for spectrum sharing in the IoVs. Therefore, it is worth and necessary to explore the various constituents of the cognitive cycles for the IoVs, which are discussed as follows.

13.5.1 Spectrum Sensing Scenarios

The fond of idle channels is an integral part of the cognitive cycle which becomes reality due to the spectrum sensing process in the conventional networks such as MANETs; however, in vehicular networks it becomes challenging to perform direct spectrum sensing techniques by the users/vehicles to find the idle channels because of high-speed movement. Therefore, we need to discuss some potential scenarios for effective detection and use of channel opportunities.

We have considered a six-lane road scenario where three roads are for going (forward road) while three are for coming (backward road) of vehicles as shown in Figure 13.4. The forward road has three lanes that are lane 1, lane 2, and lane 3, while backward road comprises the lane 4, lane 5, and lane 6. A base station (BS)/IU is assumed to be installed at the road divider so that it can cover both sides that are left and right lane.

The access regions of the BS are represented through a dotted line in addition to various links that are V2V, V2I, V2C, and V2H. The V2H is assumed to be established between the vehicle and the pedestrians walking on the footpath. The BS is assumed to be installed with continuous or sufficient power supply and computation ability so that it can support and control the multiple vehicles in its access region in addition to its communication with the internet/broadband or remote units.

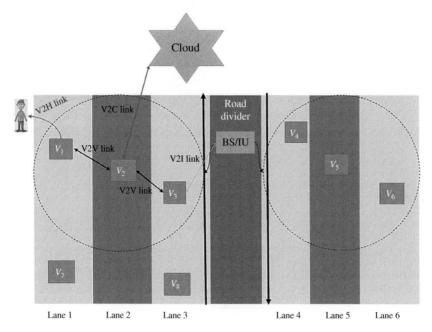

Figure 13.4 The six-lane highway scenario for internet-of-vehicles.

Scenario A: For simplicity and ease of understanding, we have considered only the forward road where vehicles V_1, V_2, and V_3 are in the lane 1, lane 2, and lane 3, respectively, in the access region of the BS on the forward road at Time T_1. Every vehicle is assumed to be equipped with the spectrum sensing and data transmission unit. The spectrum sensing unit of the vehicles V_1, V_2, and V_3 senses the channels and shares the information with the BS; however, used the idle channels allocated by the BS for their communication. At time T_2, the vehicles V_7 and V_8 enter into the access region of BS, and BS allocates the idle channels to V_7 and V_8, on the basis of spectrum sensing information provided to the BS by previous users that are V_1, V_2, and V_3. This is a most effective way to address the issue of high mobility of vehicles and demand of very fast spectrum sensing techniques in addition to the high accuracy when we implement the concept the cognitive radio on the roads.

Scenario B: The BS has a set of channels with their status that it is idle or active on the basis of historical information. As the vehicle enters into the access region of BS, the BS assigns the idle channel to the vehicles for its communication. The key issue with the database of spectrum sensing technique is the non-real time status of the channel. The database of spectrum sensing is a suitable option for vehicular networks with reference to the speed and complexity; however, it is challenging with reference to the reliability because the non-real time spectrum sensing may lead to the active status of the assigned channel that results in data collision.

Scenario C: In the scenario C, the IU/BS is assumed to be equipped with spectrum prediction, spectrum sensing, and data communication unit. IU predicts the future state of the channel and prepares a set of the idle predicted channel. When a vehicle enters into the access region of the IU, it shares the set of idle predicted channels with the vehicles (users). The vehicles perform the spectrum sensing on the highly idle predicted channels. Scenario C is more appropriate for the real-time update of the channel than the scenario A and scenario B.

13.5.2 Spectrum Sharing Scenarios

The spectrum sharing is the key constituent of the cognitive cycle and also plays an important role in the vehicular cognitive radio networks. The spectrum is shared with the PU and CU via different spectrum accessing strategies that are interweave spectrum access, underlay spectrum access, hybrid spectrum access, and overlay spectrum access; however, every technique has its pros and cons due to which the selection of the appropriate technique is a key step.

A) **Interweave spectrum access**: As discussed in Chapter 1 that the CU accesses the channels in the absence of PU and spectrum sensing is a prerequisite for this accessing technique. Therefore, it is worth mentioning that the performance of the interweave spectrum access is also a function of the spectrum sensing technique opted, in addition to the issue of the non-seamless communication. Thus, it is clear that in the vehicular networks, the interweave spectrum access is suitable only if the system demands less complexity, medium data rate (capacity), and service interruption is allowed.

B) **Underlay spectrum access:** The underlay spectrum access technique supports the seamless communication; however, the transmitted power is constrained by the PU interference tolerable limit that results in limited channel capacity. The key benefit of this technique is the no-need of spectrum sensing technique that is advantageous in terms of the time consumption and complexity. Therefore, this technique is suitable for the scenarios where low data rate with no service interruption is required while demanding the less complex systems.

C) **Hybrid spectrum access:** The hybrid spectrum access is the combination of the interweave and underlay technique and supports the communication with high and constrained power on the idle and active sensed channels, respectively. It supports the seamless communication for the users; however, the need to switch its communication from low to high and vice-versa is a major challenge with respect to the time. The switching time is a potential factor that affects the communication, and in vehicular networks a small delay can cause accident. Therefore, the hybrid spectrum access is only suitable if switching delay is tolerable for the opted network selection.

D) **Overlay spectrum access:** The overlay spectrum access is an advanced spectrum accessing technique that supports the seamless communication and transmits the data with full/high power on the idle as well as active channels, while managing the interference by using the advance interference cancelation techniques. This mode of spectrum sharing technique seems to very effective for the vehicular networks because of seamless communication and high capacity of the channel. The space complexity and energy requirements of the advance interference cancellation can be managed by vehicular networks as vehicles and IU are assumed to be equipped with sufficient power supply and cost as well as space is not an issue. However, the time consumed by the advanced interference cancellation technique is a prominent factor and needs to be tolerable limit of the delay of the network.

13.5.3 Spectrum Mobility/Handoff Scenarios

The switching of communication from one channel to another channel is known as spectrum handoff in the conventional cellular network; however, in the CRNs, it is named as spectrum mobility. In the cognitive-based vehicular networks, both the spectrum handoff and spectrum mobility are defined as different processes. It is worth mentioning that the high speed of vehicles on the roads risks both the processes of communication switching in the cognitive vehicular networks and therefore it is worth discussing both the scenarios.

A) **Spectrum Handoff:** When the vehicle enters into the access region of the new BS/IU, it needs to stop the communication on the channel of previous BS access region and have to establish its communication on the assigned channel by the current BS. As the vehicle changes its communication from one BS to other BS, this process is known as spectrum handoff. The scenario is shown in Figure 13.5, when the vehicle V_7 enters into the access region of the BS-2, from the access region of the BS-1 and this process is known as spectrum handoff.

B) **Spectrum Mobility:** Unlike the spectrum handoff, it is possible that the PU comes back on the channels used by the vehicles and vehicles have to switch their communication in the access region of the same BS, and this process is known as spectrum mobility. As shown in Figure 13.5, it is assumed that the BS is equipped with five channels which are CH1, CH2, CH3, CH4, and CH5. Initially, the V_2 establishes communication on the CH3; however, with the emergence of the PU on the CH3, the V_2 switches its communication on the

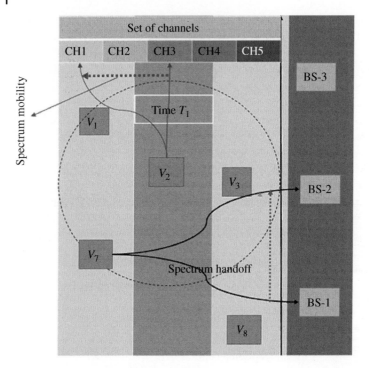

Figure 13.5 Spectrum mobility and spectrum handoff scenarios in the IoVs.

CH1 while remaining in the access region of the BS-2 and this process of switching is known as spectrum mobility.

13.6 Frameworks of Vehicular Networks with Cognitive Radio

The PHY and MAC layer challenges are well explored in the literature which reveals that the small modification in CR and IoVs networks' PHY and MAC protocols is applicable to the CR-IoVs network. However, it is worth mentioning that high speed of vehicles affects the routing path in the network which motivates the need to explore the routing protocols.

Therefore, various researchers have presented the potential way for the routing protocols in the CR-VANET [37–40]. As per Singh et al. [44], the high mobility (for instance, 100–200 km/h) of vehicles affects the topology very dynamically which results in the very short lifespan of the vehicular communication link. Moreover, the dense and sparse nature of the vehicle density on the roads introduces the fragmentation problem that leads to the network unreachability for some vehicles. In addition to this, the high speed of vehicles directly affects the signal reception quality because of the Doppler effect and fast fading scenarios.

To address the aforementioned challenges, some popular routing algorithms are opportunistic forwarding [45], trajectory-based forwarding [46], and geographic forwarding [47]. An opportunistic forwarding algorithm is a potential forwarding algorithm which is suitable for the scenarios with frequent disconnections and can be combined with other approaches that use trajectory-based or geographic forwarding. The geographic forwarding algorithms forward packets toward the destination by keeping in mind the geographical location of the receiver. The positive and negative aspects of the geographical location method are the scalable nature and non-efficient nature for handling dead ends and voids. The trajectory routing is another routing algorithm because it considers the road infrastructure as an overlay directed graph, with intersections as graph nodes and roads as graph edges allowing messages to move in predefined trajectories. The authors in [48] have proposed a SDN-based routing protocol for the CR-based IoVs in order to improve the network intelligence so that a stable route between the source and destination vehicle can be found. The proposed algorithm improves the network performance with reference to the end-to-end delay, delivery-ratio, and low overhead.

The prominent and probable network architectures for the CR-based IoVs are illustrated as follows.

13.6.1   CR-Based IoVs Networks Architecture

Similar to the conventional CR networks, the CR-VANET architecture can be decomposed into two types: centralized and distributed architecture as shown in Figure 13.6. In the centralized architecture, all the vehicles on roads in the access region of a particular BS/RSU/IU needs to be controlled by the BS. Here, the controlling means the BS is responsible for all the processes of cognitive cycle such as assignment of the idle channels, sharing common information with all the vehicles, providing channels for the spectrum mobility, instructing, guiding, and managing regarding the vehicles' speed and directions on the roads. However, in the distributed architecture, all the vehicles are allowed to perform the function of cognitive cycle and road management individually. The distributed architecture is further divided into two types namely, the cooperative and non-cooperative. In the cooperative architecture, the vehicles perform most of the functions of cognitive cycle such as the spectrum sensing and then exchange the information with other vehicles; however, the decision about the selected idle channel, accessing process, frame time, etc., completely relies on the particular vehicle. The cooperation among vehicles increases the data availability on each vehicle that increases the

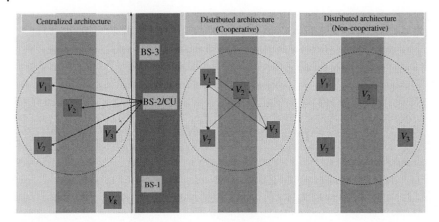

Figure 13.6 The frameworks of CR-based internet-of-vehicles.

probability of correct decision as well as avoids the issues of hidden terminal problems in the spectrum sensing and spectrum sharing processes. The key issue with this process is the time consuming nature as exchanging/waiting the information with/from all other vehicles can be more than the vehicle network time in the access region of BS.

Conversely, in the non-cooperative network architecture, each vehicle performs all the functions of cognitive cycle and road management without sharing any information with other vehicles. This is a fast-working network architecture; however, the reliability is a challenging issue due to hidden terminal problems.

13.7 Key Potentials and Research Challenges

The potentials and research challenges of the CR-based IoVs are depicted in Figure 13.7 and illustrated as follows.

13.7.1 Key Potentials

A) **Predictable mobility:** One good and positive point with the vehicular networks is the fixed road network that leads to predicted mobility of the nodes. This provides an edge for the network and topology designs while facing the challenges of road signs and traffic rules.

B) **Sufficient energy and storage:** Any vehicle on roads have to carry a load of minimum 100 kg, e.g. Scooter or Motorbike which is assumed to be equipped with a battery and a continuous power supply can be considered with reference to communication and storage devices. This continuous and sufficient power

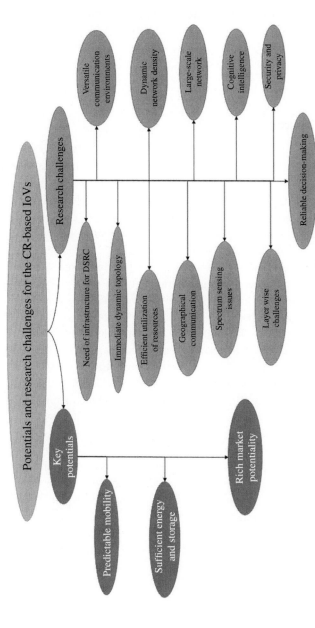

Figure 13.7 Key potentials and research challenges for the CR-based IoVs.

supply introduces a common characteristic of nodes that is ample energy and computing power (including both storage and processing), since the nodes are vehicles instead of small handheld devices. This motivates back the need of energy-efficient nature of the system; however, the reliability and speed are the prime constraints for CR-VANET.

C) **Rich market potentiality:** The market is full of non-autonomous vehicles and very rapidly moving toward the automatic nature of vehicles; the mixed population of the vehicles on the roads and a race toward automatic nature of the vehicles have opened a lot of opportunities in the market.

13.7.2 Research Challenges

A) **Need of infrastructure for DSRC:** DSRC is a potential technology for the vehicular communication that uses a public key infrastructure that distributes and manages digital certificates for vehicles. This put a bind on the vehicles to have access of RSU/IU. The deployment of such IUs on the highway seems possible up to an extent; however, the deployment, management, operating, as well as maintenance of such units in the remote and rural areas is a very challenging task due to access and cost.

B) **Immediate dynamic topology:** The topology is a potential phenomenon to decide the path of communication from the transmitting node to the receiving node. In the vehicular communication, the nodes/vehicles move with a very high speed that changes the position of vehicles very frequently which leads to very high dynamic topology. The addition of CR with vehicular networks puts extra pressure because unlike the conventional vehicular networks, less time is also required for spectrum sensing and channel selection. Therefore, the network and topology design must be on the basis of these additional challenges.

C) **Dynamic network density:** The network density in the vehicular network is directly related to the vehicles' density on the roads and it is a well-known fact that number of vehicles on the roads are very high in busy/Jam hours such as around office starting and closing timings. However, other than busy hours, the number of vehicles on the roads will be less and very less in the night hours. Therefore, the optimal performance designs need to be performed by keeping in mind the aforementioned different conditions. The dynamic nature of the network density constraints the network to first know about the status of the network density, and then proceeds for the further process. This may take some time of the network topology; therefore, the fast detection of network density and optimal design in the time less than a fraction of the network topology is a challenging task.

D) **Large-scale network:** The network size in the vehicular networks can be very large on the big highways, city center, and crowded cities. Implementing, managing, and maintenance of such large networks is a critical as well challenging task because it requires infrastructure cost, labor cost, as well as administration cost.

E) **Geographical communication:** The communication in the vehicular networks is little different when compared with the conventional networks where communication endpoints are defined by using the ID or group ID; however, in the vehicular networks, the communication relies on the geographical locations where the packets need to be forwarded (e.g. in safe-driving applications).

F) **Various communication environments:** The road scenarios for the vehicular networks also play an important role as various scenarios can be thought about the vehicular networks. The variety of scenarios such as highway in busy and idle hours, city centers, streets, buildings, and tree scenarios. Handling such large number of scenarios is a challenging task because achieving direct line of sight (LOS) will be very infrequent.

G) **Security and privacy:** The security aspects that have to be considered for vehicular services, especially for the IoVs, are highly confidential. Therefore, the data generated by the sensing devices at the vehicles need to be encrypted to protect against intrusion attacks. In case of IoVs, unlike IoTs, the size of sensing units is not a constraint that leads to relaxing point in terms of power sources and energy-efficient nature of the protocols. However, the introduction of CR in the IoVs makes it more challenging for security concerns due to spectrum sensing and accessing techniques; therefore, various security concerns arise for the cognitive-inspired IoVs such as at the spectrum sensing level, spectrum sharing, and spectrum mobility. Spectrum sensing is vulnerable to primary-user emulation attacks, where the intruder mimics the primary-user activities and restricts the CU/medical devices to find the idle channel even when it is idle [49]. Moreover, in the absence of a PU, if the CU is issuing the available spectrum, mimicking of CU activities starts the cognitive user emulation attack. The proposed solution for these attacks in conventional CR networks is to identify the intruders by differentiating legitimate users in the following ways, namely (i) key allocation based, (ii) trust based, etc. [50]. However, very high moving speed of the vehicles and their heterogeneous nature makes it a very challenging and prominent research area that is yet to be explored. Accessing techniques also introduce security challenges for the IoMTs; these are called common control channel attack and beacon falsification attack. Moreover, the cross-layer design and re-configurability with a SDN introduce further challenges [51–53]. The controlled power transmission using the spread spectrum technique via the ultrawide-band (UWB) is prominent and less vulnerable to security threats, that is why it is suitable for vehicular networks with less data rate demand. In addition to this, the key challenge is an open research issue that is yet to be explored, e.g. (i) design of the cognitive-inspired IoVs' physical parameter specifications, (ii) design of MAC protocols, and (iii) design of minimum and maximum values of transmission parameters due to very dense node implementation on the roads, especially in the office hours.

H) **Spectrum sensing issues:** The high speed of vehicles in IoVs affects the performance of spectrum sensing techniques due to signal fading and fast-changing locations. In the case of fast varying channels over time, some works focus on improving the performance of sensing [54] and signal identification [55–57]. Similarly, improved algorithms have been proposed for channel estimation and equalization [58] in high-speed environments. In [59], the authors have explored the sensing performance of the multi-antenna energy detector by using temporal signal correlation in the cognitive radio networks. Huang et al. [60] have investigated the radar sensing-throughput trade-off for radar-assisted CR-IoVs.

I) **Cognitive intelligence:** Cognitive-based IoVs enables the IoV with the more perceptive/ sensing ability, through cognition in intra-vehicle networks, inter-vehicle networks, and beyond-vehicle networks. It also enables the IoVs to achieve the little information about the inter- as well as intra-vehicle scheduling strategies to the whole transportation system.

J) **Reliable decision-making:** Decision-making is an important aspect in the vehicular communication, and CR-based IoVs are most vulnerable to the decision-making as both CR and IoVs are critical for the decision-making process. The selection of the channel for spectrum sensing, spectrum analysis and decision, and spectrum mobility are the potential aspects of the CR; however, the high-speed, heterogeneous nature of vehicles/nodes also affects the decision-making of the IoVs.

K) **Efficient utilization of resources:** With perception of network traffic status and real-time road circumstance, the decisions derived by analytic technologies such as machine learning and deep learning can help resource cognitive engine to conduct more effective control over vehicles, and to enhance information sharing efficiency within vehicular networks.

L) **Layer-wise challenges:** The physical layer challenges are illustrated as follows, especially during cooperation among vehicles. The use of relay node communication appears to be very important in the vehicular networks and selection of the relay nodes/vehicles is a very prime aspect on the physical layer. However, the use of relay nodes in the physical layer imposes the need of exchange of extra control messages at the MAC layer. On the other hand, the routing or network layer is also responsible for best path in terms of the throughput, latency, and reliability. The MAC layer is also having significant functionalities regarding the accessing decision of the channels. The MAC layer protocols in cooperation are generally classified into three types that are contention free, contention based, and hybrid. In the contention free, entire resource is divided into the multiple users by the controlling entity. For example, in the TDMA, a scheduler regulates the nodes/vehicles and defined which channel will be used by which vehicle and at what time. In the contention-based cooperative MAC protocols, one node/vehicle contends with the other neighboring vehicles that also want to access the same channel, e.g. CSMA, CSMA/CA. The Hybrid cooperative MAC protocols support the advantages of both the contention free and contention based MAC protocols. Similar to TDMA, the hybrid MAC protocol experiences less collision among two hop neighbors and attains high channel utilization under extreme contention conditions. Similar to the advantages of the CSMA/CA, the hybrid MAC protocol incurs low latency and elevates the channel utilization under low contention conditions. Therefore, in the CR-VANET, the role of hybrid MAC is very important as it provides the benefits of both the techniques. The network/routing layer is responsible for the selection of the network path and has following potential challenges. Routing is a potential phenomenon in the conventional mobile ad-hoc networks in order to select the best path from the transmitter to the receiver, in terms of the improved throughput latency and reliability. In the CR-VANET, it shows very high significance since vehicles move with very high speed and V2V framework motivates us to go for cooperative communication that leads to the demand of highly efficient routing protocols. The predicted paths provide an edge for the routing protocols by predicting the future location of the vehicle; however, very high speed of the vehicles introduces the challenge of very optimal routing path because change in location means change in routing path. This new path may be worse path compared to the other routing path. In the routing path, the relay vehicles play a very critical part and relay vehicle is allowed for any of the following modes namely, Amplify and forward, Compress and forward, Store and forward, and Decode and forward. The ultimate purpose of the various protocols is to maximize throughput, optimize the power allocation, minimize the outage transmission, improvement in reliability, maximize utilization, and minimize the reservation slot collision.

There are other potential research challenges that appear for the CR-assisted IoVs when we have to explore other technologies with CR-IoVs in order to improve its performance such as visible light communication [61], non-orthogonal multiple access [62], use of machine learning techniques [63], and deep reinforcement learning techniques [64, 65].

13.8 Summary

This chapter has explored the potential aspects of resource sharing in IoVs. We have started with the prominent technologies for the IoVs communication such as DSRC, WAVE, LTE, LTE-Advanced, 5G NR, etc., where it is perceived that in the near future, the allocated spectrums for IoVs services will be scarce; therefore, an efficient spectrum utilization such as dynamic spectrum access will play an important role. Therefore, various frameworks of dynamic spectrum access using cognitive cycle are well explored for the IoVs networks in order to achieve the spectral-efficient behavior. Further, the prominent architectures of the CR-based IoVs networks are presented. Finally, the potential aspects and research challenges are illustrated in detail to address the future research directions.

References

1 Kenney, J.B. (2011). Dedicated short-range communication (DSRC) standards in the United States. *Proceedings of the IEEE* **99** (7): 1162–1182.

2 IEEE (2020). IEEE 1609- Family of Standards for Wireless Access in Vehicular Environments (WAVE). https://www.standards.its.dot.gov/Factsheets/Factsheet/80 (accessed 2 July 2020).

3 ISO 21217 (2014). Intelligent Transport Systems – Communications Access for Land Mobiles (CALM) – Architecture. https://www.iso.org/standard/61570.html (accessed 2 July 2020).

4 Wannstrom J. LTE-Advanced. https://www.3gpp.org/technologies/keywords-acronyms/97-lte-advanced (accessed 2 July 2020).

5 5G: New Radio. https://www.qualcomm.com/invention/5g/5g-nr (accessed 2 July 2020).

6 Fidaus, I.D., Setiawan, H., and Pradana, F. (2015). Design and simulation WiMax networks 802.16d and 802.16e in Sleman. *Proceedings of International Seminar on Intelligent Technology and Its Applications* (ISITIA), Surabaya, Indonesia (20–21 May 2015), 419-422.

7 Lee J-S, Su Y-W, and Shen C-C (2008). A Comparative Study of Wireless Protocols: Bluetooth, UWB, ZigBee, and Wi-Fi. *IECON 2007 – 33rd Annual Conference of the IEEE Industrial Electronics Society*, Taipei, Taiwan (5–8 November 2007), 47–51.

8 FCC. (1999). Intelligent Transportation Services. Allocated 75 MHz at 5.850-5.925 GHz to the mobile service for use by Dedicated Short Range Communications systems operating in the Intelligent Transportation System radio service, DA/FCC #: FCC-99-305, Docket/RM: 98-95, RM-9096, FCC Record Citation: 14 FCC

Rcd 18221 (32). https://www.fcc.gov/document/intelligent-transportation-services (accessed 2 July 2020).

9 Dedicated short-range communications (DSRC) and spectrum policy. U.S. Department of Transportation. https://www.its.dot.gov/presentations/world_ congress2016/Leonard_DSRC_Spectrum2016.pdf (accessed 2 July 2020).

10 IEEE 802.11TM. Wireless Local Area Networks: The Working Group for WLAN Standards. http://www.ieee802.org/11/ (accessed 2 July 2020).

11 IEEE (1609). Family of Standards for Wireless Access in Vehicular Environments (WAVE). https://www.standards.its.dot.gov/Factsheets/Factsheet/80 (accessed 2 July 2020).

12 Dedicated Short Range Communications (DSRC) Message Set Dictionary J2735_200911. https://www.sae.org/standards/content/j2735_200911/ (accessed 2 July 2020).

13 Dedicated Short Range Communication (DSRC) Systems Engineering Process Guidance for SAE J2945/X Documents and Common Design Concepts™ J2945_201712. https://www.sae.org/standards/content/j2945_201712/ (accessed 2 July 2020).

14 Intelligent transport systems – Communications access for land mobiles (CALM) – Architecture. https://www.iso.org/obp/ui/#iso:std:iso:21217:ed-2:v1:en (accessed 2 July 2020).

15 3GPP Specification Release Numbers. https://www.electronics-notes.com/articles/ connectivity/3gpp/standards-releases.php (accessed 18 May 2020).

16 The Mobile Broadband Standard. The 3GPP release: a global initiative. https:// www.3gpp.org/about-3gpp/about-3gpp (accessed 18 May 2020).

17 Naik, G., Choudhury, B., and Park, J.-M. (2019). IEEE 802.11bd & 5G NR V2X: evolution of radio access technologies for V2X communications. *IEEE Access* **7**: 70169–70184.

18 Zeng, L., and Zhu, Y. (2012). 3G-assisted routing in vehicular networks. *Proceedings of IEEE Sensors*, Taipei, Taiwan (28–31 October 2012), 1–4.

19 Zhao, Q., Zhu, Y., Chen, C. et al. (2013). When 3G meets VANET: 3G-assisted data delivery in VANETs. *IEEE Sensors Journal* **13** (10): 3575–3584.

20 Araniti, G., Campolo, C., Condoluci, M. et al. (2013). LTE for vehicular networking: a survey. *IEEE Communications Magazine* **51** (5): 148–157.

21 Andrade, H.G.V.D., FerreiraC.C.L.L. ; Filho, A.G.D.S. (2016). Latency analysis in real lte networks for vehicular applications. *Proceedings of 6th Brazilian Symposium on Computing Systems Engineering (SBESC)*, Joao Pessoa, Brazil (1–4 November 2016), 1–6.

22 Xu, C., Huang, X., Ma, M., and Bao, H. (2018). An anonymous handover authentication scheme based on LTE-A for vehicular networks. *Wireless Communications and Mobile Computing* **2018**: 1–15.

23 Mir, Z.H. and Filali, F. (2014). LTE and IEEE 802.11p for vehicular networking: a performance evaluation. *EURASIP Journal on Wireless Communications and Networking* **114**: 1–15.

24 Gupta, S.K., Khan, J.Y., and Ngo, D.T. (2020). An LTE-Direct-Based Communication System for Safety Services in Vehicular Networks. In *Moving Broadband Mobile Communications Forward-Intelligent Technologies for 5G and Beyond*. IntechOpen, 2020. DOI: https://doi.org/10.5772/intechopen.91948.

25 Kim, W. and Lee, E.-K. (2017). LTE network enhancement for vehicular safety communication. *Mobile Information Systems* **2017**: 1–18.

26 Storck, C.R. and Figueiredo, F.D. (2019). A 5G V2X ecosystem providing internet of vehicles. *Sensors* **19** (3): 1–20.

27 Here's what you need to know about 5G and C-V2X. https://www.ericsson.com/en/news/2019/9/5g-and-v2x (accessed 18 May 2020).

28 How NR based sidelink expands 5G C-V2X to support new advanced use cases. https://www.qualcomm.com/media/documents/files/nr-c-v2x-webinar-march-2020-presentation.pdf (accessed 18 May 2020).

29 Flore, D. 5G V2X The automotive use-case for 5G. https://www.3gpp.org/ftp/information/presentations/Presentations_2017/A4Conf010_Dino%20Flore_5GAA_v1.pdf (accessed 18 May 2020).

30 Cellular V2X communications towards 5G. https://www.5gamericas.org/wp-content/uploads/2019/07/2018_5G_Americas_White_Paper_Cellular_V2X_Communications_Towards_5G__Final_for_Distribution.pdf (accessed 18 May 2020).

31 Tayyaba, S.K., Khattak, H.A., Almogren, A. et al. (2020). 5G vehicular network resource management for improving radio access through machine learning. *IEEE Access* **8**: 6792–6800.

32 Ge, X., Li, Z., and Li, S. (2017). 5G software defined vehicular networks. *IEEE Communications Magazine* **55** (7): 87–93.

33 Masini, B., Bazzi, A., and Natalizio, E. (2017). Radio access for future 5G vehicular networks. *Proceedings of IEEE 86th Vehicular Technology Conference (VTC-Fall)*, Toronto, Canada (24–27 September 2017), 1–7.

34 Putzschler, U. Building the 5G highway for connected cars. https://www.nokia.com/blog/building-5g-highway-connected-cars/.

35 Ruan, Y., Zhang, R., Li, Y. et al. (2019). Spectral-energy efficiency trade-off in cognitive satellite-vehicular networks towards beyond 5G. *IEEE Transactions on Vehicular Technology* **68** (6): 5809–5819.

36 Stine, J.A. and Portigal, D.L. (2004). Spectrum 101 An Introduction to Spectrum Management. MITRE Technical Report, Washington C3 Center McLean, Virginia. https://www.mitre.org/sites/default/files/pdf/04_0423.pdf (accessed 19 May 2020).

37 ICT Regulation Toolkit/5. Radio Spectrum Management/Spectrum Management Overview. http://www.ictregulationtoolkit.org/toolkit/5.1 (accessed 19 May 2020).

38 Spectrum Management. https://dot.gov.in/spectrum# (accessed 19 May 2020).

39 What is spectrum management?. https://www.transportation.gov/pnt/what-spectrum-management (accessed 19 May 2020).

40 Spectrum management, committed to connecting the world. https://www.itu.int/pub/R-REP-SM (accessed 19 May 2020).

41 Independent communications authority of South Africa. https://www.icasa.org.za/legislation-and-regulations/engineering-and-technology/radio-frequency-spectrum-management (accessed 20 December 2020).

42 Liang, L., Ye, H., and Li, G.Y. (2019). Spectrum sharing in vehicular networks based on multi-agent reinforcement learning. *IEEE Journal on Selected Areas in Communications* **37** (10): 2282–2292.

43 Carvalhoabc, F.B.S., Lopesac, W.T.A., Alencarac, M.S., and Filhod, J.V.S. (2015). Cognitive vehicular networks: an overview. *Procedia Computer Science* **65**: 107–114.

44 Singh, K.D., Rawat, P., and Bonnin, J.-M. (2014). Cognitive radio for vehicular ad hoc networks (CR-VANETs): Approaches and challenges. *EURASIP Journal on Wireless Communications and Networking* **49**: 1–22.

45 Chen, Z.D., Kung, H., and Vlah, D. (2001). Ad hoc relay wireless networks over moving vehicles on highways. *Proceedings of the 2nd ACM International Symposium on Mobile Ad Hoc Networking & Computing*, Long Beach. New York, 247–250.

46 Niculescu, D., and Nath, B. (2003). Trajectory based forwarding and its applications. *Proceedingsp of 9th Annual International Conference on Mobile Computing and Networking*, San Diego. New York, 260–272.

47 Bose, P., Morin, P., Stojmenovic, I., and Urrutia, J. (2001). Routing with guaranteed delivery in ad hoc wireless networks. *Wireless Networks* **7** (6): 609–616.

48 Gafoor, H. and Koo, I. (2018). CR-SDVN: a cognitive routing protocol for software-defined vehicular networks. *IEEE Sensors Journal* **18** (4): 1761–1772.

49 Chen, R., Park, J.-M., and H. Reed, J. (2008). Defense against primary user emulation attacks in cognitive radio networks. *IEEE Journal on Selected Areas in Communications* **26** (1): 25–37.

50 Li, J., Feng, Z., Feng, Z., and Zhang, P. (2015). A survey of security issues in cognitive radio networks. *China Communications* **12** (3): 132–150.

51 Mendez, D., Papapanagiotou, I., and Mena, B.Y. (2018). Internet of things: survey on security. *Information Security Journal: A Global Perspective* **27** (3): 162–182.

52 Ammar, M., Russello, G., and Crispo, B. (2018). Internet of Things: a survey on the security of IoT frameworks. *Journal of Information Security and Applications* **38**: 8–27.

53 Alaba, F.A., Othman, M., Abaker, I. et al. (2017). Internet of Things security: a survey. *Journal of Network and Computer Applications* **88** (15): 10–28.

54 Hassan, K., Gautier, R., Dayoub, I., Radoi, E., and Berbineau, M. (2012). Non-parametric multiple-antenna blind spectrum sensing by predicted eigenvalue threshold. *Proceedings of IEEE International Conference on Communications* (ICC), Ottawa, ON, Canada (10–15 June 2012), 1624–1629.

55 Kharbech, S., Dayoub, I., Simon, E., Zwingelstein-Colin, M. (2013). Blind digital modulation detector for MIMO systems over high-speed railway channels. *Proceedings of International Workshop on Communication Technologies for Vehicles*, Springer, Berlin, Heidelberg, 232–241.

56 Hassan, K., Dayoub, I., Hamouda, W., and Berbineau, M. (2010). Automatic modulation recognition using wavelet transform and neural networks in wireless systems. *EURASIP Journal on Advances in Signal Processing* https://doi.org/10.1155/2010/53289869.

57 Hassan, K., Dayoub, I., Hamouda, W. et al. (2012). Blind digital modulation identification for spatially-correlated MIMO systems. *IEEE Transactions on Wireless Communications* **11** (2): 683–693.

58 Simon, E. and Khalighi, M. (2013). Iterative Soft-Kalman channel estimation for fast time-varying MIMO-OFDM channels. *IEEE Wireless Communications Letters* **2** (6): 599–602.

59 Song, K., Liu, F., Wang, P., and Wang, C. (2020). Sensing performance of multi-antenna energy detector with temporal signal correlation in cognitive vehicular networks. *IEEE Signal Processing Letters* **27**: 1050–1054.

60 Huang, S., Jiang, N., Gao, Y. et al. (2020). Radar sensing-throughput tradeoff for radar assisted cognitive radio enabled vehicular ad-hoc networks. *IEEE Transactions on Vehicular Technology* **69** (7): 7483–7492.

61 Nauryzbayev, O.G., Abdallah, M., and Al-Dhahir, N. (2020). Outage analysis of cognitive electric vehicular networks over mixed RF/VLC channels. *IEEE Transactions on Cognitive Communications and Networking* **6** (3): 1096–1107.

62 Li, J., Dang, S., Yan, Y. et al. (2020). Generalized quadrature spatial modulation and its application to vehicular networks with NOMA. *IEEE Transactions on Intelligent Transportation Systems*: 1–10. https://doi.org/10.1109/TITS.2020.3006482.

63 Hossain, M.A., Noor, R.M., Yau, K.-L.A. et al. (2020). Comprehensive survey of machine learning approaches in cognitive radio-based vehicular ad hoc networks. *IEEE Access* **8**: 78054–78108.

64 Li, M., Gao, J., Zhao, L., and Shen, X. (2020). Deep reinforcement learning for collaborative edge computing in vehicular networks. *IEEE Transactions on Cognitive Communications and Networking* https://doi.org/10.1109/TCCN.2020.3003036.

65 Chen, X., Wu, C., Chen, T. et al. (2020). Age of information aware radio resource management in vehicular networks: A proactive deep reinforcement learning perspective. *IEEE Transactions on Wireless Communications* **19** (4): 268–2281.

Index

Spectrum Sharing in Cognitive Radio Networks: Towards Highly Connected Environments, First Edition. Prabhat Thakur and Ghanshyam Singh.
© 2021 John Wiley & Sons, Inc. Published 2021 by John Wiley & Sons, Inc.

Printed and bound by CPI Group (UK) Ltd, Croydon, CR0 4YY